Computational Biology

Volume 20

The *Computational Biology* series publishes the very latest, high-quality research devoted to specific issues in computer-assisted analysis of biological data. The main emphasis is on current scientific developments and innovative techniques in computational biology (bioinformatics), bringing to light methods from mathematics, statistics and computer science that directly address biological problems currently under investigation.

The series offers publications that present the state-of-the-art regarding the problems in question; show computational biology/bioinformatics methods at work; and finally discuss anticipated demands regarding developments in future methodology. Titles can range from focused monographs, to undergraduate and graduate textbooks, and professional text/reference works.

Author guidelines: springer.com > Authors > Author Guidelines

More information about this series at http://www.springer.com/series/5769

Marina Axelson-Fisk

Comparative Gene Finding

Models, Algorithms and Implementation

Second Edition

 Springer

Marina Axelson-Fisk
Chalmers University of Technology
Gothenburg
Sweden

ISSN 1568-2684
Computational Biology
ISBN 978-1-4471-6875-1   ISBN 978-1-4471-6693-1   (eBook)
DOI 10.1007/978-1-4471-6693-1

Springer London Heidelberg New York Dordrecht

Springer-Verlag London Ltd. is part of Springer Science+Business Media (www.springer.com)

To Anders

Preface to the Second Edition

The first edition of *Comparative Gene Finding: Models, Algorithms and Implementation* was published in March 2010. Since then a lot has happened and the field is gradually changing. A main driving force has been the ever-increasing use of next-generation sequencing (NGS) technology, which is revolutionizing a manifold of related fields. The pressure on computational methods and tools to handle these large amounts of data is greater than ever. In particular, since the "old" sequence analysis tools are not well adapted to the new situation with a huge data volume, much shorter read lengths, and an increased level of sequencing errors.

The gene prediction process these days typically involve the running of a multitude of bioinformatics tools, for repeat masking, for gene prediction, and for homology analyses of data from a variety of sources. The analysis tools are preferably gathered in an annotation pipeline that automatizes the processes and produces a consensus annotation by means of some kind of *combiner* software. Therefore, in this second edition we have chosen to add a chapter on annotation pipelines for next-generation sequencing data. The chapter gives a brief description of DNA sequencing in general, and of NGS techniques in particular, as well as a few application areas relevant to the gene prediction problem. The various issues involved in building a pipeline, is presented, with a discussion of the main steps including sequence assembly, *de novo* repeat masking, gene prediction, and genome annotation. Furthermore, Chap. 2 is extended to include a section on conditional random fields (CRF) as yet another model for computational gene finding. CRFs make a valuable contribution in the new sequencing technology era, as they allow for a more flexible inclusion of differing input formats and complex interdependencies between data.

Besides this and a few minor corrections, the second edition is largely unaltered. The intended reader and the required prerequisites stated in the former preface therefore remain unchanged.

Gothenburg, February 2015 Marina Axelson-Fisk

Preface to the First Edition

Comparative genomics is a new and emerging field, and with the explosion of available biological sequences the requests for faster, more efficient, and more robust algorithms to analyze all this data are immense. This book is meant to serve as a self-contained instruction of the state of the art of computational gene finding in general, and of comparative approaches in particular. It is meant as an overview of the various methods that have been applied in the field, and a quick introduction into how computational gene finders are built in general. A beginner to the field could use this book as a guide through to the main points to think about when constructing a gene finder, and the main algorithms that are in use. On the other hand, the more experienced gene finder should be able to use this book as a reference to the different methods and to the main components incorporated in these methods. I have focused on the main uses of the covered methods and avoided much of the technical details and general extensions of the models. In exchange I have tried to supply references to more detailed accounts of the different research areas touched upon.

The book makes no claim of being comprehensive, however. As the amount of available data has exploded, as has the literature around computational biology and comparative genomics over the past few years, and although I have attempted to leave no threads untouched, it has been impossible to include all different approaches and aspects of the field. Moreover, I am likely to have missed several important references that rightfully should have been mentioned in this text. To all of you I sincerely apologize.

The structure of the book is meant to follow the natural order in which a gene finding software is built, starting with the main models and algorithms, and then breaking them down into the intrinsic submodels that cover the various features of a gene. The book is initiated in Chap. 1 with a brief encounter of genetics, describing the various biological terms and concepts that will be used in the succeeding chapters. Here we discuss the general terms of gene structure, and discuss the problems of settling on a gene definition, before we describe the gene finding problem that we have set out to solve. The end of the chapter includes a historical overview of the algorithm development of the past few decades. Chapter 2 covers some of the algorithms most commonly used for single species gene finding. Each model section includes a theoretical encounter and illustrative examples, and is concluded with a description of an existing gene finding software that uses the model. In Chap. 3 we move on to sequence alignments. The chapter is divided into two parts. The first part describes different scoring schemes used in pairwise alignments, the application of dynamic programming, and the basic properties and statistical foundation of heuristic database searches. The second part describes the most common approaches to multiple sequence alignment, and the various attempts to deal with the increased computational complexity. In Chap. 4 we take on the main topic of the book, comparative gene finding. Here we combine the ideas in Chaps. 2 and 3 to a comparative setting, and describe how the strengths of both areas can be combined to improve the accuracy of gene finding. Again, each section is structured into a theoretical part, examples and an overview of the use of the model in an existing gene finder. Chapter 5 takes us through the gene features most commonly captured by a computational gene model, and describes the most important submodels used. A variety of different algorithms are described in detail, along with several illustrations and examples. Chapter 6 goes through the basics of parameter training, and covers a number of the different parameter estimation and optimization techniques commonly used in gene finding. In Chap. 7 we illustrate how to implement a comparative gene finder by giving the details behind the cross-species gene finder SLAM. SLAM uses a generalized hidden Markov model as main algorithm and has been used both by the Mouse Genome Sequencing Consortium to compare the initial sequence of mouse to the human genome, and by the Rat Genome Sequencing Consortium to perform a three-way analysis of human, mouse, and rat. The different steps and aspects in constructing a comparative gene finder are explained, and concluded with an encounter of various accuracy assessment measures used to debug and benchmark the resulting software.

This book covers a number of different fields, including probability theory, statistics, information theory, optimization theory, and numerical analysis. The reader is expected to have some background in bioinformatics in general, and in mathematics and mathematical statistics in particular. Basic knowledge of analysis, probability theory, and random processes will prove very valuable. The level and the structure of the book is such that it can readily be used as a course book for master level students, but it can also provide valuable insights and give a good overview to scientists wanting to get into the field quickly. Besides being specifically focused on the algorithmic details surrounding computational gene finding, it provides a good lesson on the intrinsic parts of computational biology and biological sequence analysis, as well as in giving an overview of a number of important mathematical and statistical areas applied in bioinformatics. A master-level course could very well be structured simply by following the book chapter-by-chapter, and perhaps include a smaller implementation project at the end.

Gothenburg, November 2009   Marina Axelson-Fisk

Acknowledgments

First and foremost I would like to thank my colleagues Lior Pachter at UC Berkeley and Simon Cawley at Affymetrix Inc. Thank you for inviting me to the SLAM-project in the first place, and for introducing me to the very exciting world of bioinformatics. Most of my gene finding work so far has been in collaboration with them (under my maiden name Alexandersson), and without them this book would never have come about. I really miss you guys, working with you has been a real pleasure and I only wish you could have written this book together with me.

Next I want to extend my gratitude to my colleagues at the Mathematical Sciences here in Gothenburg (Chalmers University of Technology and Gothenburg University). This is a great department to be in, and I want to thank you for all your help, support, and friendship, both professionally and personally. In particular I want to mention those of you who has taken the time off their busy schedule to review this text: Johan Jonasson, Graham Kemp (at Computer Science), Olle Häggström, Marita Olsson, and Mikael Patriksson. Thank you for all your help, your invaluable comments, and for being so darn picky.

Finally I want to acknowledge the Swedish Research Council (Vetenskapsrådet) for their generous funding, which made it possible for me to focus on such a large and time-consuming project as this.

Contents

Acronyms

CAI	Codon Adaptation Index
CBI	Codon Bias Index
CDS	CoDing Sequence
CML	Conditional Maximum Likelihood
CNS	Conserved Non-coding Sequence
CRF	Conditional Random Field
DAG	Directed Acyclic Graph
DNA	Deoxyribonucleic Acid
GHMM	Generalized Hidden Markov Model
GPHMM	Generalized Pair Hidden Markov Model
HMM	Hidden Markov Model
IMM	Interpolated Markov Model
ICM	Interpolated context model
LDA	Linear Discriminant Analysis
MCE	Minimum Classification Error
MCMC	Markov Chain Monte Carlo
MDD	Maximal Dependence Decomposition
MEA	Maximum Expected Accuracy
ML	Maximum Likelihood
MMI	Maximum Mutual Information
ORF	Open Reading Frame
PHMM	Pair Hidden Markov Model
PSSM	Position-Specific Scoring Matrix
QDA	Quadratic Discriminant Analysis
RBS	Ribosomal Binding Site
RNA	Ribonucleic Acid
SVM	Support Vector Machine
TSS	Transcription Start Site
UTR	Untranslated Region
VLMM	Variable-Length Markov Model
VOM	Variable-Order Markov Model

WAM	Weight Array Model
WMM	Weight Matrix Model
WWAM	Windowed Weight Array Model

Chapter 1
Introduction

This book is meant to serve as an introduction to the new and very exciting field of comparative gene finding. We introduce the field in its current state, and go through the process of constructing a comparative gene finder by breaking it down into its separate building blocks. But before we can dive into the algorithmic details of such a process, we begin by giving a brief introduction to the underlying biological theory. In this chapter we introduce the basic concepts of genetics needed for this book, and define the gene finding problem we have set out to solve. We round off by giving a brief account of the historical developments of approaching the gene finding problem up to where it stands today. In the last section we split the process of building a gene finder into its smaller parts, and the rest of the book is structured in the same manner.

1.1 Some Basic Genetics

Every living organism consists of cells, from just one cell as in the bacterium *Escherichia coli* (*E. coli*) to many trillions ($\sim 10^{12}$) as in human. Higher organisms also contain considerable amounts of noncellular material, such as bones and water. With a few exceptions each cell contains a complete copy of the genetic material, which is the blueprint that directs all the activities in the organism, and that contains the code for the inheritable traits that are passed from parent to offspring. The genetic material is composed of the chemical substance *deoxyribonucleic acid*, or DNA for short. A single-stranded DNA molecule is a long polymer of small subunits called *nucleotides* (or *bases*). Each nucleotide consists of a sugar molecule, a phosphoric acid molecule, and one of four *nitrogen bases*: adenine (A), thymine (T), guanine (G) or cytosine (C), giving rise to the four-letter DNA code {A,T,G,C}. Adenine and guanine belong to the class of *purines*, while cytosine and thymine belong to the *pyrimidines*. The nitrogen base in purines is slightly larger than that of pyrimidines, consisting of a six-nitrogen and a five-nitrogen ring fused together, while pyrimidines only have a single six-nitrogen ring. In living organisms, DNA usually comes in double-stranded form, where two single-stranded DNA molecules are arranged

© Springer-Verlag London 2015
M. Axelson-Fisk, *Comparative Gene Finding*, Computational Biology 20,
DOI 10.1007/978-1-4471-6693-1_1

into a long ladder, coiled to the shape of a *double helix*. The backbone of the ladder is formed by the sugar and phosphate molecules of the nucleotides, while the "rungs" of the ladder consist of the nitrogen bases joined in the middle by hydrogen bonds. In this structure, A always binds to T, and G always binds to C, in so-called *base pairs* (bp), causing the two sides of the ladder, or the two *strands* in the double helix, to mirror each other.

There are two main types of cells, correspondingly distinguishing between two main types of organisms, namely *eukaryotic* and *prokaryotic* cells. Besides the fact that eukaryotic cells are considerably more complex than prokaryotic cells, an important difference is that eukaryotic cells contain a nucleus while prokyarotic cells do not. Most of the genetic material reside in the nucleus in eukaryotes, and is carried on large, physically separate, DNA macromolecules called *chromosomes*. While the chromosomes in eukaryotes are linear, the DNA in prokaryotes is organized into circular rings. These rings are technically not chromosomes, although many tend to use the term for prokaryotes as well. Each eukaryotic specie has a characteristic number of chromosomes, which for instance is 46 in a typical human cell, 40 in mouse, and only 8 in fruit fly. With "typical" cells we mean *diploid* cells, where the chromosomes are organized in *pairs*. In each pair, one chromosome descends from the mother, and the other from the father. There are two types of chromosome pairs, the *autosomes* and the *sex chromosomes*. In an autosome pair the two individual chromosomes are of the same length, carry the same inheritable traits, and the number of copies of the chromosomes is the same in both males and females. The sex chromosomes, on the other hand, may have very different characteristics and are also the main indicator of the gender of the organism. The human genome consists of 23 chromosome pairs, including 22 autosomes and one pair of sex chromosomes, X and Y. A female carries the pair XX while males carry the pair XY, where the Y chromosome naturally always comes from the father. The *ploidy* of an organism signifies the number of copies of the unique set of chromosomes in that organism. Thus, a diploid cell contains two copies of each distinct chromosome (except for the sex chromosomes), whereas a *haploid* cell bears only one copy of each. Most cells in higher organisms are diploid, but specifically the *gametes* (the sperm and ova) are haploid. Examples of haploid organisms are fungi, wasps, and ants, and for instance, plants may switch between a haploid and a diploid, or even a polyploid state.

The genetic material on the chromosomes is organized into subunits called the *genes* of the organism. The genes are subsequences of DNA spread out along the chromosomes, and are intervened by possibly very long stretches of nonfunctional DNA. The genes provide the templates for the proteins and RNA molecules that are responsible for all activity in the organism, and are traditionally defined as the units of inheritance that control the hereditary traits passed on from parent to offspring. The *genome* of an organism is its complete set of DNA (or RNA for some viruses), including both the genes and the nonfunctional stretches of DNA. The genome sizes vary greatly between organisms; from 600,000 bp in the smallest (known) free-living organism (a bacterium) to some 3 billion bp in human. While the genome is very compact in lower organisms, with very little nonfunctional material, in higher organisms the genes are hidden in a vast sea of "junk" DNA. In human, for instance, the genes constitute only about 3 % of the human genome. If the rest of the sequence really is junk, or if it has some sort of function, direct or indirect, is still under much debate [10, 43].

Although mirroring one another, the chromosome strands run in opposite directions, and are said to be *antiparallel*. One strand is called the 5′ → 3′, the *forward*, or the *sense* strand, while the other the 3′ → 5′, the *reverse*, or the *antisense* strand. The 3′ and 5′ connotations correspond to the orientation of the 3′ and 5′ carbon atoms of the sugar rings in the nucleotides (see Fig. 1.1), and the *reading direction* (i.e., the direction genes are transcribed) goes from 5′ to 3′. That is, the transcription starts at the 5′ end of the gene and ends in the 3′ end. Sequences appearing before or after a gene (in the reading direction) is commonly referred to as *upstream* or *downstream* of the gene respectively. Faced with a novel sequence to analyze, which strand that gets assigned to be the forward or the reverse strand is of course arbitrary.

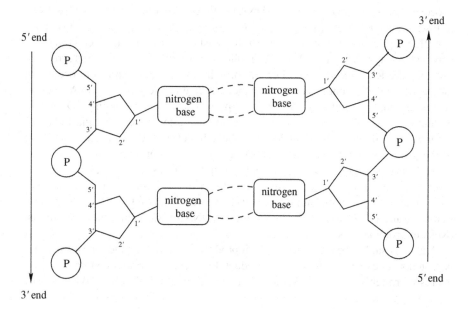

Fig. 1.1 Each nucleotide consists of a phosphoric acid molecule (P), a sugar ring, and a nitrogen base. The antiparallel strands of the DNA double helix run in opposite directions. The direction is given using the 3′ and 5′ carbon atoms of the sugar rings in the nucleotides. Reading the sequences from left to right, one will have the 5′ atom to the left of the 3′, while in the antiparallel strand the situation is reversed

1.2 The Central Dogma

The *genotype* of an organism is the set of genes that the genome contains. The *phenotype*, on the other hand, is the set of observable characteristics of that organism, such as size, structure, number of cells, tissues, and organs and, also, behavior and function of the organism. The *Central Dogma in Molecular Biology* formulated by Sir Francis Crick, first in 1958, and then re-stated in 1970 [27], describes how the genetic information contained in genes is transferred, giving the connection between genotypes and phenotypes. Moreover, the central dogma states that protein sequences are never translated back to DNA, RNA or a new protein, DNA is never created out of RNA (with the exception of retroviruses), and DNA is never directly translated into protein. Crick classified the $3 \times 3 = 9$ possible information transfers between DNA, RNA, and protein into *general transfers*, *special transfers*, and *unknown transfers* (see Fig. 1.2). General transfers (solid arrows in Fig. 1.2) are those that are believed to occur naturally in all cells, such as DNA to DNA (DNA replication), DNA to RNA (transcription), RNA to RNA (RNA replication), and RNA to protein (translation). The RNA replication, which may seem as the least natural of all the general transfers, may be the way in which RNA viruses replicate. Special transfers (dashed arrows in Fig. 1.2), the RNA to DNA and DNA to protein transfers, are known to occur, but only under specific conditions. For instance, reverse transcription of RNA to DNA occurs in some retroviruses and retrotransposons. Direct translation of DNA into protein has been observed, but only artificially using test tubes. The third class of the transfers, the unknown transfers, are those believed never to occur, consisting of the protein to DNA, protein to RNA, and protein to protein transfers, and are not shown in the figure. Potential exceptions to these rules may exist, and while the central dogma has been adjusted since it first was stated, it remains the backbone of molecular biology. For the purposes of this book we are only interested in the general transfers, and of those only the transcription (DNA to RNA) and the translational (RNA to protein) ones.

A gene is said to be *expressed* when it produces a product, which for most genes is proteins but can end with an RNA product. In essence, a protein-coding gene gets expressed roughly in two steps; the *transcription* and the *translation*. When a gene

Fig. 1.2 The central dogma. The *solid arrows* signify general transfers, the *dashed arrows* signify special transfers, while unknown transfers are those absent with respect to a complete graph

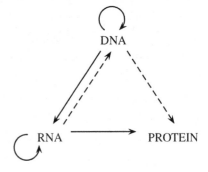

is transcribed, the entire gene sequence, sometimes extending over many thousands of DNA bases, is copied into another kind of nucleic acid, called *ribonucleic acid* (RNA). The RNA molecule is a *complementary* copy of the DNA template, meaning that a T in the DNA template results in an A in the RNA molecule, a C in a G, and vice versa. The only exception is that the nucleotide thymine (T) in DNA is replaced by the molecule *uracil* (U) in RNA. Thus, in the copying process, an A in the DNA template results in a U in the RNA molecule.

The resulting RNA molecule is called a *primary transcript*, and is often further processed before it is passed on for translation. One such process, particularly common in eukaryotes, is called *splicing*, in which nonfunctional parts of the transcribed sequence are excised out of the molecule. For some genes the processed RNA is the final product, while for most genes the RNA molecule is passed on to the translation step. The processed RNA molecule is then called a *messenger* RNA (mRNA) because it carries the genetic information from the DNA sequence to the protein-synthesis machinery of the cell. In this machinery, the mRNA molecule is translated into protein by letting each triplet of the RNA sequence code for specific a *amino acid*. These triplets are then called *codons* and the specific mapping between RNA triplets and amino acids is often referred to as the *genetic code* (see Fig. 1.3). The resulting protein is finally folded into a specific three-dimensional structure and transported out into the organism to whatever place it is supposed to be.

Second base in codon

		T	C	A	G	
First base in codon	T	TTT ⌉ Phe TTC ⌋ TTA ⌉ Leu TTG ⌋	TCT ⌉ TCC TCA Ser TCG ⌋	TAT ⌉ Tyr TAC ⌋ TAA Stop TAG Stop	TGT ⌉ Cys TGC ⌋ TGA Stop TGG Trp	T C A G
	C	CTT ⌉ CTC CTA Leu CTT ⌋	CCT ⌉ CCC CCA Pro CCG ⌋	CAT ⌉ His CAC ⌋ CAA ⌉ Gln CAG ⌋	CGT ⌉ CGC CGA Arg CGG ⌋	T C A G
	A	ATT ⌉ ATC Ile ATA ⌋ ATG Met	ACT ⌉ ACC ACA Thr ACG ⌋	AAT ⌉ Asn AAC ⌋ AAA ⌉ Lys AAG ⌋	AGT ⌉ Ser AGC ⌋ AGA ⌉ Arg AGG ⌋	T C A G
	G	GTT ⌉ GTC Val GTA GTG ⌋	GCT ⌉ GCA GCC Ala GCG ⌋	GAT ⌉ Asp GAC ⌋ GAA ⌉ Glu GAG ⌋	GGT ⌉ GGC GGA Gly GGG ⌋	T C A G

Third base in codon

Fig. 1.3 The genetic code

1.3 The Structure of a Gene

Most genes contain information for making specific proteins, which then perform a
wide variety of activities in the cell. Other genes, called noncoding genes, encode
functional RNA molecules often involved in the regulation of gene expression and
protein synthesis. These genes are not translated into proteins, and lack the typical
sequence constraints of coding sequences, something that makes them hard to detect
by traditional gene finding programs. Protein-coding genes can vary a lot in size and
organization, but they share several conserved features. Therefore it is common to
ignore the noncoding genes in gene finding algorithms, and, likewise, we will here
concentrate on the identification of protein-coding genes, and henceforth, when we
write "gene" we mean a protein-coding gene (Fig. 1.4).

The boundaries of a gene are often defined as the beginning and end of the transcription, and the core of the gene is the coding region consisting of the DNA sequence
that eventually gets translated into protein. Furthermore, the genes in higher organisms are not contiguous, but are often split into alternating coding and noncoding
segments. The coding segments, called *exons*, constitute the template sequences for
the amino acid sequence, and very often a separate exon corresponds to a discrete
functional or structural unit of the protein. The exons are interspersed by noncoding
regions of highly variable lengths called *introns*. When the gene is expressed, the
entire region surrounding the gene and sometimes extending over many thousands
of bases is transcribed into an RNA molecule using the DNA as a template. The
introns are then cut out in a process called *splicing*, and the exon sequences are
"glued" together again and translated into the corresponding protein. In the translation the combined exon sequence, often called the coding sequence or CDS, is
divided into triplets, or three-letter words, called *codons*, that each code for one of
20 possible *amino acids* (except for the three terminal codons TAA, TAG, and TGA).

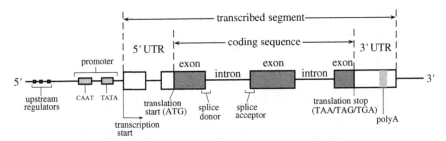

Fig. 1.4 The structure of a eukaryotic gene. The gene is defined as the segment that is transcribed
into RNA. The coding sequence consists of exons that get translated into amino acids, and intervening introns that get spliced out before translation and do not encode for protein. The 5′UTR
and 3′UTR are untranslated regions flanking the coding sequence and not involved in the protein
synthesis. The promoter contains binding sites (such as CAAT and TATA) for enzymes involved in
the transcription. Additional regulatory elements, such as enhancers and suppressors, may reside
further upstream of the gene

The amino acids corresponding to the codon sequence are linked together in a long linear string called a *polypeptide*, and proteins are composed of one or several large such polypeptides.

Although the end product after splicing will consist of even base triplets (i.e., the length is divisible by 3), the splice signal can occur in the middle of a codon. If a codon is to be spliced, one part appears at the end of one exon in the underlying DNA sequence, and the other at the beginning of the next exon. The intron in between the exons is said to have *phase* 0, 1 or 2 depending on whether it falls between two complete codons, after the first base of a codon, or after the second base of a codon, respectively. Similarly, a gene can have either one of three possible *reading frames* (six if we include the reverse strand in the count) corresponding to the location of the coding sequence of the gene with respect to the beginning of the entire input sequence. For instance, if the coding region of the gene starts k bases into the sequence, the reading frame of the gene is $k \bmod 3$. An *open reading frame* (ORF) is a string of codons occurring between an initial codon ATG and ending with one of the three stop codons TAA, TAG, or TGA, but without any other interrupting stop codons in between.

Not all genes are active in all types of cells or at all times. Some genes are almost always active, though, and these are called *housekeeping genes*. Others are more differentially expressed and might be active only in specific cell types or at particular stages of development of the organism, or they may be activated only when necessary by specific processes in the cell. Such differential expression is achieved by regulating the transcription and the translation of the genes in various ways, and besides consisting of exons and introns the gene is comprised of a number of regulating components, such as the promoter, the UTRs, and the polyadenylation signal (polyA) to mention a few. The UTRs, or *untranslated regions*, are regions of untranslated exons both upstream and downstream of the coding region. The promoter is the region surrounding the transcription start of the gene and regulates the binding of transcription factors to the gene, and the polyA signal resides at the end of the transcript and is part of the process that prepares the mRNA for translation.

The problem of gene finding is to accurately predict the gene structure, in particular to locate the different gene features and predict the resulting polypeptide. It is not obvious how to define a gene, something that we discuss further in Sect. 1.5. For our purposes, however, we will say that a gene constitutes a contiguous segment of DNA, composed by a number of features needed to generate the final protein. These features include:

- Upstream non-transcribed regulatory elements, including the promoter.
- Transcription start site.
- 5'UTR.
- Coding region, including coding exons, splice sites, and introns.
- 3'UTR, including the polyA-signal.

1.4 How Many Genes Do We Have?

In May 2000, with the DNA sequence of the human genome near completion, Dr. Ewan Birney, a senior scientist at the European Bioinformatics Institute (EBI), organized a sweepstake, where he invited researchers to bet on the total number of genes in the human genome. The winner was to be announced at a Cold Spring Harbor conference in 2003 [105], and for the fairness of the sweepstake the pot was to be split between the nearest bidders in each year 2000–2002. Dr. Birney was convinced that his annotators at Ensembl [35] would have a final answer well in time, but as the 2003 conference drew nearer it became evident that the human gene number was still far from being resolved. Still, the rules of the sweepstake stated that a winner had to be announced, and therefore the most recent estimate of EBI resulted in 21,000 genes as the most probable "final answer". As it turned out, all bets placed were well above the "final answer", and thus the lowest bet in each year, no matter how far off it may seem, became the winning bet. The winner of year 2000 bet 27,462 genes, and was asked how he came up with such a low number in a year when popular estimates were considerably higher, closer to 50,000 in fact. His explanation was that he had been in a bar drinking at the time. It had been late at night and at that point the behavior of the other bar visitors had not seemed much more sophisticated than that of a fruit fly, which, at that time was thought to have around 13,500 genes. Thus, he simply doubled this number, and used his birth date, April 27, 1962, as his final (and winning) bet.

There has been a huge interest in the total number of genes in the human genome in recent years, which may seem as a drift away from more important questions. However, while the gene number merely is a product of the efforts to identify and characterize all functional units in a genome, the number itself may be of interest. For instance, the fact that the number of genes in human is much less than originally thought, has appeared very enticing to some and raised many questions, both of biological and philosophical nature. How is it possible that such a biologically complex organism such as ourselves have a gene number only about a third larger than the nematode worm *Caenorhabditis elegans*, and just about five times than that of the bacterium *Pseudomonas aeruginosa* [25]? Or, to quote Comings in 1972 [26], "The lowly liverwort has 18 times as much DNA as we, and the slimy, dull salamander known as Amphiuma has 26 times our complement of DNA." The question of the apparent lack of relationship between genome size and biological complexity has been named the *C-value paradox*, where the *C*-value signifies the amount of DNA in a haploid eukaryotic cell. The term *C-value* was coined by Hewson Swift already in 1950 [102], in reference to a "remarkable constancy in the nuclear DNA content of all the cells in all the individuals within a given animal species" reported by Vendrely and Vendrely [44, 104]. This constancy was taken as evidence that it was the DNA, and not the proteins, that served as the hereditary material. However, the combination of a fairly constant DNA content within species, and huge variations of genome sizes between species, seemed 'paradoxical'. Although the *C*-value seemed related to morphological complexity in lower organisms, the variation was profound among higher organisms, with numerous plants, insects, and amphibians having much larger genomes than humans, something that appeared very provocative at the time. However, the basic paradox was that the DNA amount should be constant, since it harbors the genes, and yet was much larger than expected based on the presumed number of genes in any given organism. The solution to the paradox came with the realization that most DNA is noncoding, so that the size of the genome has little to do with the number of genes it holds. Moreover, the now well-known fact that the genes are not aligned along the genome as simple pearls on a string, but exhibit complex dependencies both in sequence residence and function, appeared to resolve the lack of relation between genome size and biological complexity.

This solution raised a number of questions on its own, however, including the purpose, evolution, and discontinuous distribution among species of this noncoding DNA [44]. Gregory therefore suggested an update of the C-value paradox to the C-value enigma [43]. The genome size variation is randomly distributed over species. Although large among lower organisms including amphibians, plants, and unicells, the size range is rather constrained in mammals, birds, and prokaryotes. In addition, the C-values of related species appears discontinuous, as multiples of some basal value even, unrelated to shifts in chromosome number. Moreover, there appears to be a strong connection between cell size and nuclear size, and a strong *negative* correlation to cell division rates; large genome sizes are much more prevalent in species with large, slowly dividing cells. Gregory suggested that any solution to the C-value enigma had to consider not just one, but three questions: (1) The variation of genome sizes, continuous or discontinuous, over species, (2) the nonrandom distribution of variation over different groups of species, where some vary greatly and others appear constrained, and (3) the strong connection between the C-value and the cell volume, as well as the negative correlation with division rates.

Claverie coins a similar paradox, the *N-value paradox* [25], for the lack of relation between gene content and organism sophistication. The question is how to define biological complexity. Diversity of cell types? Brain circuitry? Cultural achievements? Claverie suggests the number of "theoretical transcriptome states" that the genome of an organism can give rise to as a measure of complexity, where the *transcriptome* is the set of all transcripts (mRNA molecules) in an organism. If we, as a simplified model, assume that each gene in the genome only has two possible states, it is either ON or OFF, the human genome could theoretically give rise to $2^{30,000}$ different "transcriptome states". Compared to about $2^{20,000}$ for the nematode worm, humans appear to be 10^{3000} times more complex. These numbers have to be decreased due to co-regulation of genes, and because some of these hypothetical transcriptome states will be lethal to the organism, but this is true for all organisms. Besides, the genes exhibit more than two states (alternatively spliced forms) on average in human, while this rate is less than two in worm [58]. The conclusion is that already a small number of genes can generate tremendous complexity, by using sophisticated mechanisms of gene regulation, rather than just increasing the number of genes [25].

While on the subject of paradoxes, Harrison et al. [49] made an interesting discovery when comparing homology matches of the protein coding genes in yeast (*Saccharomyces cerevisiae*), worm (*Caenorhabditis elegans*), and fruit fly

(*Drosophila melanogaster*). The authors scanned all three genomes as well as human chromosomes 21 and 22 against three different subsets of the Swiss-Prot database [14]: bacterial proteins, all other phyla, and all other organisms than those analyzed. Although the worm had substantially more annotated protein-coding genes than both fly and yeast, the amount of homology detected was greater in both fly and yeast, regardless of what subset of Swiss-Prot was used. The authors suggested to name the tendency of a stable ratio of homology between worm and fly the *H-value paradox*, where the *H*-value represents the total amount of detected protein homology measured in bases. While this discrepancy could be due to overpredictions in worm, another plausible explanation is that the worm might have undergone a contraction in gene number, followed by an expansion of organism-specific genes. Yet another explanation could be differences in genome annotation, since the number of annotated genes have been modified several times in both fly and yeast.

Returning to the actual question of how many genes there are, the gene number in human has varied a great deal over the years. The original predictions, based on the estimated number of gene products in a cell, started out well over 100,000. But when the draft sequence of the human genome was published in 2001, the initial analysis came to an estimate of about 30,000–40,000 protein-coding genes [53]. Ensembl [35], which has become the golden standard for human genome annotation, currently reports 22,258 protein-coding genes and an additional 6,411 RNA genes. Although significant variations are still reported every now and then, the general scientific community seems to stabilize more and more to about 23,000 protein-coding genes, and about 27,000 genes when including RNA-genes, a number that is remarkably close to that of the winning bet in the gene sweepstake. Settling on a final number seems difficult even for such a well-analyzed genome as the yeast *S. cerevisiae*. Although completed already over a decade ago and having undergone hundreds of (published) genome-wide analyses, the total number of genes still varies between sources. Table 1.1 was reported in 2004 [4], and checking current gene numbers we

Table 1.1 Estimated gene numbers in *S. cerevisiae*

Reference	Year	Gene number
Goffeau et al. [42]	1996	5885
Cebrat et al. [21]	1997	~4800
Kowalczuk et al. [61]	1999	>4800
Blandin et al. [13]	2000	5651
Zhang and Wang [113]	2000	5645
Wood et al. [108]	2001	<5570
Mackiewicz et al. [71]	2002	5322
Kumar et al. [68]	2002	~6000
Kellis et al. [57]	2003	5726
SGD [24]	2005	5888
MIPS [76]	2005	6335

get from the Saccharomyces genome database (SGD) a total of 6607 ORFs of which 4825 are verified, 971 are uncharacterized, and 811 are dubious [24].

One reason for the variations in gene number in yeast has been due to the detection of "small ORFs" (smORFs). While the initial annotation of yeast only included ORFs of at least 100 amino acids, it has been discovered that many genes are shorter than that. For instance, Kellis et al. [57] performed comparative analysis between *S. cerevisiae* and three other fungi, which resulted in the inclusion of 188 smORFs constituting about 3 % of the total gene count. A similar proportion of missed smORFs in human would result in an additional 900 or so genes. Moreover, Southan listed a number of reasons as to why we can expect the gene number to rise back to over 30,000 again [98]:

1. Model eukaryotes tend to show a post-completion rise in gene. number.
2. The human genome is still incomplete.
3. Gene finding softwares have a significant false-negative rate.
4. Automated gene annotation pipelines are conservative.
5. Transcript coverage by mRNA and EST sequences is incomplete.
6. Sampling experiments tend to reveal new genes.
7. A fraction of rapidly evolving small proteins remain undetected.

Ultimately, before trying to establish the actual number of genes in a genome, we need a clear definition of the concept 'gene', which has turned out to be easier said than done.

1.5 Problems of Gene Definitions

The definition of a gene has been revised several times during the past century, and the search for a comprehensive formulation is still not settled. Some even argue that the concept should be declared dead, to leave place for new, more suitable definitions. The concept "gene" has gone from being an abstract unit of inheritance, to carrying specific traits passed on from parents to offspring, to being directly associated with enzymes and other proteins, to becoming physical molecules lined up along the chromosomes. The notion that it is the proteins that carry out all the activities in the cell is up for debate as well. An increasing amount of evidence postulates that RNA molecules are more important than originally thought. Moreover, the findings of new kinds of RNA molecules emphasize their importance even more [32, 87].

At the beginning of the twentieth century an English physician named Archibald Garrod proposed that a *gene* is the object responsible for the production of a specific protein. Although ignored for decades, his work was affirmed in 1941 by George Beadle and Edward Tatum, who showed that genes affect heredity by determining enzyme structures [8]. Their "one gene/one enzyme"-hypothesis, awarding them the Nobel prize in 1958, has subsequently been refined. Now we know that not all genes code for enzymes but for structural and regulatory proteins as well. Also, many proteins, such as hemoglobin, consist of several polypeptide chains, each controlled by a different gene. Today, the Beadle and Tatum hypothesis translates, more accurately, into "one gene/one polypeptide," but this concept is still questionable.

There are numerous problems with, and exceptions and contradictions to any gene definition one has come up with. For instance, some definitions may want to include the regulation in what is called a gene. However, regulatory sites exist that affect every step of the gene expression process, including mRNA degradation, and post-modifications to the resulting protein. Besides being very difficult to find since these sites can be located virtually anywhere, some sites reside very far away from the gene they regulate, something that would make the defined gene sequence very long. Other problems involve *trans*-splicing : separate mRNA molecules may be spliced together, both from "genes" residing on opposite strands of each other, or even on separate chromosomes. For instance, the major peptide of the enzyme Glucose 6-phosphate dehydrogenase is encoded from information from two different chromosomes [56]. Moreover, posttranslational events contradict the sequence view where the DNA is thought to encode a corresponding RNA molecule. Proteins may be spliced into two separate proteins, thus the start and end of a protein is not determined by the DNA sequence. Similarly, separate proteins may be joined after translation, or the protein may be modified into a completely different structure and function after translation.

The Human Gene Nomenclature, which is responsible for the naming of genes, defines a gene as "a DNA segment that contributes to phenotype/function. In the absence of demonstrated function a gene may be characterized by sequence, transcription or homology" [106]. This is a fairly unspecific definition, as it includes all kinds of transcripts. Since most gene finding methods only deal with protein-coding genes, a narrower definition could be "chromosome-derived transcripts giving rise to one or more protein forms with shared sequence identity that assign them as products of a single genomic locus and strand orientation" [98]. However, since gene finding algorithms are unable to resolve multiple translation starts and alternative splice forms within the same transcript sequence, for this purpose we need to be more specific.

The Human Gene Nomenclature definition results in a baseline number of protein-producing loci in the genome. But for the purpose of gene finding, we need a definition that unravels the resulting protein sequence. That is, in the presence of alternative splice forms, for instance, we would like to be able to identify each form separately.

This is a huge challenge for automatic gene finding algorithms, which tend to be confused by alternatively spliced transcripts. Regardless, we need a slightly different definition from the one above, as we aim to identify the underlying sequence to a specific protein. If we try to keep the eye on the money, the purpose of gene finding is to identify functional regions, and, if possible, the correct boundaries giving rise to a specific transcript/protein. This is a difficult task, however, as it may produce an artificial definition that has little to do with reality. Therefore, perhaps the most satisfactory definition for our purposes is the one produced by the Encyclopedia of DNA elements (ENCODE), a project whose goal is to find all functional elements in the genome: "the gene is a union of genomic sequences encoding a coherent set of potentially overlapping functional products" [40].

1.6 The Gene Finding Problem

There is a wide range of approaches to the problem of gene finding, addressing different types of issues. Roughly, the approaches can divided into ab initio methods, *similarity-based* methods, and *comparative* methods, and various combinations of these categories. The most direct approach is the similarity based, or evidence based, method, where known mRNA, cDNA, or protein sequences are matched against the input sequence, and where high similarities are strong indicators for homologous genes. The advantage is that such matches have a high reliability, and also give a good clue to the function of the new gene. The disadvantage is when no match is found in the database. Nothing can be said, and no sequence can be ruled out as noncoding. Ab initio methods, also called de novo methods, use the statistical patterns and known consensus sequences of signals to detect novel genes. The advantage is that there is no need for a homologous genome or a database of known genes. All that is needed is a fair parameter representation of what types of patterns one wants to look for. For a completely unknown genome, however, this method may pose a problem, since there are no known patterns yet. But using the pattern of a related sequence may still help the search forward, and then the parameter settings can be updated as one goes along. One disadvantage is that novel genes with unusual patterns that do not resemble any known genes are likely to be missed. Comparative methods use two or more evolutionary related sequences to identify novel genes, either in one sequence at a time or simultaneously in all sequences. These methods utilize the strengths in ab initio gene finding, as well as strengthening the signal of potential genes by means of homology. Genes with unusual patterns may still be correctly predicted if the sequence similarity is high enough.

The identification of genes is complicated for several reasons. The fact that the regions coding for functional units only comprise 3 % of the human genome poses a combinatorial difficulty. Another difficulty is that of overlapping or nested genes.

That is, genes may overlap one another, or be nested within one another, both on the same strand or on opposite strands. Such events are very difficult to model, and often the best-case scenario is that the algorithm predicts one of the genes correctly, but more likely the outcome will be a mish-mash of both genes. Alternative splicing is a similar problem. Alternative splicing means that the RNA transcript can be spliced in different ways and thereby result in different gene products. This is a common process in eukaryotes, and explains the discrepancy between the number of genes in an organism and the number of RNA and protein products active in the organism. Usually, the difference between different splice forms correspond to different combinations of the exons in the transcript, but more intrinsic splicing processes are known to exist. Regardless, alternative splicing poses a big problem for gene prediction models. Gene prediction softwares usually only predict a single most likely parse of the input sequence. Allowing for suboptimal parses is one possible solution around this.

Other challenges to gene finding include frameshifts and sequencing errors, but the latter become less and less of an issue, as the sequencing process has been greatly improved over the past decade. Nevertheless, if a sequencing error occurs that disrupts the codon sequence of the gene, the resulting gene prediction will most often be incorrect as well. Moreover, the presence of *pseudogenes*, is likely to confuse the gene predictor, especially the ab initio kind. Pseudogenes are nonfunctional sequences that still resemble genes. They may for instance be artifacts of previously active genes that during evolution have been "turned off" one way or another. Comparative methods tend to avoid pseudogenes, as they need to appear in both sequences at a fairly high sequence similarity to be detected.

We have seen in Sect. 1.3 that the gene is a complicated object with many different components. However, due to incomplete knowledge and the lack of sufficient training data, and due to computational complexity, the gene models are often simplified to become feasible. The most common assumptions and simplifications include:

- Only model protein coding genes.
- Only model the coding part of the genes.
- Length constraints placed on the different gene features.
- Predict one optimal parse only.

As gene prediction softwares evolve and the biological knowledge and data increases, more and more gene components can be included in the models. This can help increase the accuracy of the predictions, and shed light not only on which genes an organism contains, but how the genome is regulated.

1.7 Comparative Gene Finding

All living things on earth are genetically related. The beginning of life starts with one common ancestor, and then through evolution new species develop. The evolution occurs through random modifications, called *mutations*, to the genetic code. While such modifications occurring in a functional region of the genome are most often harmful, and therefore selected against, every now and then a modification makes for an improvement to the fitness of the organism, and is kept. In this manner, the organisms slowly accumulate genetic differences, until eventually they may become separate species. *Phylogeny* is the study of the genetic relationship between organisms, and is aiming at understanding the course of evolution and the similarities and dissimilarities between organisms.

There are two main driving forces to evolution: *natural selection* and *genetic drift*. Natural selection is the process in which species gradually adapt to their surroundings. In this process, positive mutations are accumulated over generations, while negative mutations are gradually sifted out. The selection occurs naturally, as individuals with a higher fitness become more successful in the population in terms of survival and reproduction. Genetic drift is the process in which fluctuations between different genotypes occur. With different genotypes we mean different variants of the same gene, called *alleles*, resulting in different phenotypes. For instance, different eye colors between individuals are regulated by different alleles of the same corresponding gene(s). While natural selection is a nonrandom process that sees to the fitness of the population, genetic drift is a completely random process free of selectional pressure. For instance, even if two alleles of the same genes are equally fit, their frequencies will still fluctuate over time due to genetic drift, and sometimes one variant will eventually vanish from the population altogether.

The main mechanism of natural selection is that of mutations. Mutations are randomly occurring, permanent changes to the nucleotide sequence in the genome. These are typically acquired, either by copying errors of the DNA prior to cell division, or by exposure to environmental agents, called *mutagens*, that harm the DNA. Examples of mutagens are ultraviolet light, radiation, or certain chemicals. We say that a mutation that results in an improvement in the organism is a *positive* mutation, while a harmful mutation is called *negative*, and a mutation that has no effect is called *neutral*. The basic types of mutations are *substitutions*, *insertions*, *deletions* and *frameshifts*. In a substitution, one nucleotide is replaced by another. If this occurs in the protein coding part of a gene, it may cause a change in the amino acid in the encoded protein, which in turn may result in a completely different structure and function of the protein. We speak of a *silent* substitution, if it causes no change to the amino acid, a *missense* substitution if it alters the amino acid, and a *nonsense* substitution if the codon is turned into a stop codon and the corresponding protein sequence gets truncated. In an insertion, extra bases are inserted into the sequence, while in a deletion bases are removed from the sequence. From a comparative genomics point of view, insertions and deletions are mirroring processes of one another, and are sometimes jointly referred to as *indels*. Frameshifts are actually the result of an insertion or a deletion in a coding region, in which the codon triplets are "shifted" to encode an entirely different amino acid sequence.

Comparative genomics is built on the observation that the DNA sequences of functional elements evolve at a much lower rate than in nonfunctional DNA. The reason is that while a mutation in a functional region is most often harmful, and selected against, mutations in nonfunctional regions is under no selectional pressure and therefore is kept to a higher extent. Sequence alignments of evolutionary-related organisms can be used to highlight important functional regions, such as protein-coding segments or regulatory signals, as well as understand the genome evolution and the development of the organisms. Comparative gene finding utilizes the evolutionary relationships by using the similarities and dissimilarities between sequences to strengthen the signal of functional elements in the sequence. It improves the accuracy of gene finding immensely, in particular by bringing down the false prediction rate, and by identifying the exact gene structures more accurately.

1.8 History of Algorithm Development

One of the first gene finding papers that presented a computer method for identifying protein coding regions in genomic DNA was published in 1982 by Staden and McLachlan [100]. It had been noticed by several others already, however, that coding regions exhibit particular statistical patterns in their base composition that may be utilized for gene finding. Staden and McLachlan identified the following criteria as useful sensors for coding sequence: open reading frames (ORFs), start codons, codon usage, ribosomal binding sites, and splice sites. That the authors were true pioneers in the field is shown by the following quote: "An ideal method would use all these criteria to give the probability that a section was coding in a particular reading frame" [100]. The method presented was a first step toward this, as it analyzed the codon usage in ORFs. Their method was later developed further by integrating both content sensors and signal sensors for splice sites and ribosomal binding sites in a software called ANALYSEQ [99]. Although it early seemed clear that the base composition in coding regions follow a distinct pattern, it was Fickett who showed that this pattern does not appear in noncoding DNA [33]. In his paper the author presented a test algorithm, called TESTCODE, that just as the Staden and McLachlan method analyzed the coding potential of open reading frames (ORF), but that also utilized the difference in pattern between coding and noncoding regions.

In 1990 came the first automatized gene finder, called GeneModeler (gm), which was able to integrate several coding sensors at once [34]. The flow of the program is to pass the sequence through a series of tests, where the potential gene model is passed through to the next step if it scores above a given threshold associated with the specific gene feature. The program starts off by identifying all ORFs and their corresponding frame in the input sequence, and moves on, using various weight matrices, to identify potential translation initiation sites and splice sites. Gene models on the form ORF—intron—ORF are constructed and both ORFs and introns are scored for sequence composition. Each gene model is then extended to include further introns and ORFs if possible, and all possible gene models that make it through the tests are reported, even those that are in conflict with one another.

Somewhere around this time the development of different splice site detectors started to emerge. Gelfand constructed a gene finder that uses discriminant analysis for splice site detection and the TESTCODE algorithm for coding potential, and his gene finder reported only the best variant of spliced mRNA emerging from the model [37]. Brunak and colleagues applied neural networks to the splice site detection of vertebrate pre-mRNA sequences [17]. Neural networks had been applied to the problem of splice site detection earlier, but in this paper the prediction accuracy was greatly improved by training the model on both false and true splice sites. Both the discriminant analysis approach and the neural network approach showed great improvements over using weight matrices to predict splice sites.

Following in the footsteps of GeneModeler, Guigó et al. constructed GeneID, which is a hierarchically structured rule-based gene finder. A novelty with GeneID was that it used profiles, constructed from weighted multiple alignments of various vertebrate sequences, for the detection of start and stop codons and splice sites. The input sequence is processed in several rounds, starting by the detection of potential exon boundaries, and then by processing the candidate exons further using discriminant analysis, until finally constructing the highest scoring gene model from the set of candidate exons.

At this point, the idea of using external homology information, such as from known EST, cDNA, or mRNA sequences, to improve gene prediction had awakened. The three main approaches to similarity-based gene finding can be summarized as follows. GRAIL [110, 111], described in more detail in Sect. 2.4.4, combines the scores of a number of different sensors in a neural network, including information from an EST database. ESTs are generally too short to determine the gene structure alone, but can still provide useful information about the beginning and end of a gene, sort out false positives, identify missed exons, and improve boundary prediction. GeneParser [95, 96] incorporates external homology information in the gene finding algorithm by invoking BLAST scores from matching the target sequence to a protein database. Procrustes [38] uses a combinatorial approach, where potential exon blocks in the input sequence are concatenated and matched to a known protein. While this approach had been attempted before [39, 95] the *spliced alignment* algorithm of Procrustes reduced computational complexity enough to make a complete search of all possible exon assemblies feasible.

The earliest gene finders would analyze only the input sequence, disregarding the mirroring complementary strand. One problem with this approach is that genes appearing on the opposite strand will signal high coding potential on the input strand as well. With GENMARK [15] came the novelty of performing gene finding on both strands simultaneously, and thereby reducing the number of false positives considerably. GENMARK uses both homogeneous and nonhomogeneous Markov chains as sequence composition models in introns and exons. A variant to the Markov chain approach in GENMARK, which addresses the issue with small training sets, is the *interpolated* Markov model (IMM), which has been implemented in the microbial gene finder GLIMMER [30, 90]. Instead of using basic counts of codons and hexamers in a training set, the probabilities are estimated using an interpolation of different lengthed context sequences, depending on their reliability in the training set. The IMM approach and its use in GLIMMER is described in more detail in Sect. 2.3.

One of the first *hidden* Markov model-based (HMM) gene finders, named Eco-Parse [65], was developed for gene finding in *E. coli*. The method is similar to the profile HMMs presented in Sect. 3.2.9 [64], but with particular focus on the codon structure in genes. With EcoParse a flora of HMM-based gene finders, using dynamic programming and the Viterbi algorithm to parse a sequence, emerged. The success of using HMMs for gene finding spread, and the method has been implemented in numerous softwares since, including HMMgene [62, 63] and VEIL [50]. The very successful extension of the standard HMM, the *generalized* HMM (GHMM), was applied to the gene finding problem a few years later, first in Genie [66], and then followed by one of the most popular gene finders of all times, Genscan [18], which is described in detail in Sect. 2.2.4. The generalization of the standard HMM involves allowing for generalized length distributions of the exons, something that has proven to improve accuracy immensely (see Sect. 5.2 for a discussion on this).

The development of gene finding continued in the search for better and more efficient algorithms for the task. More and more gene finders chose to invoke external homology one way or the other, early examples being GeneBuilder [79], CEM [6] and GeneWise [12]. Meanwhile, the Human Genome Project [53] had started and was up and running, generating more and more reliable human sequences for every day. The idea of comparing longer contiguous sequence between homologous organisms, and in particular comparing human and mouse, grew stronger. The first true comparative gene finder to enter the stage was ROSETTA [7], which is described in Sect. 4.2.1. ROSETTA combined the gene finding task with that of aligning long homologous sequences by predicting genes in human and mouse simultaneously. The method provided a proof of concept that comparative gene finding was both feasible and very successful. Since then, numerous comparative methods have been introduced, all with varying strengths and weaknesses, and with various success. After the Human Genome Project published the initial analysis of the human sequence in 2001 [53], a new project was launched, the Mouse Genome Sequencing Consortium [81], which aimed at sequencing the mouse genome with the main purpose to provide thorough comparative analyses between human and mouse. Three comparative softwares were used in this study: SLAM [1], SGP-2 [85] and Twinscan [60]. SGP-2 is a heuristic comparative gene finder that combines the GeneID algorithm [47] with TBLASTX [41] alignments. Twinscan is a semi-comparative approach that extends the Genscan GHMM-algorithm by boosting the predictions using the alignment to a homologous informant sequence. Twinscan is described a little more in Sect. 4.1.2. SLAM uses a generalized pair HMM (GPHMM), which is a merging of the GHMM-algorithm used in Genscan, and the PHMM often used for sequence alignment, and is described briefly in Sect. 4.4.2 and from an implementary point of view in Chap. 7.

A different approach that utilizes the increasing amount of well-annotated sequences, is that of gene mapping, most notably applied in Projector [78] and GeneMapper [22]. While gene mapping traditionally means the mapping of DNA sequences onto their corresponding chromosomes, in gene finding terminology it means the mapping, or projection, of annotated genes of one organism onto a new homologous sequence of another organism. Projector uses a version of the PHMM-algorithm in DoubleScan [77], but instead of searching for the optimal parse among all possible, it is limited to find a parse that agrees with the provided annotated sequence. GeneMapper is similar, but can be used both in a pairwise and a multiple setting. While the pairwise version uses regular dynamic programming to map the genes, the multiple version works with profiles. Both Projector and GeneMapper are discussed further in Sect. 4.5.

While most comparative gene finders are performing pairwise comparisons, a natural next step would be to make multiple comparisons. This has turned out to be very complex computationally, however, such that no direct extension of the pairwise methods to three-ways or higher has been possible yet. Attempts along these lines have been made, though, for instance in N-SCAN [46] and DOGFISH [19], where instead of using a single informant sequence to boost the predictions, a multiple alignment of several informant sequences are used. N-SCAN is described further in Sect. 4.6.1. Neither N-SCAN or DOGFISH are truly comparative in the sense that they do not perform gene prediction in multiple sequences simultaneously, but merely uses multiple alignments of homologous sequences as informants. An interesting future development would be to be able to make full-fledged multiple sequence annotations. Perhaps with more clever algorithms and even more powerful computers this can be made possible not too far off in the future.

Table 1.2 lists a large bulk of the gene finding softwares that have been introduced over the years. Unnamed algorithms have been excluded, and for all the softwares out there that have been forgotten in this list, I sincerely apologize.

Table 1.2 Gene finding softwares

Year	Software	Description
1982	TESTCODE [33]	Statistical pattern analysis of ORFs
1984	ANALYSEQ [99]	Combining statistical pattern sensors with basic signal sensors
1990	GeneModeler (gm) [34]	Rule-based gene finder
1991	GeneID [47]	Hierarchically structured rule-based gene finder
	NetGene [17]	Splice site detection in pre-mRNA using neural networks
1992	GRAIL [110]	Neural network-based gene finder
	SORFIND [54]	Prediction of internal exons in human using statistical patterns
1993	GENMARK [15]	Markov chain-based gene finder
	GeneParser [95, 96]	Neural network-based gene finder integrating external homology
	GREAT [39]	Vector dynamic programming-based gene finder
1994	EcoParse [65]	HMM-based gene finder for *E. coli*
	FGENEH [97]	Discriminant analysis-based gene finder for human
	GenLang [31]	Generative grammar-based gene finder
1996	Genie [66, 67]	GHMM-based gene finder
	Procrustes [38]	Similarity-based gene finder using spliced alignments
1997	GeneWise [12]	Similarity-based HMM gene finder
	Genscan [18]	GHMM-based gene finder
	HMMgene [62, 63]	HMM-based gene finder
	MZEF [114]	Quadratic discriminant analysis-based gene finder
	VEIL [50]	HMM-based gene finder
1998	GLIMMER [30, 90]	Interpolated Markov model-based microbial gene finder
	MORGAN [89]	Decision tree-based gene finder
	OPRHEUS [36]	Similarity-based gene finder for bacterial genomes
	Pombe [23]	Discriminant analysis-based gene finder for fission yeast
	SelfID [3]	Iterative Markov model-based microbial gene finder
1999	CRITICA [5]	Similarity-based prokaryotic gene finder
	GeneBuilder [79]	Similarity-based gene finder
2000	CEM [6]	Similarity-based gene finder
	FGENESH [88]	HMM-based gene finder for human
	ROSETTA [7]	Comparative heuristic gene finder
2001	EuGene [91]	Eukaryotic gene finder that integrates arbitrary sources using a directed acyclic graph
	GeneMarkS [9]	HMM-based prokaryotic gene finder
	Phat [20]	GHMM-based gene finder for Plasmodium falciparum
	Pro-Frame [80]	Similarity-based gene finder using spliced alignments

(continued)

Table 1.2 (continued)

Year	Software	Description
	Pro-Gen [83]	Similarity-based gene finder using spliced alignments
	SGP-1 [107]	Similarity-based gene finder
	Twinscan [60]	GHMM-based comparative gene finder
2002	DoubleScan [77]	PHMM-based comparative gene finder
	GAZE [51]	Heuristic gene finder integrating arbitrary number of sensors
2003	AGenDA [103]	Similarity-based gene finder
	Augustus [101]	HMM-based gene finder
	DIGIT [112]	Bayesian-based gene finder combining the results of several gene finder
	EasyGene [69]	HMM-based gene finder with dynamical state space generation
	EvoGen [86]	Comparative HMM-based gene finder that models both gene structure and evolution
	Exonomy [72]	GHMM-based gene finder
	GlimmerM [72]	Gene finder using interpolated Markov models and decision trees
	SGP-2 [85]	Comparative gene finder based on GeneID and TBLASTX
	SLAM [1]	GPHMM-based comparative gene finder
	Unveil [72]	HMM-based gene finder
	ZCURVE [48]	Prokaryotic gene finder that uses the Z-curve representation of DNA
2004	EGPred [55]	Similarity-based gene finder
	Ensembl [28, 35]	Similarity-based gene annotation system
	eShadow [84]	Comparative gene finder using phylogenetic shadowing
	GenomeWise [11]	Similarity-based gene finder
	GlimmerHMM [74]	HMM- and IMM-based gene finder
	Phylo-HMM [93]	Comparative gene finder using phylogenetic HMMs
	Projector [78]	Similarity-based gene finder using gene mapping
	SNAP [59]	HMM-based gene finder
	TigrScan [74]	GHMM-based gene finder
2005	ExonHunter [16]	Similarity-based HMM gene finder
	GeneMark.HMM-E [70]	HMM-based eukaryotic gene finder using non-supervised learning
	GenomeThreader [45]	Similarity-based gene finder using spliced alignments
	JIGSAW [2]	Heuristic gene finder combining multiple sources of evidence
	N-SCAN [46]	Extension of Twinscan to multiple informant sequences
	TWAIN [73, 75]	GPHMM-based comparative gene finder

(continued)

Table 1.2 (continued)

Year	Software	Description
2006	Agene [82]	GHMM-based gene finder using ADPH state length distributions
	DOGFISH [19]	Similarity-based gene finder using multiple informant sequences
	GeneAlign [52]	Similarity-based gene finder
	GeneMapper [22]	Similarity-based gene finder using gene mapping
	shortHMM [109]	GHMM-based exon prediction
2007	Conrad [29]	Comparative gene finder using semi-Markov conditional random fields
2009	mGene [92]	Gene finder based on GHMMs and support vector machines

1.9 To Build a Gene Finder

The main steps in constructing a gene finding algorithm can be summarized as
follows:

1. Choose a basic mathematical model, such as a hidden Markov model, neural
   network, or decision tree. This model will then serve as an umbrella that integrates
   a number of subcomponents of the gene into a final prediction. Such models used
   for single species gene finding are presented in Chap. 2, and those for comparative
   gene finding in Chap. 4.
2. Determine a gene model, including all subcomponents it should contain. In this
   step we construct the state space of our mathematical model. It may seem that
   this step should come first, but the choice of the main mathematical model will
   affect the appearance of the state space.
3. Choose submodels for the different components in the state space, such as exon
   and intron models, splice site detectors, etc. Different submodels are presented
   in Chap. 5.
4. Train the entire model, including all submodels. For this we collect a training
   set of known genes and estimate the parameters of our model. Various training
   algorithms are presented in Chap. 6.

Faced with a novel sequence it is common to preprocess it before plugging it into the
gene finder. At the very least it is useful to *repeat mask* the sequence. The universally
used software for this is called RepeatMasker [94]. The program compares the input
sequence to a library of repetitive elements and returns an output sequence where
repeats and regions of low complexity have been "masked" by replacing these regions
by N's in the input sequence. The altered stretches signify regions unlikely to harbour
a gene, and the masked output sequence is ready for use by the gene finder. If the
gene finder is comparative, taking two homologous sequences as input, an additional
possible preprocessing step is to limit the search space in some manner. SLAM [1],

for instance, performs an initial approximate alignment that limits the search space to a region in the dynamic programming matrix most likely to contain the optimal path. Other efforts to reduce the computational complexity appear in Chap. 3 in the context of multiple alignments.

Once the gene finder is constructed and trained and the input sequences have gone through the preprocessing, it is ready to perform the gene prediction. Typically we begin by running the model on a test set, where we know the true answer, and then assess the accuracy of our model. The final outcome is a *parse* of the input sequence, meaning an ordered list of states predicted for the sequence. The development steps 1–5 above are likely to be iterated several times, as the accuracy assessment will help detect problems in the algorithm or areas of the model that have room for improvement. Various measures of accuracy assessment are described in Sect. 4.4.2.

References

1. Alexandersson, M., Cawley, S., Pachter, L.: SLAM: cross-species gene finding and alignment with a generalized pair hidden Markov model. Genome Res. **13**, 496–502 (2003)
2. Allen, J.E., Salzberg, S.L.: JIGSAW: integration of multiple sources of evidence for gene prediction. Bioinformatics **21**, 3596–3603 (2005)
3. Audic, S., Claverie, J.-M.: Self-identification of protein-coding regions in microbial genomes. Proc. Natl. Acad. Sci. USA **95**, 10026–10031 (1998)
4. Axelson-Fisk, M., Sunnerhagen, P.: Comparative genomics and gene finding in fungi. In: Sunnerhagen, P., Piskur, J. (eds.) Topics in Current Genetics: Comparative Genomics Using Fungi as Models, pp. 1–28. Springer, Berlin (2005)
5. Badger, J.H., Olsen, G.J.: CRITICA: coding region identification tool invoking comparative analysis. Mol. Biol. Evol. **16**, 512–524 (1999)
6. Bafna, V., Huson, D.H.: The conserved exon method for gene finding. Int. Conf. Intell. Syst. Mol. Biol. **8**, 3–12 (2000)
7. Batzoglou, S., Pachter, L., Mesirov, J., Berger, B., Lander, E.S.: Human and mouse gene structure: comparative analysis and application to exon prediction. Genome Res. **10**, 950–958 (2000)
8. Beadle, G., Tatum, E.: Genetic control of biochemical reactions in Neurospora. Proc. Natl. Acad. Sci. USA **27**, 499–506 (1941)
9. Besemer, J., Lomsadze, A., Borodovsky, M.: GeneMarkS: a self-training method for prediction of gene starts in microbial genomes. Implications for finding sequence motifs in regulatory regions. Nucleic Acids Res. **29**, 2607–2618 (2001)
10. Biémont, C., Vieira, C.: Junk DNA as an evolutionary force. Nature **443**, 521–524 (2006)
11. Birney, E., Clamp, M., Durbin, R.: GeneWise and GenomeWise. Genome Res. **14**, 988–995 (2004)
12. Birney, E., Durbin, R.: Dynamite: a flexible code generating system for dynamic programming methods used in sequence comparison. Proc. Int. Conf. Intell. Syst. Mol. Biol. **5**, 56–64 (1997)
13. Blandin, G., Durrens, P., Tekaia, F., Aigle, M., Bolotin-Fukuhara, M., Bon, E., Casarégola, S., de Montigny, J., Gaillardin, C., Lépingle, A., Llorente, B., Malpertuy, A., Neuvéglise, C., Ozier-Kalogeropoulus, O., Perrin, A., Potier, S., Souciet, J.-L., Talla, E., Toffano-Nioche, C., Wésolowski-Louvel, M., Marck, C., Dujon, B.: Genomic exploration of the hemiascomycetous yeasts: 4. The genome of Saccharomyces cerevisiae revisited. FEBS Lett. **487**, 31–36 (2000)
14. Boeckmann, B., Bairoch, A., Apweiler, R., Blatter, M.C., Estreicher, A., Gasteiger, E., Martin, M.J., Michoud, K., O'Donovan, C., Phan, I., Pilbout, S., Schneider, M.: The SWISS-PROT

protein knowledgebase and its supplement TrEMBL in 2003. Nucleic Acids Res. **31**, 365–370 (2003)

15. Borodovsky, M., McIninch, J.: GENMARK: parallel gene recognition for both DNA strands. Comput. Chem. **17**, 123–133 (1993)

16. Brejova, B., Brown, D.G., Li, M., Vinar, T.: ExonHunter: a comprehensive approach to gene finding. Bioinformatics **21**, i57–i65 (2005)

17. Brunak, S., Engelbrecht, J., Knudsen, S.: Prediction of human mRNA donor and acceptor sites from the DNA sequence. J. Mol. Biol. **220**, 49–65 (1991)

18. Burge, C., Karlin, S.: Prediction of complete gene structures in human genomic DNA. J. Mol. Biol. **268**, 78–94 (1997)

19. Carter, D., Durbin, R.: Vertebrate gene finding from multiple-species alignments using a two-level strategy. Genome Biol. **7**, S6.1–S6.12 (2006)

20. Cawley, S.E., Wirth, A.I., Speed, T.P.: Phat—a gene finding program for Plasmodium falciparum. Mol. Biochem. Parasitol. **118**, 167–174 (2001)

21. Cebrat, S., Dudek, M.R., Machiewicz, P., Kowalczuk, M., Fita, M.: Asymmetry of coding versus noncoding strand in coding sequences of different genomes. Microb. Comp. Genomics **2**, 259–268 (1997)

22. Chatterji, S., Pachter, L.: Reference based annotation with GeneMapper. Genome Biol. **7**, R29 (2006)

23. Chen, T., Zhang, M.Q.: Pombe: a gene-finding and exon-intron structure prediction system for fission yeast. Yeast **14**, 701–710 (1998)

24. Cherry, J.M., Adler, C., Ball, C., Chervitz, S.A., Dwight, S.S., Hester, E.T., Jia, Y., Juvik, G., Roe, T., Schroeder, M., Weng, S., Botstein, D.: SGD: saccharomyces genome database. Nucleic Acids Res. **26**, 73–79 (1998)

25. Claverie, J.M.: Gene number: what if there are only 30,000 human genes? Science **291**, 1255–1257 (2001)

26. Comings, D.E.: The structure and function of chromatin. Adv. Hum. Genet. **3**, 237–431 (1972)

27. Crick, F.: Cetnral dogma of molecular biology. Nature **227**, 561–563 (1970)

28. Curwen, V., Eyras, E., Andrews, T.D., Clarke, L., Mongin, E., Searle, S.M.J., Clamp, M.: The ensembl automatic gene annotation system. Genome Res. **14**, 942–950 (2004)

29. DeCaprio, D., Vinson, J.P., Pearson, M.D., Montgomery, P., Doherty, M., Galagan, J.E.: Conrad: gene prediction using conditional random fields. Genome Res. **17**, 1389–1398 (2007)

30. Delcher, A.L., Harmon, D., Kasif, S., White, O., Salzberg, S.L.: Improved microbial gene identification with GLIMMER. Nucleic Acids Res. **27**, 4636–4641 (1999)

31. Dong, S., Searls, D.B.: Gene structure prediction by linguistic models. Genomics **23**, 540–551 (1994)

32. The FANTOM consortium and RIKEN genome exploration research group and genome science group (genome network project core group). Science **309**, 1559–1563 (2005)

33. Fickett, J.W.: Recognition of protein coding regions in DNA sequences. Nucleic Acids Res. **10**, 5303–5318 (1982)

34. Fields, C.A., Söderlund, C.A.: GM: a practical tool for automating DNA sequence analysis. Comput. Appl. Biosci. **6**, 263–270 (1990)

35. Flicek, P., Aken, B.L., Beal, K., Ballester, B., Caccamo, M., Chen, Y., Clarke, L., Coates, G., Cunningham, F., Cutts, T., Down, T., Dyer, S.C., Eyre, T., Fitzgerald, S., Fernandez-Banet, J., Grf, S., Haider, S., Hammond, M., Holland, R., Howe, K.L., Howe, K., Johnson, N., Jenkinson, A., Khri, A., Keefe, D., Kokocinski, F., Kulesha, E., Lawson, D., Longden, I., Megy, K., Meidl, P., Overduin, B., Parker, A., Pritchard, B., Prlic, A., Rice, S., Rios, D., Schuster, M., Sealy, I., Slater, G., Smedley, D., Spudich, G., Trevanion, S., Vilella, A.J., Vogel, J., White, S., Wood, M., Birney, E., Cox, T., Curwen, V., Durbin, R., Fernandez-Suarez, X.M., Herrero, J., Hubbard, T.J., Kasprzyk, A., Proctor, G., Smith, J., Ureta-Vidal, A., Searle, S.: Ensembl 2008. Nucleic Acids Res. **36**, D707–D714 (2008)

36. Frishman, D., Mironov, A., Mewes, H.-W., Gelfand, M.: Combining diverse evidence for gene recognition in completely sequenced bacterial genomes. Nucleic Acids Res. **26**, 2941–2947 (1998)

37. Gelfand, M.S.: Computer prediction of the exon-intron structure of mammalian pre-mRNAs. Nucleic Acids Res. **18**, 5865–5869 (1990)
38. Gelfand, M.S., Mironov, A.A., Pevzner, P.A.: Gene recognition via spliced sequence alignment. Proc. Natl. Acad. Sci. USA **93**, 9061–9066 (1996)
39. Gelfand, M.S., Roytberg, M.A.: Prediction of the exon-intron structure by a dynamic programming approach. BioSystems **30**, 173–182 (1993)
40. Gerstein, M.B., Bruce, C., Rozowsky, J.S., Zheng, D., Du, J., Korbel, J.O., Emanuelsson, O., Zhang, Z.D., Wiessman, S., Snyder, M.: What is a gene, post-ENCODE? History and updated definition. Genome Res. **17**, 669–681 (2007)
41. Gish, W., States, D.J.: Identification of protein coding regions by database similarity search. Nat. Genet. **3**, 266–272 (1993)
42. Goffeau, A., Barrell, B.G., Bussey, H., Davis, R.W., Dujon, B., Feldmann, H., Galibert, F., Hoheisel, J.D., Jacq, C., Johnston, M., Louis, E.J., Mewes, H.W., Murakami, Y., Philippsen, P., Tettelin, H., Oliver, S.G.: Life with 6000 genes. Science **274**, 563–567 (1996)
43. Gregory, T.R.: Coincidence, coevolution, or causation? DNA content, cell size, and the C-value enigma. Biol. Rev. **76**, 65–101 (2001)
44. Gregory, T.R.: The C-value enigma in plants and animals: a review of parallels and an appeal for partnership. Ann. Bot. **95**, 133–146 (2005)
45. Gremme, G., Brendel, V., Sparks, M.E., Kurtz, S.: Engineering a software tool for gene structure prediction in higher organisms. Inf. Softw. Tech. **47**, 965–978 (2005)
46. Gross, S.S., Brent, M.R.: Using multiple alignments to improve gene prediction. J. Comput. Biol. **13**, 379–393 (2006)
47. Guigó, R., Knudsen, S., Drake, N., Smith, T.: Prediction of gene structure. J. Mol. Biol. **226**, 141–157 (1992)
48. Guo, F.-B., Ou, H.-Y., Zhang, C.-T.: ZCURVE: a new system for recognizing protein-coding genes in bacterial and archaeal genomes. Nucleic Acids Res. **31**, 1780–1789 (2003)
49. Harrison, P.M., Kumar, A., Lang, N., Snyder, M., Gerstein, M.: A question of size: the eukaryotic proteome and the problems in defining it. Nucleic Acids Res. **30**, 1083–1090 (2002)
50. Henderson, J., Salzberg, S., Fasman, K.H.: Finding genes in DNA with a hidden Markov model. J. Comput. Biol. **4**, 127–141 (1997)
51. Howe, K.L., Chothia, T., Durbin, R.: GAZE: a generic framework for the integration of gene-prediction data by dynamic programming. Genome Res. **12**, 1418–1427 (2002)
52. Hsieh, S.J., Lin, C.Y., Liu, N.H., Chow, W.Y., Tang, C.Y.: GeneAlign: a coding exon prediction tool based on phylogenetical comparisons. Nucleic Acids Res. **34**, W280–W284 (2006)
53. Human genome sequencing consortium: initial sequencing and analysis of the human genome. Nature **409**, 745–964 (2002)
54. Hutchinson, G.B., Hayden, M.R.: The prediction of exons through an analysis of spliceable open reading frames. Nucleic Acids Res. **20**, 3453–3462 (1992)
55. Issac, B., Raghava, G.P.S.: EGPred: prediction of eukaryotic genes uisng ab initio methods after combining with sequence similarity approaches. Genome Res. **14**, 1756–1766 (2004)
56. Kanno, H., Huang, I.-Y., Kan, Y.W., Yoshida, A.: Two structural genes on different chromosomes are required for encoding the major subunit of human red cell glucose-6-phosphate dehydrogenase. Cell **58**, 595–606 (1989)
57. Kellis, M., Patterson, N., Endrizzi, M., Birren, B., Lander, E.S.: Sequencing and comparison of yeast species to identify genes and regulatory elements. Nature **423**, 241–254 (2003)
58. Kim, H., Klein, R., Majewski, J., Ott, J.: Estimating rates of alternative splicing in mammals and invertebrates. Nat. Genet. **36**, 915–917 (2004)
59. Korf, I.: Gene finding in novel genomes. BMC Bioinform. **5**, 59 (2004)
60. Korf, I., Flicek, P., Duan, D., Brent, M.R.: Integrating genomic homology into gene structure prediction. Bioinformatics **17**, S140–S148 (2001)
61. Kowalczuk, M., Mackiewicz, P., Gierlik, A., Dudek, M.R., Cebrat, S.: Total number of coding open reading frames in the yeast genome. Yeast **15**, 1031–1034 (1999)
62. Krogh, A.: Two methods for improving performance of an HMM and their application for gene finding. Proc. Int. Conf. Intell. Syst. Mol. Biol. **5**, 179–186 (1997)

63. Krogh, A.: Using database matches with HMMGene for automated gene detection in Drosophila. Genome Res. **10**, 523–528 (2000)
64. Krogh, A., Brown, M., Mian, I.S., Sjölander, K., Haussler, D.: Hidden Markov models in computational biology: applications to protein modeling. J. Mol. Biol. **235**, 1501–1531 (2002)
65. Krogh, A., Mian, I.S., Haussler, D.: A hidden Markov model that finds genes in E.coli DNA. Nucleic Acids Res. **22**, 4768–4778 (1994)
66. Kulp, D., Haussler, D., Reese, M.G., Eeckman, F.H.: A generalized hidden Markov model for the recognition of human genes in DNA. Proc. Int. Conf. Intell. Syst. Mol. Biol. **4**, 134–142 (1996)
67. Kulp, D., Haussler, D., Reese, M.G., Eeckman, F.H.: Integrating database homology in a probabilistic gene structure model. Pac. Symp. Biocomput. **2**, 232–244 (1997)
68. Kumar, A., Harrison, P.M., Cheung, K.-H., Lan, N., Echols, N., Bertone, P., Miller, P., Gerstein, M.B., Snyder, M.: An integrated approach for finding overlooked genes in yeast. Nat. Biotech. **20**, 58–63 (2002)
69. Larsen, T.S., Krogh, A.: Easy-Gene—a prokaryotic gene finder that ranks ORFs by statistical significance. BMC Bioinform. **4**, 21–35 (2003)
70. Lomsadze, A., Ter-Hovhannisyan, V., Chernoff, Y.O., Borodovsky, M.: Gene identification in novel eukaryotic genomes by self-traning algorithm. Nucleic Acids Res. **33**, 6494–6506 (2005)
71. Mackiewicz, P., Kowalczuk, M., Mackiewicz, D., Nowicka, A., Dudkiewicz, M., Laszkiewicz, A., Dudek, M.R., Cebrat, S.: How many protein-coding genes are there in the Saccharomyces cerevisiae genome? Yeast **19**, 619–629 (2002)
72. Majoros, W.H., Pertea, M., Antonescu, C., Salzberg, S.L.: GlimmerM, Exonomy and Unveil: three ab initio eukaryotic gene finders. Nucleic Acids Res. **31**, 3601–3604 (2003)
73. Majoros, W.H., Pertea, M., Delcher, A.L., Salzberg, S.L.: Efficient decoding algorithms for generalized hidden Markov model gene finders. BMC Bioinform. **6**, 16–28 (2005)
74. Majoros, W.H., Pertea, M., Salzberg, S.L.: TigrScan and GlimmerHMM: two open source ab initio eukaryotic gene finders. Bioinformatics **20**, 2878–2879 (2004)
75. Majoros, W.H., Pertea, M., Salzberg, S.L.: Efficient implementation of a generalized pair hidden Markov model for comparative gene finding. Bioinformatics **21**, 1782–1788 (2005)
76. Mewes, H.W., Heumann, K., Kaps, A., Mayer, K., Pfeiffer, F., Stocker, S., Frishman, D.: MIPS: a database for genomes and protein sequences. Nucleic Acids Res. **27**, 44–48 (1999)
77. Meyer, I.M., Durbin, R.: Comparative ab initio prediction of gene structures using pair HMMs. Bioinformatics **18**, 1309–1318 (2002)
78. Meyer, I.M., Durbin, R.: Gene structure conservation aids similarity based gene prediction. Nucleic Acids Res. **32**, 776–783 (2004)
79. Milanesi, L., D'Angelo, D., Rogozin, I.B.: GeneBuilder: interactive in silico prediction of gene structure. Bioinformatics **15**, 612–621 (1999)
80. Mironov, A.A., Noivchkov, P.S., Gelfand, M.S.: Pro-Frame: similarity-based gene recognition in eukaryotic DNA sequences with errors. Bioinformatics **17**, 13–15 (2001)
81. Mouse Genome Sequencing Consortium: Initial sequencing and comparative analysis of the mouse genome. Nature **420**, 520–562 (2002)
82. Munch, K., Krogh, A.: Automatic generation of gene finders for euakryotic species. BMC Bioinform. **7**, 263–274 (2006)
83. Novichkov, P.S., Gelfand, M.S., Mironov, A.A.: Gene recognition in eukaryotic DNA by comparison of genomic sequences. Bioinformatics **17**, 1011–1018 (2001)
84. Ovcharenko, I., Boffelli, D., Loots, G.G.: eShadow: a tool for comparing closely related sequences. Genome Res. **14**, 1191–1198 (2004)
85. Parra, G., Agarwal, P., Abril, J.F., Wiehe, T., Fickett, J.W., Guigó, R.: Comparative Gene Prediction in Human and Mouse. Genome Res. **13**, 108–117 (2003)
86. Pedersen, J.S., Hein, J.: Gene finding with a hidden Markov model of genome structure and evolution. Bioinformatics **19**, 219–227 (2003)
87. RIKEN genome exploration research group and genome science group (genome network project core group) and the FANTOM consortium. Science **309**, 1564–1566 (2005)

88. Salamov, A.A., Solovyev, V.V.: Ab initio gene finding in Drosophila genomic DNA. Genome Res. **10**, 516–522 (2000)
89. Salzberg, S.L., Delcher, A.L., Fasman, K.H., Henderson, J.: A decision tree system for finding genes in DNA. J. Comput. Biol. **5**, 667–680 (1998)
90. Salzberg, S.L., Delcher, A.L., Kasif, S., White, O.: Microbial gene identification using interpolated Markov models. Nucleic Acids Res. **26**, 544–548 (1998)
91. Schiex, T., Moisan, A., Rouzé, P.: EuGene: an eucaryotic gene finder that combines several sources of evidenc. In: Gascuel, O., Sagot, M.-F. (eds.) Computational Biology, pp. 111–125. Springer, Berlin (2001)
92. Schweikert, G., Zien, A., Zeller, G., Behr, J., Dieteric, C., Ong, C.S., Philips, P., De Bona, F., Hartmann, L., Bohlen, A., Krüger, N., Sonnenburg, S., Rätsch, G.: mGene: accurate SVM-based gene finding with an application to nematode genomes. Genome Res. June 29 Epub (2009)
93. Siepel, A., Haussler, D.: Computational identification of evolutionary conserved exons. RECOMB **8**, 177–186 (2004)
94. Smit, A.F.A., Hubley, R., Green, P.: RepeatMasker. http://www.repeatmasker.org
95. Snyder, E.E., Stormo, G.D.: Identification of coding regions in genomic DNA sequences: an application of dynamic programming and neural networks. Nucleic Acids Res. **21**, 607–613 (1993)
96. Snyder, E.E., Stormo, G.D.: Identification of protein coding regions in genomic DNA. J. Mol. Biol. **248**, 1–18 (1995)
97. Solovyev, V.V., Salamov, A.A., Lawrence, C.B.: Predicting internal exons by oligonucleotide composition and discrimant analysis of spliceable open reading frames. Nucleic Acids Res. **22**, 5156–5163 (1994)
98. Southan, C.: Has the yo-yo stopped? an assessment of human protein-coding gene number. Proteomics **4**, 1712–1726 (2004)
99. Staden, R.: Computer methods to locate signals in nucleic acid sequences. Nucleic Acids Res. **12**, 505–519 (1984)
100. Staden, R., McLachlan, A.D.: Codon preference and its use in identifying protein coding regions in long DNA sequences. Nucleic Acids Res. **10**, 141–156 (1982)
101. Stanke, M., Waack, S.: Gene prediction with a hidden Markov model and a new intron sub-model. Bioinformatics **19**, ii215–ii225 (2003)
102. Swift, H.: The constancy of desoxyribose nucleic acid in plant nuclei. Proc. Natl. Acad. Sci. USA **36**, 643–654 (1950)
103. Taher, L., Rinner, O., Garg, S., Sczyrba, A., Brudno, M., Batzoglou, S., Morgenstern, B.: AGenDA: homology-based gene prediction. Bioinformatics **19**, 1575–1577 (2003)
104. Vendrely, R., Vendrely, C.: La teneur du noyau cellulaire en acide désoxyribonucléique à travers les organes, les individus et les espéces animales : Techniques et premiers résultats. Experientia **4**, 434–436 (1948)
105. Wade, N.: Gene sweepstakes end, but winner may well be wrong. New York Times, 3 June 2003
106. Wain, H.M., Bruford, E.A., Lovering, E.C., Lush, M.J., Wright, M.W., Povey, S.: Guidelines for human gene nomenclature. Genomics **79**, 464–470 (2002)
107. Wiehe, T., Gebauer-Jung, S., Mitchell-Olds, T., Guigó, R.: SGP-1: prediction and validation of homologous genes based on sequence alignments. Genome Res. **11**, 1574–1583 (2001)
108. Wood, V., Rutherford, K.M., Ivens, A., Rajandream, M.-A., Barrell, B.: A re-annotation of the Saccharomyces cerevisiae genome. Comp. Funct. Genomics **2**, 143–154 (2001)
109. Wu, J., Haussler, D.: Coding exon detection using comparative sequences. J. Comput. Biol. **13**, 1148–1164 (2006)
110. Xu, Y., Mural, R.J., Einstein, J.R., Shah, M.B., Uberbacher, E.C.: GRAIL: a multi-agent neural network system for gene identification. Proc. IEEE **84**, 1544–1552 (1996)
111. Xu, Y., Uberbacher, E.C.: In: Salzberg, S.L., Searls, D.B., Kasif, S. (eds.) Computational Methods in Molecular Biology, pp. 109–128. Elsevier Science B.V., Amsterdam (1998)

112. Yada, T., Takagi, T., Totoki, Y., Sakaki, Y., Takaeda, Y.: DIGIT: a novel gene finding program by combining gene-finders. Pac. Symp. Biocomput. **8**, 375–387 (2003)
113. Zhang, C.-T., Wang, J.: Recognition of protein coding genes in the yeast genome at better than 95 % accuracy based on the Z curve. Nucleic Acids Res. **28**, 2804–2814 (2000)
114. Zhang, M.Q.: Identification of protein coding regions in the human genome by quadratic discriminant analysis. Proc. Natl. Acad. Sci. USA **94**, 565–568 (1997)

Chapter 2
Single Species Gene Finding

A gene finding model usually consists of a main algorithm that serves as a kind of "umbrella" algorithm for a large number of rather complex submodels. The submodels represent various features of a gene, such as exons, introns, and splice site models. Each submodel scores the probability, or likelihood, that each given sequence region constitutes the corresponding gene feature, and then these scores are passed on up to the main algorithm. The main algorithm uses these scores as foundation for parsing the entire input sequence into complete gene structures that adhere to the biological rules implemented in the model. This chapter covers a range of mathematical models commonly used as main algorithms in single species gene finding. Similar models used for comparative gene finding are presented in Chap. 4, while the various kinds of submodels used for specific gene features are presented in Chap. 5.

2.1 Hidden Markov Models (HMMs)

One reason for the popularity of Markov models is that, due to their flexibility, most processes can be approximated by a Markov chain. Markov theory is a well-studied technique and includes a machinery of powerful algorithms to be used in data analysis. The word "chain" may indicate that the random process generates a discrete chain of events, but a Markov chain can evolve both in discrete and continuous time, and have either a discrete or continuous state space. The Markov chains we will consider here, however, will all have a discrete, finite state space. Moreover, since most Markov models presented in this book will be of discrete-time type, this will be our main focus in this section. But since continuous-time Markov models will be mentioned in connection with substitution models in pairwise alignments (Sect. 3.1), we give a brief account of that theory as well. For more details, see [16].

A powerful extension of the Markov theory are the hidden Markov models (HMMs). HMMs were originally developed for speech recognition, with one of the best references being the introduction by Rabiner [30]. Nowadays, HMMs have become an integral part of bioinformatics with applications including modeling the

© Springer-Verlag London 2015
M. Axelson-Fisk, *Comparative Gene Finding*, Computational Biology 20,
DOI 10.1007/978-1-4471-6693-1_2

periodic patterns occurring in a gene, the sequence alignment pairing of nucleotides and amino acids, and the point mutation process of sequence evolution. For a deeper and more general description of HMMs applied to bioinformatics, see [18].

2.1.1 Markov Chains

A *random process*, also called a *stochastic process*, is basically the evolution in time of some random variable. For instance, the mutation process in evolution can be seen as a random process. What makes the process random is that it jumps randomly between different *states* in a *state space*. A *Markov chain* is a random process which is "memoryless" in the sense that the next jump only depend on the current state, and not the past of the process. This property, called the *Markov property*, is described in more detail below.

We typically write a random process as a sequence of indexed random variables $(X_1, X_2, \ldots)$, where X_t is the *state* of the random process at time index $t \in T$. If the index set T is finite or countable, such as the integers, we call the process a *discrete-time* random process. If the indices come from a continuous set, such as an interval on the real line, the process is a *continuous-time* random process. The process evolves by jumping between the states in a state space S. Just as with the time index, the state space can be finite, countable, or continuous. Note that there is no initial assumption about independence between the random variables in the process. Different settings on the index set T, the state space S, and various interdependencies between the indices in the process make up a wide variety of random processes. Markov chains are thus a special case of this.

Discrete-Time Markov Chains

Consider a physical process that at any instant in time will reside in one of N possible states, call them $S = \{s_1, \ldots, s_N\}$. Assume that the process jumps between states at discrete time points $t = 1, 2, 3, \ldots$, and let X_t denote the state at time t. Using the definition of conditional probabilities, the probability of any sequence of random variables $(X_1, \ldots, X_T)$ can be for states $i_1, \ldots, i_T \in S$ be decomposed as

$$\mathbf{P}(X_1 = i_1, \ldots, X_T = i_T) = \tag{2.1}$$
$$= \mathbf{P}(X_T = i_T | X_{T-1} = i_{T-1}, X_{T-2} = i_{T-2}, \ldots, X_1 = i_1)$$
$$\cdot \mathbf{P}(X_{T-1} = i_{T-1} | X_{T-2} = i_{T-2}, \ldots, X_1 = i_1)$$
$$\cdots$$
$$\cdot \mathbf{P}(X_2 = i_2 | X_1 = i_1)$$
$$\cdot \mathbf{P}(X_1 = i_1).$$

The conditional probabilities in (2.1) represent the probabilities to jump from state X_t to X_{t+1}, possibly conditioning on all the past states. What characterizes a Markov chain, however, is that it is "memoryless". That is, given the current state, the future and the past of the process are independent. This feature, called the *Markov property*, can be formalized as follows:

Definition 2.1 The process $(X_1, X_2, \ldots)$ is a *Markov chain* if it for $i, j, i_1, \ldots, i_{t-2} \in S$ satisfies the *Markov property*

$$\mathbf{P}(X_t = j | X_{t-1} = i, X_{t-2} = i_{t-2}, \ldots, X_1 = i_1) = \mathbf{P}(X_t = j | X_{t-1} = i). \quad (2.2)$$

The probability of a sequence $(X_1, \ldots, X_T)$, generated by a Markov chain, thus becomes

$$\mathbf{P}(X_1 = i_1, \ldots, X_T = i_T) = \mathbf{P}(X_1 = i_1) \prod_{t=2}^{T} \mathbf{P}(X_t = i_t | X_{t-1} = i_{t-1}). \quad (2.3)$$

Definition 2.2 The probability of the first state X_1 is determined by the *initial distribution* $\boldsymbol{\pi} = \{\pi_1, \ldots, \pi_N\}$, where

$$\pi_i = \mathbf{P}(X_1 = i), \quad i \in S \quad \sum_{i=1}^{N} \pi_i = 1. \quad (2.4)$$

The chain proceeds according to the *transition matrix* $\mathbf{A} = (a_{ij})_{i,j \in S}$, which is an $(N \times N)$-matrix consisting of *transition probabilities*

$$a_{ij} = \mathbf{P}(X_t = j | X_{t-1} = i), \quad i, j \in S. \quad (2.5)$$

The transition matrix is *stochastic*, meaning that all entries are nonnegative $a_{ij} \geq 0$, and each row adds up to one

$$\sum_{j=1}^{N} a_{ij} = 1. \quad (2.6)$$

A Markov chain with transition probabilities as in (2.5) is said to be of *first order*, due to its dependency on only the previous state. This can be generalized, however, such that each state depends on several of the previous states. For instance, a *second*-order Markov chain depends on the previous two states, and has transition probabilities on the form

$$a_{ijk}^{(2)} = \mathbf{P}(X_t = k | X_{t-1} = j, X_{t-2} = i), \quad i, j, k \in S. \quad (2.7)$$

Just as in the first-order case, the transition probabilities are nonnegative and the rows sum up to one

$$\sum_{k=1}^{n} a_{ijk} = 1. \qquad (2.8)$$

To generalize further, a kth-order Markov chain depends on the k previous states and is defined as

$$a_{ij}^{(k)} = \mathbf{P}(X_t = j | X_{t-1} = i_1, X_{t-2} = i_2, \ldots, X_{t-k} = i_k). \qquad (2.9)$$

where $\mathbf{i} = (i_1, \ldots, i_k)$ and $i_1, \ldots, i_k, j \in S$. The sequence $X_{t-1}, \ldots, X_{t-k}$ is sometimes referred to as the *context* of X_t. Note that while a first-order Markov chain of N states has an $(N \times N)$ transition matrix, a kth-order Markov chain has an $(N^k \times N)$ transition matrix, with one row for each of the N^k possible contexts. A Markov chain of zeroth-order has no context and only consists of independent state frequencies π_i.

Example 2.1 Building a Markov chain from data
As a toy-example, consider a machine that generates a DNA sequence according to a first-order Markov chain. The state space $S=\{A,C,G,T\}$ is illustrated in Fig. 2.1 where the states are represented as circles, and the arrows between them represent the transitions. The machine starts in some state according to the initial distribution $\pi = \{\pi_A, \pi_C, \pi_G, \pi_T\}$, generates the corresponding DNA base, and then jumps to a new state according to the transition probabilities $a_{ij}, i, j \in S$.

Assume that the machine generated a DNA sequence of length $T = 24$,

$$\text{CCTCCCGGACCCTGGGCTCGGGAC}$$

By noting that

$$a_{ij} = \mathbf{P}(X_t = j | X_{t-1} = i) = \frac{\mathbf{P}(X_{t-1} = i, X_t = j)}{\mathbf{P}(X_{t-1} = i)}, \qquad (2.10)$$

we can deduce that a first-order Markov chain on $S=\{A,C,G,T\}$ models dinucleotide frequencies $\{AA, AC, AG, AT,\ldots, TG, TT\}$. Thus, by counting the number of times nucleotide i is followed by nucleotide j for all $i, j \in S$ in our sequence, we can estimate the model parameters by

$$\hat{\pi}_i = \frac{c_i}{\sum_k c_k} \qquad \hat{a}_{ij} = \frac{c_{ij}}{c_i}, \qquad (2.11)$$

Fig. 2.1 The state space of a DNA sequence generating machine

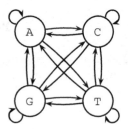

Table 2.1 The frequency counts and estimated model parameters

c_{ij}		To (j)				
		A	C	G	T	c_i
From (i)	A	0	2	0	0	2
	C	0	5	2	3	11
	G	2	1	5	0	8
	T	0	2	1	0	3

where c_i is the frequency count of the single residue i, and c_{ij} is the frequency count of the dinucleotide $\{ij\}$ for $i, j \in S$. The dinucleotide frequency counts of the observed sequence above are shown in Table 2.1. Note that since the sequence ends with a C, the C-row will not add up.

The estimated model parameters thus become

$$\hat{\pi} = (0.08, 0.46, 0.33, 0.13), \qquad \hat{\mathbf{A}} = \begin{pmatrix} 0.00 \ 1.00 \ 0.00 \ 0.00 \\ 0.00 \ 0.56 \ 0.22 \ 0.22 \\ 0.25 \ 0.12 \ 0.62 \ 0.00 \\ 0.00 \ 0.67 \ 0.33 \ 0.00 \end{pmatrix}. \tag{2.12}$$

Using the estimated model we can predict the next nucleotide in the sequence. For instance, given that $X_T = C$, there is an estimated 56 % chance that the next symbol is 'C'. Similarly, an entire new sequence can be scored based on this model. For instance

$$p(\text{CCTG}) = \tag{2.13}$$
$$= \mathbf{P}(X_1 = C)\mathbf{P}(X_2 = C|X_1 = C)\mathbf{P}(X_3 = T|X_2 = C)\mathbf{P}(X_4 = G|X_3 = T)$$
$$= \pi_C \cdot a_{CC} \cdot a_{CT} \cdot a_{TG}$$
$$= 0.0167.$$

Such scoring can be used to examine how characteristic a new sequence is to the given model and, for instance, to distinguish a coding sequence from a noncoding sequence. This is discussed further in Example 2.2. Probabilities of indices at longer distances in the process can be determined similarly by using

$$\mathbf{P}(X_{T+2} = C|X_T = C) = \tag{2.14}$$
$$= \sum_{k \in S} \mathbf{P}(X_{T+2} = C|X_{T+1} = k) \, \mathbf{P}(X_{T+1} = k|X_T = C)$$
$$= \sum_{k \in S} a_{Ck} \, a_{kC}$$
$$= 0.00 \cdot 0.00 + 0.56 \cdot 0.56 + 0.22 \cdot 0.12 + 0.22 \cdot 0.67$$
$$= 0.49. \qquad\qquad \square$$

In general, the *n-step transition matrix* $\mathbf{A}^{(n)} = (a_{ij}(n))_{i,j \in S}$, corresponding to the nth power of $\mathbf{A}$ represents the transitions from i to j in n steps, where

$$a_{ij}(n) = \sum_{k \in S} a_{kj} a_{ik}(n-1). \tag{2.15}$$

The previous example is an example of a Markov chain that does not vary over time. The transition probabilities are the same regardless of where we are in the sequence, that is, X_t is independent of how long the process has run.

Definition 2.3 A Markov chain is said to be *time-homogeneous* (or just *homogeneous*) if the following condition holds

$$\mathbf{P}(X_t = j | X_{t-1} = i) = \mathbf{P}(X_h = j | X_{h-1} = i) \text{ for } t \neq h. \tag{2.16}$$

and *inhomogeneous* otherwise.

Example 2.2 Markov chain classification of E. coli
The single most powerful method of discriminating between coding and noncoding sequences is to use the statistical differences in sequence patterns. We use the same model as in Example 2.1 with the state space shown in Fig. 2.1.

Assume that we want to use this model to discriminate between coding and noncoding sequences in the bacteria *Escherichia coli*. First, we use a training set of known coding and noncoding sequences to estimate the model parameters. Table 2.2 shows the dinucleotide frequencies and base counts for coding and noncoding sequences in the *E. coli* strain O157:H7 [26].
The probability of a new sequence $(X_1, \ldots, X_T)$ is given by

$$\mathbf{P}(X_1 = i_1, \ldots, X_T = i_T) = \pi_{i_1} \prod_{t=1}^{T-1} a_{i_t, i_{t+1}}. \tag{2.17}$$

The probabilities π and a_{ij} can be estimated as in (2.11) using the frequency counts in Table 2.2. Now, in order to test if the given sequence is coding or not, we can

Table 2.2 The dinucleotide frequency counts in *E. coli* O157:H7 coding and noncoding sequences

		Coding to (j)						Noncoding to (j)				
	c_{ij}	A	C	G	T	c_i	c_{ij}	A	C	G	T	c_i
From (i)	A	0.310	0.224	0.199	0.268	0.245	A	0.321	0.204	0.200	0.275	0.262
	C	0.251	0.215	0.313	0.221	0.243	C	0.282	0.233	0.269	0.215	0.239
	G	0.236	0.308	0.249	0.207	0.273	G	0.236	0.305	0.235	0.225	0.240
	T	0.178	0.217	0.338	0.267	0.239	T	0.207	0.219	0.259	0.314	0.259

calculate the probability in (2.17) for two different models, coding and noncoding, using the corresponding frequency counts in Table 2.2.

The two probabilities are then compared using a *likelihood-ratio test*, or a *log-odds ratio* decision rule

$$S(X) = \log \frac{\mathbf{P}_C(X_1 = i_1, \ldots, X_T = i_T)}{\mathbf{P}_N(X_1 = i_1, \ldots, X_T = i_T)} \begin{cases} > \eta & \Rightarrow \text{ coding,} \\ < \eta & \Rightarrow \text{ noncoding,} \end{cases} \tag{2.18}$$

where $\mathbf{P}_C$ is the probability when the parameters have been estimated using coding frequencies, and $\mathbf{P}_N$ the corresponding probability using noncoding frequencies. The threshold value η is chosen to satisfy a desired significance level (e.g., $\alpha = 0.05$). It is customary in sequence analysis to use logarithms of the probabilities to prevent the probabilities of long sequences from falling below computer precision and become numerically unstable. As a positive side effect products are transformed into sums, which results in a more efficient computation.

The decision rule in (2.18) is of course very crude. The (length-normalized) log-odds scores of coding versus noncoding sequences in *E. coli* are illustrated in Fig. 2.2. We see that while the peaks of the two distributions are separated, which is necessary in order to discriminate between the models, the overlap is significant, making it hard to separate coding sequence from noncoding sequence based on this score alone. Several improvements to the decision rule would be possible already at this early stage. For one thing, a more sensitive approach would utilize the fact that coding sequences are organized in codons. Thus, a quick fix would be to upgrade the above model to a second-order Markov chain, using transition probabilities trained on triplets rather than on dinucleotides.

Moreover, it is a known fact that the probability of a triplet in a coding region depends on its position with respect to the reading frame of the sequence. Thus, an even better model would be an *inhomogeneous* second-order Markov chain. We would then train four different Markov chains, one for each coding frame and one for noncoding sequences. An example of this is given in Sect. 5.3.5. □

Fig. 2.2 Distribution of log-odds ratio scores of length-normalized coding (*dark gray*) and noncoding (*light gray*) sequences in *E. coli*

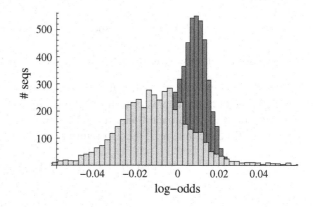

Example 2.2 illustrates a strategy for classifying an unknown DNA sequence into coding or noncoding. What we really want, however, is to extract one or several coding regions from a longer sequence consisting of intermediate stretches of non-coding regions. Furthermore, in organisms where splicing may occur, we would like to combine the coding regions into complete gene structures if possible. The Hidden Markov Model (HMM) theory, presented in the next section, provides a suitable framework for this.

Stationarity and Reversibility

An important question of Markov theory is the limit behavior of the chain. What are the characteristics of a process that has run for a long time? Although the chain itself will never converge toward a specific state (unless $a_{ii} = 1$ for some $i \in S$), the state distribution may still stabilize. More specifically, what is the probability of state i occurring when time goes to infinity? Will the behavior of the chain converge? We call a distribution over the state space $\tau = \{\tau_1, \ldots, \tau_N\}$ a *stationary distribution* if

(a) $\tau_i \geq 0$ for all i, and $\sum_i \tau_i = 1$.
(b) $\tau = \tau \mathbf{A}$, which is to say that $\tau_j = \sum_{i=1}^{N} \tau_i a_{ij}$ for all $j \in S$.

The stationary distribution is sometimes called the *invariant, equilibrium,* or *steady state* distribution. The concept of stationarity is central in Markov theory, since convergence toward a stationary distribution somehow guarantees that the process is well-behaved in some respect. The stationary distribution may or may not exist, and even if it exists, the process may or may not ever reach it. We need a couple of more concepts before we can state the requirement for a stationary distribution to exist.

Definition 2.4 We say that state $i \in S$ *communicates* with state $j \in S$, writing $i \rightarrow j$, if, starting from i, the probability of ever reaching state j is positive. That is, if $a_{ij}(m) > 0$ for some $m \geq 0$. We say that i and j *intercommunicates* if $i \rightarrow j$ and $j \rightarrow i$. Furthermore, we say that the state space is *irreducible* if all its states intercommunicate.

Definition 2.5 We call a state *recurrent* if the probability of eventually returning is 1. That is, if

$$\mathbf{P}(X_t = i \text{ for some } t > 1 | X_1 = i) = 1. \tag{2.19}$$

If this probability is strictly less than 1, we say that the state is *transient*.

Note that although we will return to a recurrent state with probability one, the expected time of return may very well be infinite. To make sure the expected return time is finite, we need an additional restriction on the recurrence. Starting in state $X_1 = i$, let T_i be the time until the first return to state i

$$T_i = \min\{t > 1 : X_t = i | X_1 = i\}. \tag{2.20}$$

Definition 2.6 We say that a recurrent state is *positive* if the expected time of return is finite $E[T_i] < \infty$.

Now we can state the following important result:

Theorem 2.1 *An irreducible chain has a stationary distribution τ if and only if all states are positive recurrent. In that case, τ is unique and is given by $\tau_i = 1/E[T_i]$.*

However, just because the stationary distribution exists, it is not guaranteed that the chain ever reaches it. For this we need an extra condition.

Definition 2.7 A state i is said to have *period $d(i)$* if any return to the state must occur in multiples of $d(i)$ time steps. Formally, the period of state i is defined as

$$d(i) = \gcd\{n : a_{ii}(n) > 0\}, \qquad (2.21)$$

where 'gcd' stands for the 'greatest common divisor'. We say that state i is *periodic* if $d(i) > 1$ and *aperiodic* otherwise. That is to say that $a_{ii}(n) = 0$ unless n is a multiple of $d(i)$. Furthermore, a Markov chain is said to be aperiodic if at least one of its states is aperiodic.

Theorem 2.2 *If the chain is irreducible and aperiodic, then for all $i, j \in S$*

$$a_{ij}(n) \rightarrow \frac{1}{E[T_i]} \quad as \ n \rightarrow \infty. \qquad (2.22)$$

Note that the limit in Eq. 2.22 is what gives the stationary distribution in Theorem 2.1. Thus, if the chain is irreducible and aperiodic with positive recurrent states, the transition probabilities converge to the stationary distribution.

Another useful property is *time reversibility*. Let $X_1, \ldots, X_T$ be an irreducible, positive recurrent Markov chain with initial probabilities π and transition matrix $\mathbf{A}$. Let $Y_1, \ldots, Y_T$ be the chain running in reverse, that is

$$Y_t = X_{T-t}. \qquad (2.23)$$

Then Y is a Markov chain as well, with transition probabilities b_{ij} say.

Definition 2.8 We say that X is *time eversible* if $a_{ij} = b_{ij}$ for all $i, j \in S$.

Thus, since

$$
\begin{aligned}
b_{ij} &= \mathbf{P}(Y_t = j | Y_{t-1} = i) \\
&= \mathbf{P}(X_{T-t} = j | X_{T-(t-1)} = i) \\
&= \frac{\mathbf{P}(X_{T-(t-1)} = i | X_{T-t} = j)\mathbf{P}(X_{T-t} = j)}{\mathbf{P}(X_{T-(t-1)} = i)} \\
&= a_{ji}\frac{\pi_j}{\pi_i}, \qquad (2.24)
\end{aligned}
$$

it holds that X is time reversible if and only if $\pi_i a_{ij} = \pi_j a_{ji}$.

Theorem 2.3 *For an irreducible chain, if there exists a distribution π such that*

$$0 \le \pi_i \le 1, \quad \sum_i \pi_i = 1, \quad \pi_i a_{ij} = \pi_j a_{ji} \text{ for all } i, j,$$

then the chain is time reversible, positive recurrent, and stationary with stationary distribution π.

The interpretation of time reversibility is that it is not possible to determine the direction of the process, or the order of states, just by observing the state sequence. This is a very useful property for substitution models in sequence alignment (see Sect. 3.1) as it allows us to model the distance between two evolutionary-related sequences by analyzing the process of evolving one into the other, rather than making inferences about the distance to some unknown common ancestor in between.

Continuous-Time Markov Chains

A continuous-time Markov chain is very similar to its discrete counterpart. It jumps between states in a state space $S = \{s_1, \ldots, s_N\}$ and is parametrized by its initial distribution and transition probabilities. The main difference is that instead of making the jumps at discrete time points, the chain makes a transition after having spent a continuous amount of time in the state. The time spent in a state is called the *holding time*, or *waiting time*. The holding time in discrete-time chains is thus always equal to 1, while for continuous-time processes the holding time is a continuous random variable.

Let $\{X(t) : t \ge 0\}$ be a continuous random process, indexed by the positive real numbers, and with a discrete state space $S = \{s_1, \ldots, s_N\}$. The process is Markov if it satisfies the *Markov property*, which for continuous-time processes translates to

$$\mathbf{P}(X(t_n) = j | X(t_0) = i_0, \ldots, X(t_{n-1}) = i_{n-1}) = \mathbf{P}(X(t_n) = j | X(t_{n-1}) = i_{n-1}),$$
$$(2.25)$$

for a sequence of times $t_1 < t_2 < \cdots t_n$ and for all $j, i_0, i_1, \ldots, i_{n-1} \in S$. Just as in the discrete case, the first state $X(t_0)$, $(t_0 = 0)$, is given by an initial distribution $\pi = \{\pi_1, \ldots, \pi_N\}$, but the transition probabilities now need to be parametrized by time as well. We denote the probability of making a transition from state i to state j between time points s and t, where $s < t$, as follows

$$a_{ij}(s, t) = \mathbf{P}(X(t) = j | X(s) = i), \quad s < t. \tag{2.26}$$

When the transition probabilities are independent of how long the process has run, we call the chain *time-homogeneous* (or just *homogeneous*). That is, for a homogeneous Markov process it holds that

$$a_{ij}(s, t) = a_{ij}(0, t - s) \text{ for all } i, j, s < t. \tag{2.27}$$

Henceforth, we write $a_{ij}(t) = a_{ij}(0, t)$ for the transition probability of a homogeneous chain over time period t, and let $\mathbf{A}(t) = \left(a_{ij}(t)\right)_{i,j \in S}$ denote the transition matrix for this time period. As in the discrete-time case the rows of the transition matrix sum to one

$$\sum_{j=1}^{N} a_{ij}(t) = 1, \tag{2.28}$$

and a time interval can be split up in smaller segments by

$$a_{ij}(s + t) = \sum_{k=1}^{N} a_{ik}(s) a_{kj}(t) = \sum_{k=1}^{N} a_{ik}(t) a_{kj}(s) \text{ if } s, t \geq 0. \tag{2.29}$$

If we assume that the transition probabilities $a_{ij}(t)$ are continuous functions of t, we can assume that for an infinitely small time interval "nothing happens." That is, as $h \downarrow 0$

$$a_{ij}(h) \to \begin{cases} 1 & \text{if } i = j, \\ 0 & \text{if } i \neq j, \end{cases} \tag{2.30}$$

and the transition matrix reduces to the identity matrix

$$\mathbf{A}(t) \to \mathbb{I} \text{ as } t \downarrow 0. \tag{2.31}$$

A difficulty that arises with continuous-time Markov chains is that we no longer have a clear notion of the rates of change. In the discrete-time theory the transition probabilities both represent the changes over unit times, as well as the rates of change between states. In a continuous setting, however, a time interval can be divided into infinitely many subintervals, such that while the transition probability $a_{ij}(t)$ gives us the probability of changing from state i to state j in time t, it does not tell us how *many* changes that have occurred in between.

We need some notion of the "instantaneous" rates of change. That is, assuming that $X(t) = i$, we would like to know the behavior of the process in a small time interval $(t, t + h)$, where $h > 0$ is very close to 0. Various things may happen during that time, but for a small enough h the events reduce to one of the two possibilities:

- Nothing happened with probability $a_{ii}(h) + o(h)$, implying that the state is the same at time t as at $t + h$.
- The chain made a single move to a new state with probability $a_{ij}(h) + o(h)$.

The $o(h)$ (little-o) is an error term that accounts for any extra, unobserved, transitions during the time interval. The term $o(h)$ basically states that for small enough h the probability of any extra events becomes negligible, and the probability of a particular transition is approximately proportional to h.

That is, there exist constants $\{\mu_{ij} : i, j \in S\}$ such that

$$a_{ij}(h) \approx \begin{cases} \mu_{ij}h & \text{if } i \neq j, \\ 1 + \mu_{ii}h & \text{if } i = j. \end{cases} \tag{2.32}$$

The matrix $\mathbf{Q} = (\mu_{ij})_{i,j}$ is called the *transition rate matrix*, also known as the *generator* of the transition matrix $\mathbf{A}(t)$. Note that $\mu_{ij} \geq 0$ if $i \neq j$, and $\mu_{ii} \leq 0$. The elements μ_{ij} for $i \neq j$ models the rate at which the chain enters state j from i, while $-\mu_{ii}$ models the rate at which the chain leaves state i. Moreover, when the chain leaves state i (with rate $-\mu_{ii}$), it must enter one of the other states, giving

$$\mu_{ii} = -\sum_{j \neq i} \mu_{ij}, \tag{2.33}$$

with the result that the rows of $\mathbf{Q}$ sum to 0. The relation between the rate matrix $\mathbf{Q}$ and the transition matrix $\mathbf{A}(t)$ can be deduced using the *forward equations*

$$\frac{da_{ij}(t)}{dt} = \sum_{k \in S} a_{ik}(t)q_{kj}, \tag{2.34}$$

or, similarly, using the *backward equations*

$$\frac{da_{ij}(t)}{dt} = \sum_{k \in S} q_{ik}a_{kj}(t). \tag{2.35}$$

Subject to the boundary condition $\mathbf{A}(0) = \mathbb{I}$, where $\mathbb{I}$ is the identity matrix, the forward and backward equations are given by

$$\mathbf{A}(t) = e^{\mathbf{Q}t} = \sum_{n=0}^{\infty} \frac{t^n}{n!} \mathbf{Q}^n, \tag{2.36}$$

where $\mathbf{Q}^n$ is the nth power of $\mathbf{Q}$. The properties of the transition rate matrix can be summarized as follows:

- The non-diagonal elements q_{ij} correspond to the probability per unit time of jumping from state i to state j.
- The row sums of the non-diagonal elements $q_i = -q_{ii}$ correspond to the total transition rate out of state i.
- The total transition rate q_i is also the rate at which the time to the next jump decreases. That is, the holding time of state i is exponentially distributed with parameter q_i.
- The number of jumps in a time interval is Poisson distributed with parameter q_i.

The transitions of a continuous-time Markov process can be viewed as an embedded discrete-time Markov chain, also known as the *jump process*.

The transition probability a_{ij} of the jump process, is the conditional probability of jumping from state i to state j, given that a transition occurs, and is given by

$$a_{ij} = \begin{cases} \dfrac{q_{ij}}{q_i} & \text{if } i \neq j, \\ 0 & \text{if } i = j. \end{cases} \tag{2.37}$$

Analogously to the discrete case, a distribution $\tau = \{\tau_1, \ldots, \tau_N\}$ on the state space is a *stationary distribution* if $\tau_i \geq 0$, $\sum_i \tau_i = 1$, and $\tau = \tau A(t)$ for all $t \geq 0$. In terms of the rate matrix, τ is a stationary distribution if and only if

$$\tau Q = 0. \tag{2.38}$$

Theorem 2.4 *Let X be an irreducible Markov chain with transition matrix* $A(t)$.

(a) *If there exists a stationary distribution* τ, *it is unique and* $a_{ij}(t) \to \tau_j$ *as* $t \to \infty$.
(b) *If there is no stationary distribution then* $a_{ij}(t) \to 0$ *as* $t \to \infty$.

2.1.2 Hidden Markov Models

While the observed output in a standard Markov model is simply the sequence of states, a *hidden* Markov model (HMM) is comprised of two interrelated random processes, a hidden process and an observed process. The hidden process is a Markov chain jumping between states as before, but it can only be observed via the observed process. The observed process generates output through random functions associated with the underlying hidden states, and is generally not Markov. In other words, given the current state the hidden process is independent of the observed process. The observed process, however, typically depends both on its previous outputs and on the hidden process. For our purposes we only need to treat HMMs with a finite state space and a discrete, finite-valued observed process producing a finite output sequence, but it may be noted that the theory is applicable to more general situations.

Example 2.3 A simple HMM
Assume we have two dice, A and B. Die A has six sides and generates numbers between 1 and 6, while die B only has four sides and generates numbers between 1 and 4 (Fig. 2.3). Assume that we roll the dice, one at a time, and switch between the dice according to a Markov chain. The state space of the Markov chain is thus $S = \{A, B\}$, and the observed outputs are the numbers we produce by rolling the die. Die A emits each number with probability $1/6$, and die B emits each number with probability $1/4$. We generate numbers from this model as follows:

Fig. 2.3 A two-state HMM, where the hidden states are the dice, and the observed outputs are the roll outcomes

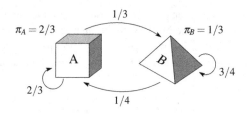

1. Choose initial die according to distribution $\{\pi_A, \pi_B\} = \{2/3, 1/3\}$.
2. Roll the die and observe the outcome.
3. Choose next die according to the transition probabilities a_{ij}, where i is the row and j the column index in the table below.

a_{ij}	A	B
A	2/3	1/3
B	1/4	3/4

4. Continue from 2.

Assume now that we only know the observed sequence of numbers, and know nothing about in which sequence the dice were rolled. Thus, the die sequence (state sequence) is *hidden* from us, and the die numbers are our observed sequence generated through random functions depending on the hidden state. The HMM algorithms can help us determine the most likely state sequence for the observed sequence, given our model.   □

We let $\{X_t\}_{t=1}^{T}$ denote the Markov process as before with state space $S = \{s_1, \ldots, s_N\}$, initial probabilities $\pi = \{\pi_1, \ldots, \pi_N\}$, and transition probabilities a_{ij}, $i, j \in S$. At each step t, the process emits an observation Y_t, where $\{Y_t\}_{t=1}^{T}$ denotes the observed process, taking values in some symbols set $V = \{v_1, \ldots, v_M\}$. Each variable Y_t depends on the current (hidden) state X_t, and possibly on the previous outputs $Y_1, \ldots, Y_{t-1}$. For simplicity we will use the shorthand $Y_a^b = Y_a, \ldots, Y_b$ for a subsequence between time indices a and b. We denote the *emission distribution* of Y as

$$b_j(Y_t | Y_1^{t-1}) = \mathbf{P}(Y_t | Y_1^{t-1}, X_t = j). \tag{2.39}$$

To summarize, our HMM is characterized by the state space S, the emission alphabet V and the initial, transition and emission probabilities $\{\pi_i, a_{ij}, b_j : i, j \in S\}$.

One way to more easily understand the procedure of a Markov model is to view it as a "sequence generating machine", by which the observed sequence could be generated as in Algorithm 1.

Algorithm 1 Generating output from a standard HMM

$t = 1$
Choose X_t according to π
while $t < T$ **do**
  Emit Y_t according to $b_{X_t}(Y_t|Y_1^{t-1})$
  Jump to state X_{t+1} according to $a_{X_t,X_{t+1}}$
  $t = t + 1$
end while

The joint probability of the hidden and the observed process is determined by noting the following:

$$\mathbf{P}(X_t = j, Y_t|X_{t-1} = i, X_1^{t-2}, Y_1^{t-1}) =$$
$$= \mathbf{P}(X_t = j|X_{t-1} = i)\mathbf{P}(Y_t|X_t, Y_1^{t-1})$$
$$= a_{ij}b_j(Y_t|Y_1^{t-1}). \tag{2.40}$$

Using the same notation as Rabiner in [30], we let $\theta = \{\pi, A, B\}$ denote the model, where π is the initial distribution, and A and B represent the transition and emission probabilities, respectively. Then the joint probability of the entire hidden and observed sequence, under the model can be written as

$$\mathbf{P}(X_1^T, Y_1^T|\theta) = \pi_{X_1}b_{X_1}(Y_1)\prod_{t=2}^{T}a_{X_{t-1},X_t}b_{X_t}(Y_t|Y_1^{t-1}). \tag{2.41}$$

In what follows, while always conditioning on the model, we omit θ in the notation.

Thus far we have described the model, generating the hidden state sequence, that is underlying our observations. In gene finding, or in any other situation of classification, we are sitting at the other end. Typically, we are faced with an observed sequence that we would like to assign state labels to. In other words, we would like to classify, or *parse*, the observed sequence. In order to do so in the framework of HMMs, we need means to:

1. Estimate the parameters of the model.
2. Validate the model.
3. Use the model as a predictive tool.

In the first step, called the *training* step, we build our model by estimating its parameters from a set of training data. That is, we use a set of sequences that are representative for the patterns we are looking for, and where we know the state labels for each observed symbol. The second step, called the *evaluation* step, is a check that our model is a reasonable representation of reality. In this step we calculate the probability that the observed data was produced by our model.

The main difficulty, when going from a standard Markov chain to HMMs, is that there is no unique correspondence between the state sequence and the observed sequence. An observed sequence could be achieved by many different state sequences, or many different *paths* through the state space. The goal of the third step, called the *parsing* step, is to determine the most likely state sequence that could have generated the observed sequence. We say that we *parse*, *classify*, *annotate*, or *decode* the observed sequence by attaching state labels to each observed symbol. The resulting state sequence then corresponds to the most probable path through the model.

By means of *dynamic programming*, described below, we can efficiently solve these problems. The corresponding HMM algorithms, utilizing the dynamic programming method, are called the *forward algorithm*, the *backward algorithm*, and the *Viterbi algorithm*. The evaluation and the decoding steps are solved directly using these algorithms. The solution to the training problem is more complicated and involves using a variant of the *expectation–maximization* (EM) algorithm called the *Baum–Welch algorithm*, described in Sect. 6.5.

2.1.3 Dynamic Programming

Many optimization problems have a recursive structure, or an *optimal substructure*, where the optimal solution can be divided into subproblems, which themselves have optimal solutions. Example 2.4 illustrates the concept of breaking a recursive structure into substructures.

Example 2.4   Fibonacci numbers
As an example of a recursive structure, consider the Fibonacci numbers, where each subsequent number in the series is the sum of the previous two numbers:

$$f(n) = \begin{cases} 0 & \text{if } n = 0, \\ 1 & \text{if } n = 1, \\ f(n-1) + f(n-2) & \text{if } n > 1. \end{cases} \tag{2.42}$$

Although not an optimization problem, it illustrates how redundant a naive implementation of $f(n)$ would be. If we, for instance, were to compute $f(5)$, using a direct (top-down) approach, we would call $f(2)$ three times, and $f(3)$ two times (see Fig. 2.4), and the number of computations needed to calculate $f(n)$ would grow exponentially with n. Such a problem, where the recursive solution contains a relatively small number of distinct subproblems repeated many times, is said to have *overlapping subproblems*. By representing each subproblem by one node only, we get a *directed acyclic graph* (DAG) (see Fig. 2.5), instead of the much redundant tree representation. Instead of having an exponentially growing number of computations, the problem grows linearly in n, since we only have to calculate each $f(n)$ once.

□

An efficient solution to problems such as that in Example 2.4 is the *dynamic programming* algorithm, which has attained a central role in computational sequence analysis [13]. Dynamic programming is a general recursive decomposition technique

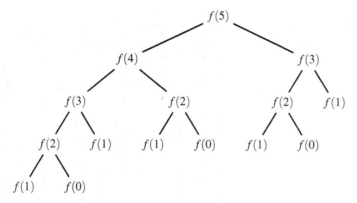

Fig. 2.4 A recursive tree illustrating the sub-calculations needed to determine $f(5)$

Fig. 2.5 A directed acyclic
graph of the computation of
the Fibonacci number $f(5)$

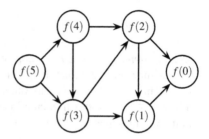

for global optimization problems, exhibiting the properties of optimal substructures
and overlapping subproblems. The word 'programming' does not refer to the process
of coding up a computer program for the purpose, but to the tabular mode of the computation. The trick used in dynamic programming is to store (or cache) and reuse the
solutions to the subproblems, an approach called *memoization* (not memorization)
in computing. The standard dynamic programming approach has three components:

1. The *recurrence relation*.
2. The *tabular computation*.
3. The *traceback*.

In the recurrence relation we establish the recursive relationships between the variables, such as in (2.42). In the tabular computation the calculations are organized in
a table that is filled in one column at a time (see Fig. 2.6). There are in general two
approaches to do this:

- In a *top-down approach* the problem is broken down into subproblems, which
  are calculated the first time they are called and then stored for further calls. This
  approach combines recursion and memoization. Figure 2.4 illustrates a top-down
  calculation of the Fibonacci algorithm. The contribution of dynamic programming
  is that the calculation of each subproblem is stored and a map function is used to
  keep track of which subproblems already have been calculated.

Fig. 2.6 The tabular
computation goes through
the table column by column

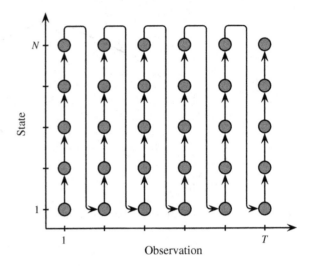

- In a *bottom-up approach* all subproblems are calculated and stored in advance. This is more efficient than the previous, but is less intuitive as it may be difficult in certain applications to figure out all subproblems needed for the calculation in advance. A bottom-down calculation of $f(5)$ would simply calculate all Fibonacci numbers subsequently, $f(0)$, $f(1)$, $f(2)$, $f(3)$, $f(4)$, $f(5)$.

The bottom-up approach is what is commonly used in HMM algorithms and sequence alignment, and is what we will consider from now on. Once the table of subproblems have been filled (bottom-up), we traceback through the table to obtain the optimal solution. In the Fibonacci example, the calculation is finished already in the tabular calculation, but in other situations such as in sequence alignment we still need to figure out the optimal solution to the global problem using the table of subproblem solutions. The easiest way to facilitate the traceback is to store pointers during the tabular computation from each cell in the table to the optimal previous position. These pointers are then followed in the traceback to determine the optimal *path* through the table. In the following sections we will show how dynamic programming is employed in HMMs.

2.1.3.1 Silent Begin and End States

Before we proceed to describe the HMM algorithms, we need to explain the notion of *silent states*. A silent state is a state with no output. Since the first state of a Markov chain follows a special initial distribution, adding a silent begin state X_0 to the model will simplify the formula in (2.41)

$$\mathbf{P}(X_1^T, Y_1^T) = \prod_{t=1}^{T} a_{X_{t-1},X_t} b_{X_t}(Y_t | Y_1^{t-1}), \qquad (2.43)$$

where now

$$a_{X_0, X_1} = \pi_{X_1}. \tag{2.44}$$

Similarly, we can model the end of the sequence by adding a silent end state X_{T+1}, such that

$$\mathbf{P}(X_{T+1} | X_T) = a_{X_T, X_{T+1}}. \tag{2.45}$$

The end state is usually not included in a general Markov chain, where the length of the chain may be undetermined and the end can occur anywhere in the sequence [11]. But since we will deal exclusively with finite sequences, adding an end state will enable the modeling of the sequence length distribution. Moreover, as we will see in Sect. 2.2.4, the inclusion of a silent begin and end state can become a valuable means to reduce computational complexity.

2.1.4 The Forward Algorithm

The *forward algorithm* is used to calculate the probability (or likelihood) of the observed data under the given model. The recurrence relation in dynamic programming is represented by the *forward variables*, defined as the joint probability of the hidden state at time $t = 1, \ldots, T$ and the observed sequence up to that time,

$$
\begin{aligned}
\alpha_i(t) &= \mathbf{P}(Y_1^t, X_t = i) \\
&= \sum_{j \in S} \mathbf{P}(Y_1^t, X_t = i, X_{t-1} = j) \\
&= \sum_{j \in S} \mathbf{P}(Y_t, X_t = i | Y_1^{t-1}, X_{t-1} = j) \mathbf{P}(Y_1^{t-1}, X_{t-1} = j) \\
&= \sum_{j \in S} \mathbf{P}(X_t = i | X_{t-1} = j) \mathbf{P}(Y_t | Y_1^{t-1}, X_t = i) \mathbf{P}(Y_1^{t-1}, X_{t-1} = j) \\
&= \sum_{j \in S} a_{ji} b_i(Y_t | Y_1^{t-1}) \alpha_j(t-1).
\end{aligned} \tag{2.46}
$$

For initialization we add a silent state X_0, where

$$\alpha_i(0) = \pi_i, \quad i \in S, \tag{2.47}$$

and for termination we add a silent end state X_{T+1}, where

$$\alpha_i(T+1) = \mathbf{P}(Y_1^T, X_{T+1} = i) = \sum_{j \in S} \alpha_j(T) a_{ji}. \tag{2.48}$$

Fig. 2.7 Each node is a sum
of the forward variables at
the previous position

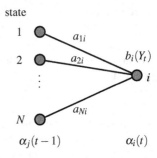

The desired probability of the observed data, given the model, is then given by

$$\mathbf{P}(Y_1^T) = \sum_{i \in S} \mathbf{P}(Y_1^T, X_{T+1} = i) = \sum_{i \in S} \alpha_i(T+1). \qquad (2.49)$$

The forward variables are efficiently calculated using the tabular computation in
dynamic programming. The calculations are organized in a table as in Fig. 2.6, that
is filled in one column at a time for increasing state and time indices. The name
forward in the forward algorithm comes from the fact that we move forward through
the data. That is, each variable $\alpha_i(t)$ at time t is a (weighted) sum over all variables
at time $t - 1$ (see Fig. 2.7).

The implementation of the forward algorithm is illustrated in Algorithm 2.

Algorithm 2 The forward algorithm

$t = 1$
Choose X_1 according to π
while $t < T$ **do**
   Emit Y_t according to $b_{X_t}(Y_t | Y_1^{t-1})$
   Jump to state X_{t+1} according to $a_{X_t, X_{t+1}}$
   $t = t + 1$
end while

2.1.5 The Backward Algorithm

There is a useful HMM algorithm closely related to the forward, called the *backward
algorithm*, which is used in particular when solving the training problem in Sect. 6.5.
As the recursion in the forward algorithm proceeds in a forward direction with respect
to time, the recursion for the backward variables goes in the opposite direction. The
backward variable $\beta_i(t)$ is the probability of all observed data after time t, given the
observed data up to this time and given that the state at time t is $X_t = i$. As with the

forward, we initialize by using a silent state, but since we are going backwards the initial state of the algorithm is the end state X_{T+1} of the chain

$$\beta_i(T+1) = 1, \quad i = 1, \ldots, N. \tag{2.50}$$

For $t = T, T-1, \ldots, 1$ we define the backward variables as

$$
\begin{aligned}
\beta_i(t) &= \mathbf{P}(Y_{t+1}^T | Y_1^t, X_t = i) \\
&= \sum_{j \in S} \mathbf{P}(Y_{t+1}^T, X_{t+1} = j | Y_1^t, X_t = i) \\
&= \sum_{j \in S} \mathbf{P}(X_{t+1} = j | X_t = i) \mathbf{P}(Y_{t+1} | Y_1^t, X_{t+1} = j) \mathbf{P}(Y_{t+2}^T | Y_1^{t+1}, X_{t+1} = j) \\
&= \sum_{j \in S} a_{ij} b_j(Y_{t+1} | Y_1^t) \beta_j(t+1).
\end{aligned}
\tag{2.51}
$$

We finish in the silent begin state X_0, but this does not need special treatment for the backward algorithm. The algorithm simply terminates upon calculation of $\beta_i(0)$, for $1 \le i \le N$. Note that, similarly to the forward algorithm, we can calculate the probability of the observed sequence using the backward algorithm as well.

$$
\begin{aligned}
\mathbf{P}(Y_1^T) &= \sum_{i \in S} \mathbf{P}(Y_1^T, X_1 = i) \\
&= \sum_{i \in S} \mathbf{P}(Y_2^T | Y_1, X_1 = i) \mathbf{P}(Y_1 | X_1 = i) \mathbf{P}(X_1 = i) \\
&= \sum_{i \in S} \pi_i b_i(Y_1) \beta_i(1).
\end{aligned}
\tag{2.52}
$$

2.1.6 The Viterbi Algorithm

The purpose of using HMMs in biological sequence analysis is to utilize their strengths as a predictive tool. That is, given that we have a model and have trained its parameters, we would like to use it to classify, or *decode*, an unlabeled sequence of observations. In other words, we would like to find the *optimal* state sequence for the given observations and the given model. However, the solution to this problem depends on our definition of "optimal". As discussed in [30], depending on the optimality criterion chosen, the solution might not even be valid. For instance, one natural criterion would be to choose the sequence of states that are individually most likely, a method commonly referred to as *posterior decoding* and discussed further in Sect. 2.1.7.1. This approach maximizes the number of correct individual states, but as soon as some state transitions in the state space have probability zero, we stand the risk of ending up with a state sequence that is indeed optimal in the sense that it reaches the highest likelihood, but that is impossible to achieve. In the end, what we would like to find is the single best state sequence among all *valid* ones. The HMM procedure that achieves this is called the *Viterbi algorithm*. The Viterbi algorithm formulation is essentially the same as the forward algorithm, except sums are replaced by maxima, and we need a little extra bookkeeping to track the maximizing terms.

We would like to optimize the probability of the hidden state sequence, given the observed data $\mathbf{P}(X_1^T|Y_1^T)$. Note, however, that this probability is maximized at the same point as the joint probability $\mathbf{P}(X_1^T, Y_1^T)$. Therefore, we define the Viterbi variables as the joint probability of hidden and observed data up to time t, maximized over all valid state sequences. The initial conditions for the recurrence relation are the same as for the forward algorithm. The Viterbi variables for the initial silent state X_0 are given by

$$\delta_i(0) = \pi_i, \quad i = 1, \ldots, N. \tag{2.53}$$

The tabular computation proceeds for $t = 1, \ldots, T$ using the recurrence relation

$$\delta_i(t) = \max_{X_1^{t-1}} \mathbf{P}(Y_1^t, X_1^{t-1}, X_t = i)$$

$$= \max_{X_1^{t-2}, j} \mathbf{P}(Y_1^t, X_1^{t-2}, X_{t-1} = j, X_t = i)$$

$$= \max_{X_1^{t-2}, j} \mathbf{P}(Y_1^{t-1}, X_1^{t-2}, X_{t-1} = j)\mathbf{P}(X_t = i|X_{t-1} = j)\mathbf{P}(Y_t|Y_1^{t-1}, X_t = i)$$

$$= \max_{1 \le j \le N} \delta_j(t-1)a_{ji}b_i(Y_t|Y_1^{t-1}). \tag{2.54}$$

As a result, each $\delta_i(t)$ represents the highest probability of all paths up to time t, ending in state i. To facilitate the traceback we store pointers from the current position to the optimal previous position,

$$\psi_i(t) = \operatorname*{argmax}_{1 \le j \le N} a_{ji}b_i(Y_t|Y_1^{t-1})\delta_j(t-1). \tag{2.55}$$

These pointers will be used to retrieve the optimal path through the state space. As for the forward algorithm, the computation is terminated by calculating the Viterbi variables for the silent end state X_{T+1}

$$\delta_i(T+1) = \max_{X_1^T} \mathbf{P}(Y_1^T, X_1^T, X_{T+1} = i)$$

$$= \max_{1 \le j \le N} a_{ji}\delta_j(T). \tag{2.56}$$

The probability of the most likely state sequence is then given by

$$\mathbf{P}(\text{most likely state sequence}) = \max_{1 \le i \le N} \delta_i(T+1). \tag{2.57}$$

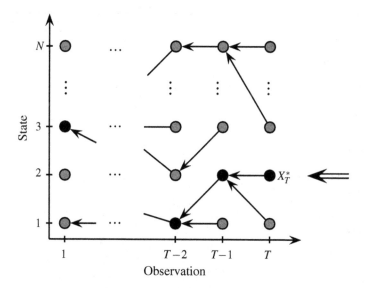

Fig. 2.8 The traceback starts in X_T^* and moves back through the state space, following the stored pointers $\psi_i(t)$

To extract the actual state sequence giving rise to this probability, we start the traceback in the silent end state giving rise to the highest value on its Viterbi variable,

$$X_{T+1}^* = \operatorname*{argmax}_{1 \le i \le N} \delta_i(T+1), \tag{2.58}$$

and backtrack recursively through the dynamic programming table (see Fig. 2.8) using

$$X_t^* = \psi_{X_{t+1}^*}(t+1), \quad t = T, T-1, \ldots, 0. \tag{2.59}$$

The resulting state sequence, $X_0^*, X_1^*, \ldots, X_{T+1}^*$ is then the optimal, or most probable, state sequence for the given observations and given model, and represents a *parse* or an *annotation* of the observed sequence. The implementation of the Viterbi algorithm is illustrated in Algorithm 3.

While the Viterbi algorithm is very efficient at finding the single best path, and is suitable to use when one path clearly dominates, it is less effective when several paths have similar near-optimal probabilities. In such cases posterior decoding might work better, even though it is not guaranteed to produce a valid solution. Posterior decoding is discussed further in Sect. 2.1.7.1.

Algorithm 3 The Viterbi algorithm

/* Initialize */

for $i = 1$ to N **do**
   Initialize: $\delta_i(0) = \pi_i$
end for

/* The tabular computation */

for $t = 1$ to T **do**
   for $i = 1$ to N **do**
      $\delta_i(t) = 0$
      for $j = 1$ to N **do**
         if $a_{ji}\, \delta_j(t-1) > \delta_i(t)$ **then**
            $\delta_i(t) = \delta_j(t-1)\, a_{ji}$
            $\psi_t(i) = j$
         end if
      end for
      $\delta_i(t) = b_i(Y_t|Y_1^{t-1})\,\delta_i(t)$
   end for
end for

2.1.7 EasyGene: A Prokaryotic Gene Finder

The gene finding problem in prokaryotes is quite different from that in eukaryotes. In particular, the prokaryotic genomes are much more dense, with much less intergenic regions and with rarely any splicing. As a result, while eukaryote genomes may contain less than 10 % of coding sequence, prokaryotes tend to be very gene rich with as much as 90 % of the sequence being coding. Moreover, the prokaryotic binding sites are usually located in direct vicinity of the protein-coding regions, and can be included in the model and thereby strengthen the gene signal. In contrast, eukaryotes binding sites can be located long distances from the actual gene and are often difficult to associate with the corresponding genes. However, although much simpler, the gene finding task is far from trivial even in prokaryotes, and is complicated by several issues.

Gene finding in prokaryotes is usually conducted by looking for *open reading frames* (ORFs). That is, long stretches of potentially coding sequences surrounded by a pair of candidate in-frame start and stop codons, but void of in-frame stop codons in between. The issue that arises in such an approach is that of separating real genes from 'spurious', or random, ORFs. The shorter the sequences considered, the more difficult this task becomes. Therefore, it is common to apply a minimum length threshold on the ORFs considered in these searches. Sharp and Cowe [37] suggested a threshold of 100 amino acids as a good trade-off between the number of missed short genes and the number of predicted spurious ORFs. It turns out, however, that such a threshold is very crude. Prokaryotic genomes contain plenty of spurious ORFs above that size, and a significant amount of true genes below it [40]. Another issue, much more prevalent in prokaryotes than in eukaryotes, is the

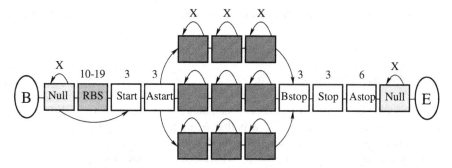

Fig. 2.9 The EasyGene gene model. The numbers above the boxes represent the number of bases modeled by that submodel, where 'X' indicates a variable number. B and E are begin and end states, the NULL-states cover intergenic region before and after the gene, the RBS state include the RBS and the spacer bases to the next state, Start and Stop model the start and stop codons, Astart, Bstop and Astop explicitly model the codons directly after the start codon and surrounding the stop codon

problem of overlapping genes, which makes the accurate detection of translation start sites notoriously difficult.

EasyGene [20] is an HMM-based prokaryotic gene finder, that attempts to address these issues. EasyGene is fully automated in that it extracts training data from a raw genomic sequence, and estimates the states for coding regions and ribosomal binding sites (RBS) used to score potential ORFs. The EasyGene state space is illustrated in Fig. 2.9. The B and E states are silent begin and end states of the HMM, and the NULL-states model everything that is not part of the gene, and not in direct vicinity of the gene. The RBS state includes the RBS as well as the bases between the RBS and the next state, and the START and STOP states correspond to the start and stop codons of the gene, respectively. While eukaryotic genes almost always start with ATG, prokaryotes use a number of alternative start codons. *E. coli* (K-12 strain), for instance, uses ATG in about 83 % of its genes, GTG in 14 % and TTG in 3 % of the cases, and an additional one or two very rare variants [4]. The stop codons are typically TAA, TAG, and TGA in both eukaryotes and prokaryotes, even if alternatives are known to exist [15]. The codons directly after the start codon and the codons surrounding the stop codon tend to follow a distribution different from the rest of the gene [38], a feature that can be used to strengthen the start and stop signals. This feature is explicitly modeled in the ASTART, BSTOP, and ASTOP states.

The model for the internal codons consists of 3 parallel submodels, allowing the HMM to keep separate statistics for atypical genes. Each submodel, consists of a series of 3 codon models, where each codon model is a 4th-order Markov model consisting of three states, one for each DNA base of the codon, capturing the codon position dependency of coding sequences. As a result the length distribution becomes negative binomial with parameters (n, p), where n is the number of serial codon models, and p the probability of transitioning out of the specific codon model. This model allows for more general length distributions than the geometric, which would be the result of using one codon model alone (this issue is discussed further in Sect. 2.2).

2.1.7.1 Posterior Decoding

As a consequence of using duplicated codon states, the length of an ORF is only realized as the sum over many HMM paths. While the Viterbi algorithm is a very efficient decoding algorithm when one path dominates, it is not appropriate when several paths have similar probabilities. Therefore, EasyGene uses posterior coding instead, also known as the *forward–backward algorithm* [30], where the *individually* most likely sequence of states is computed. The details on the forward–backward algorithm are given in Sect. 6.5, but in short we use the probabilities of being in a given state $s_i \in S$ at time t, given the observed sequence to determine the individually most likely state sequence. The forward–backward variables are defined as

$$\gamma_t(i) = \mathbf{P}(X_t = s_i | Y_1^T) = \frac{\alpha_i(t)\beta_i(t)}{\mathbf{P}(Y_1^T)}, \tag{2.60}$$

and the resulting optimal state sequence is given by

$$X_t^* = \underset{s_i \in S}{\operatorname{argmax}}\ \gamma_t(i), \quad 1 \le t \le T. \tag{2.61}$$

Assuming that there are no frameshifts or sequencing errors in the sequence, there is exactly one stop codon for each start codon, and, thus, the probability of a gene is equivalent to the posterior probability of its gene start. As a consequence, we can easily extract all possible start codons for a gene in the case of several similar scores. Moreover, in the case of overlapping genes the Viterbi algorithm would only report the highest scoring, while using posterior decoding each gene is scored and reported separately. However, posterior decoding merely bunches together independent underlying states, without checking that the parse is valid. Although this is not a big problem in prokaryotic gene finding, we still need to be careful when interpreting the output of posterior decoding.

2.1.7.2 Statistical Significance of Predictions

Along with its gene predictions, EasyGene reports a measure of statistical significance for each ORF. The measure is based on a comparison of the predicted ORFs to the expected number of ORFs in a random sequence. The random sequence model, called the NULL-model (different from the NULL-states in the main model), is a third-order Markov chain using the same base frequencies as the overall genome in question. It consists of a third-order state for intergenic regions, and a reverse codon model to capture genes on the reverse strand (see Fig. 2.10). The significance

Fig. 2.10 The NULL-model in EasyGene is used to model random sequence with the same background statistics as the overall genome. The state space consists of a third-order intergene state and a reverse codon model to capture genes on the reverse strand

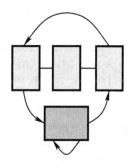

measure is based on a log-odds score between the model and the NULL-model,

$$\beta = \log \frac{\mathbf{P}(Y|M)}{\mathbf{P}(Y|N)}, \tag{2.62}$$

where $\mathbf{P}(Y|M)$ is the probability that sequence Y contains an ORF under the model, and $\mathbf{P}(Y|N)$ the same probability for the NULL-model. This score is different from that reported by Genscan, which reports posterior exon probabilities based on the HMM (see Sect. 2.2.4.3).

2.2 Generalized Hidden Markov Models (GHMMs)

A major problem with standard HMM is the intrinsic modeling of state duration. Outputting exactly one observation per jump leads to a length distribution that is exponentially decaying, something that often is unsuitable for many applications. The solution to this problem is called a hidden *semi*-Markov model, or a *generalized* HMM (GHMM). The word 'generalized' comes from this fact, that instead of having geometric length distributions we can use a length distribution of our choice.

2.2.1 Preliminaries

A standard HMM makes a transition at each time unit t, such that the *transition time* is always equal to one, and the observed output is exactly one symbol per unit. However, by making a number of *self-transitions* into the same state, we can observe a coherent subsequence emitted from the same state. We call the length of such a sequence the *duration* of the state. Due to the Markov property, rendering the process memoryless, the duration follows a geometric distribution (see Sect. 5.2.1). Hence, using a standard HMM for the purpose of gene finding, for instance, would impose a geometric distribution on each state. It has been noted, however, that exons in particular tend to follow a length distribution that is statistically very different from the geometric distribution. Thus forcing such a model on the sequence data may lead to bad predictions.

A *semi*-Markov model, has the same structure as a standard HMM, except that the process stays in each state a random amount of time before its next transition, as opposed to standard HMMs, where the time in each state always equals one (note that we are only considering discrete-time processes here). More formally, a semi-Markov process is a process W whose state sequence is generated by a Markov process X as before, but whose transition times are given by another process ζ that may depend on both the current state and the next. Thus, since properties of ζ may depend on the next state in X, the process W is in general not Markovian. The associated process (X, ζ), however, *is* a Markov process, hence the name semi-Markov.

In a *hidden* semi-Markov model, or a *generalized* hidden Markov model as we will call it henceforth, there is a duration distribution associated with each state. When a state emits output, first a duration is chosen from the duration distribution, and then the corresponding length of output is generated. This generalization can improve performance by allowing for more accurate modeling of the typical duration of each particular state. As a result, the indices of the hidden and the observed process will start to differ as soon as a state has a duration that is longer than 1. For example, assume that the model generated the following output:

Hidden: X_1: X_2: X_3:
Observed: $Y_1 Y_2 Y_3$ $Y_4 Y_5$ $Y_6 Y_7 Y_8 Y_9$

In order to handle this, we separate the time index notation in the state sequence and in the observed sequence as follows. Given that the hidden process is in state X_l, let d_l denote the duration of X_l chosen from a length distribution $f_{X_l}(d_l)$. To keep track of the indices in the hidden versus the observed sequences, we introduce partial sums for the number of emitted symbols up to (and including) state X_l

$$p_l = \sum_{k=1}^{l} d_k, \quad \text{and } p_0 = 0. \tag{2.63}$$

We let L denote the length of the Markov process, X_1^L, and T the length of the observed process, Y_1^T. For simplicity we assume that $p_L = T$, meaning that all of the observed output generated in the final state X_L is included in the observed sequence. Now the state sequence X_1^L, the duration sequence d_1^L, and the length of the state sequence L, are hidden from the observer, and the observed data remains to be the observation sequence Y_1^T. The joint probability of the hidden and observed data becomes

$$\mathbf{P}(Y_1^T, X_1^L, d_1^L) = \prod_{l=1}^{L} a_{X_{l-1}, X_l} f_{X_l}(d_l) b_{X_l}(Y_{p_{l-1}+1}^{p_l} | Y_1^{p_{l-1}}), \tag{2.64}$$

where X_0 is the silent begin state as before, with

$$a_{X_0,X_1} = \pi_{X_1}. \tag{2.65}$$

One drawback with GHMMs is that statistical inference is harder than for standard HMMs. In particular, the Baum–Welch algorithm for parameter training is not applicable. The Baum–Welch algorithm is a generalized EM-algorithm (expectation–maximization), that uses counts of transition–emission pairs to update the expectation part of the algorithm. Details on the Baum–Welch algorithm, and on how to train GHMMs, can be found in Chap. 6.

2.2.2 The Forward and Backward Algorithms

One of the attractive features of using a generalized HMM for gene finding is that it provides a natural way of computing the posterior probability of a predicted generalized state, given the observed data. How this is done is described in Sect. 2.2.4.3, in the framework of the gene finding software Genscan [7]. First, we need to adjust the forward and backward algorithms in (2.46) and (2.51), respectively, to fit the GHMM framework.

The Forward Variables

Recall that the forward variables are defined as the joint probability of observed sequence up to time t, and the hidden state at time t. In a GHMM, however, we need to adjust the definition slightly, since each state can have variable durations. We define the forward variables $\alpha_i(t)$ as the probability of the observed data, and that the hidden state i at time t actually *ended* at time t. This is to say that $X_l = i$ and $p_l = t$ for some $1 \leq l \leq L$. In what follows we let D be the maximum duration of a state.

$$\alpha_i(t) = \mathbf{P}\left(Y_1^t, \{\text{some hidden state } i \text{ ends at } t\}\right)$$

$$= \mathbf{P}\left(Y_1^t, \bigcup_{l=1}^{L} (X_l = i, p_l = t)\right)$$

$$= \sum_{j\in S}\sum_{d=1}^{D} \mathbf{P}\left(Y_1^t, \bigcup_{l=1}^{L} (X_l = i, p_l = t, d_l = d), \bigcup_{l=1}^{L} (X_l = j, p_l = t - d)\right)$$

$$= \sum_{j\in S}\sum_{d=1}^{D} \left[\mathbf{P}\left(Y_1^{t-d}, \bigcup_{l=1}^{L} (X_l = j, p_l = t - d)\right)\right. \tag{2.66a}$$

$$\left. \cdot \mathbf{P}\left(\bigcup_{l=1}^{L} (X_l = i, p_l = t, d_l = d) \middle| \bigcup_{l=1}^{L} (X_l = j, p_l = t - d)\right)\right. \tag{2.66b}$$

$$\cdot \, \mathbf{P}\!\left(Y_{t-d+1}^t \middle| Y_1^{t-d}, \bigcup_{l=1}^L (X_l = i, \, p_l = t, \, d_l = d), \, \bigcup_{l=1}^L (X_l = j, \, p_l = t - d)\right)\right] \quad (2.66\text{c})$$

The first term (2.66a) is simply $\alpha_j(t - d)$. To handle the conditioning on unions in the second (2.66b) and third (2.66c) terms, we make use of the following two lemmas.

Lemma 2.1 *If sets A, B, and C satisfy* $B \cap C = \emptyset$ *and* $\mathbf{P}(A|B) = \mathbf{P}(A|C)$, *then* $\mathbf{P}(A|B \cup C) = \mathbf{P}(A|B)$.

Lemma 2.2 *If for set A and disjoint sets* $B_1, \ldots, B_n$ *we have that* $\mathbf{P}(A|B_i) = \mathbf{P}(A|B_1)$ *for all* $1 \leq i \leq n$, *then* $\mathbf{P}(A| \bigcup_{i=1}^n B_i) = \mathbf{P}(A|B_1)$.

As a result, for the second term (2.66b) we get

$$\mathbf{P}\!\left(\bigcup_{l=1}^L (X_l = i, \, p_l = t, \, d_l = d) \middle| \bigcup_{l=1}^L (X_l = j, \, p_l = t - d)\right)$$

$$= \mathbf{P}\!\left(\bigcup_{l=1}^L (X_l = i, \, p_l = t, \, d_l = d) \middle| X_1 = j, \, p_1 = t - d\right)$$

$$= \mathbf{P}(X_2 = i, \, p_2 = t, \, d_2 = d | X_1 = j, \, p_1 = t - d)$$

$$= a_{ji} \, f_i(d). \quad (2.67)$$

Similarly, the third term (2.66c) becomes

$$\mathbf{P}\!\left(Y_{t-d+1}^t \middle| Y_1^{t-d}, \bigcup_{l=1}^L (X_l = i, \, p_l = t, \, d_l = d,), \, \bigcup_{l=1}^L (X_l = j, \, p_l = t - d)\right)$$

$$= \mathbf{P}(Y_{t-d+1}^t | Y_1^{t-d}, \, p_1 = t - d, \, d_2 = d, \, X_2 = i).$$

$$= b_i(Y_{t-d+1}^t | Y_1^{t-d}). \quad (2.68)$$

Thus, the forward algorithm results in

$$\alpha_i(t) = \sum_{j \in S} \sum_{d=1}^D a_{ji} \, f_i(d) \, b_i(Y_{t-d+1}^t | Y_1^{t-d}) \alpha_j(t - d). \quad (2.69)$$

We initialize and terminate as before with

$$\alpha_i(0) = \pi_i, \quad (2.70)$$

$$\alpha_i(T + 1) = \mathbf{P}(Y_1^T, X_1^L, X_{L+1} = i) = \sum_{j \in S} \alpha_j(T) a_{ji}. \quad (2.71)$$

Just as in the non-generalized case, the probability of the observed sequence, given the model, is given by summing over the terminal forward variables

$$P(Y_1^T) = \sum_{i \in S} \alpha_i (T + 1).$$ (2.72)

The Backward Variables

The backward variable $\beta_i(t)$ denotes the probability of all the observed data after time t, given the observed data up to time t and given that the hidden state i ended at time t. Skipping the details, the backward variables for GHMMs are given by

$$\beta_i(t) = P\left(Y_{t+1}^T \mid \bigcup_{l=1}^{L} (X_l = i, p_l = t)\right)$$

$$= \sum_{j \in S} \sum_{d=1}^{D} a_{ij} f_j(d) b_j(Y_{t+1}^{t+d} \mid Y_1^t) \beta_j(t + d).$$ (2.73)

The backward algorithm is initiated as before, using $\beta_i(T + 1) = 1$ for all $i \in S$, and is terminated upon calculation of $\beta_i(0)$.

2.2.3 The Viterbi Algorithm

Now that we know what the GHMM forward algorithm looks like, adjusting the Viterbi algorithm of the standard HMM as straightforward. Recall from Sect. 2.1.6 that the conditional probability $P(X_1^T \mid Y_1^T)$ of the state sequence given the observed data is maximized by the same state sequence as the joint probability $P(Y_1^T, X_1^T)$. The same holds in the GHMM situation; the state sequence and the associated durations that maximizes $P(X_1^L, d_1^L \mid Y_1^T)$ also maximizes the joint probability $P(Y_1^T, X_1^L, d_1^L)$ given in (2.64). As for the standard HMMs, the Viterbi algorithm only differs from the forward algorithm in that the sums are replaced by maxima. Therefore, skipping the technical details, the tabular computation of the GHMM Viterbi algorithm for $t = 1, \ldots, T$ becomes

$$\delta_i(t) = \max_{l, X_1^{l-1}, d_1^l} P(Y_1^t, X_1^{l-1}, X_l = i, p_l = t)$$

$$= \max_{j, d} \delta_j(t - d) a_{ji} f_i(d) b_i(Y_{t-d+1}^t \mid Y_1^{t-d}).$$ (2.74)

The Viterbi algorithm is initiated and terminated by

$$\delta_i(0) = \pi_i, \tag{2.75}$$

$$\delta_i(T+1) = \max_j \delta_j(T-1)a_{ji}. \tag{2.76}$$

The probability of the most likely sequence of states and durations is given by

$$\mathbf{P}(\text{most likely sequence of states and durations}) = \max_{1 \le i \le N} \delta_i(T+1). \tag{2.77}$$

As we evaluate $\delta_i(t)$ we record the values of the optimal previous position in the dynamic programming table, which now includes *two* values. In addition to knowing the most likely previous state $\psi_i(t)$ we need to know the most likely duration of that state, in order to jump back to the right previous cell in the table. That is, we record the previous state and its duration in the pair of variables

$$\big(\psi_i(t), \phi_i(t)\big) = \operatorname*{argmax}_{j,d} \ \delta_j(t-d)a_{ji}f_i(d)b_i(Y_{t-d+1}^t | Y_1^{t-d}), \tag{2.78}$$

where now $\psi_i(t)$ corresponds to the maximizing previous state j, and $\phi_i(t)$ to the maximizing duration d of that state. The most probable state sequence thus ends in a state i^* which has duration $\phi_{i^*}(T+1)$ and is preceded by state $\psi_{i^*}(T+1)$, and the whole sequence is unraveled by backtracking using the ϕ's and the ψ's.

2.2.4 Genscan: A GHMM-Based Gene Finder

Genscan [7] is probably one of the most popular single species gene finders of all times, and included several novel improvements when it first was published. The novel features include:

- The ability to predict multiple genes in a sequence.
- The ability to predict partial genes at the end of the sequence.
- The ability to predict genes on both strands simultaneously.
- Binning the parameter set into several submodels, depending on the G+C content of the input sequence.
- Modeling long-range internal dependencies in splice sites using Maximal Dependence Decomposition (MDD).

While most gene predictors up to that point assumed the existence of exactly one complete gene in the sequence to be analyzed, Genscan allows the recognition of both multiple and partial genes. Moreover, the gene prediction is efficiently performed on both strands simultaneously, by simply adding a mirror image of the state space, connected via the intergenic state (see Fig. 2.11). Other improvements include G+C dependent model parameters. Typically, the gene density is higher and the gene

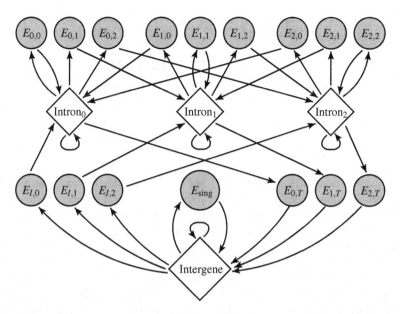

Fig. 2.11 A simplified version of the Genscan state space, modeling only the forward strand and consisting of E-states and I-states alone. The *diamond-shaped* I-states are non-generalized states emitting one symbol at a time, while the *circular* E-states are complex submodels with generalized length distributions and including the splice site models

structure is more compact in regions of higher G+C content (see Sect. 5.3.1). Since this directly affects the model parameters, Genscan separates the training sequences into four sets; sequences of less than 43 % GC-content, 43–51%, 51–57%, and more than 57 % G+C content. When a sequence is to be analyzed, the G+C content is first calculated, and the corresponding set of parameters is applied in the prediction process. Another novel feature in Genscan is the splice site predictor. Signal sequences in general, and splice sites in particular exhibit significant internal dependencies between nonadjacent positions, something that is not easily captured by common weight matrices, or even with higher order Markov models. Genscan uses the Maximal Dependence Decomposition (MDD) model for modeling splice sites. The MDD breaks down the splice site sequence into its specific positions and creates a decision tree arranged by the position dependencies in the order of importance. The MDD is described in more detail in Sect. 5.4.3. The improvements introduced in Genscan were quickly adapted by the gene finding community, and now constitute an integral part of most modern gene finders.

A simplified version of the Genscan state space is illustrated in Fig. 2.11. The figure only shows the forward strand and only includes exons (E-states), introns, and intergene (I-states). The full state space consists of 27 separate states, 13 for each strand, and a joint intergene state. The additional states include promoters, UTR-states, and polyA-signals as well. However, the identification of such regions

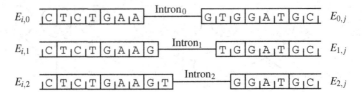

Fig. 2.12 Illustrating the notion of exon and intron phase. Intron$_j$ comes between exon $E_{i,j}$ and exon $E_{j,k}$, $j = 0, 1, 2$

is significantly more difficult than the E- and I-states, leading to much less reliable predictions in comparison to the protein-coding portion of the genes. The reverse strand can be included in the state space by adding a mirror image of the forward states, joined by a common intergenic state. The diamond-shaped I-states are simple, non-generalized states with geometrically distributed durations. The circular E-states are more complex: they include codon compositions, a generalized length distribution, and exon boundary models (splice sites, start or stop). Four types of exons are defined, depending on which boundaries that contain them: single exons (start to stop), initial exons (start to donor), internal exons (acceptor to donor), and terminal exons (acceptor to stop). The introns and internal exons are separated into three groups, corresponding to the phase of the surrounding exons. Exons can be spliced anywhere in the reading frame; exactly between two complete codons, after the first base of the codon, or after the second base (see Fig. 2.12). If an internal intron appears exactly between two complete codons, it has phase 0, while if it occurs after the first or second nucleotide, it has phase 1 or 2, respectively. The internal exons are indexed correspondingly, with $E_{i,j}$ signifying that the previous exon ended with i extra bases (and the preceding intron has phase i), and the current exon ends with j extra bases. The notion of phase is discussed further in Sect. 5.1.1. At a first glance it may seem unnecessary to include such a large number of exon and intron states, but it turns out to be computationally efficient, as it allows us to keep track of the phase without requiring anything more than a first-order Markov chain.

It is typically the exon states that need to be generalized in the gene finding state space, since their length distributions differ significantly from the geometric distribution. Also, the length distribution tend to differ between different exon types as well (see Sect. 5.2 for details). Therefore, Genscan uses separate empirical length distributions for single, initial, internal, and terminal exons. The introns and intergenic regions, however, seem to follow the geometric length distribution fairly well provided that a certain minimum threshold is exceeded. Also, the 5'UTR and 3'UTR state lengths are modeled using geometric distribution.

2.2.4.1 Sequence Generation Algorithm

Consider generating a genomic sequence from the Genscan model. Say that we start off in the intergenic state and do a number of self-transitions. The state duration of the intergene and the introns in each step is always $d_l \equiv 1$. Thus, each time we visit the intergenic state, a single-base Y_t is generated according to some distribution $b_{IG}(Y_t|Y_1^{t-1})$. The output may depend on all of the preceding sequence, but most models only include contexts of a few bases back. Eventually, after some geometrically distributed time t we make a transition into a different state, for instance the $E_{I,2}$ state. The $E_{I,2}$ state represents an initial exon including the two first bases of its last codon. An exon duration d is generated according to some generalized length distribution $f_{E_{I,2}}(d)$, where $f_{E_{I,2}}(d) = 0$ if $(d \mod 3) \neq 2$. We generate d bases Y_{t+1}^{t+d} according to emission distribution $b_{E_{I,2}}(Y_{t+1}^{t+d}|Y_1^t)$, and jump with probability 1 to intron I_2. Note that while the exon sequence may depend on the entire previous sequence as well, it is typically modeled as independent of the preceding I-state. In intron I_2 we continue as in the intergene state, executing a geometric number of self-transitions and emitting a state-specific base in each step, before jumping to one of the internal exons $E_{2,i}$ or the terminal exon $E_{2,T}$. The exon sequence will finish off the last codon of the previous exon, before continuing to generate complete codons. It should be clear by now that the length of each specific exon state is fixed mod 3.

The algorithm for generating a sequence of predetermined length from the Genscan model is summarized in Algorithm 4. The last state of the algorithm is truncated appropriately to ensure that $p_L = T$, meaning that the observed sequence ends exactly at the last base of the last state.

Note that we left out any mentioning of the splice sites surrounding the introns. While the I-states are simple, non-generalized states, the exons are themselves complex submodels. Each exon submodel consists of its length distribution, the model for the coding sequence as well as the exon boundaries (start, stop, donor, and acceptor signals). Details on the exon submodels are given in Chap. 5.

Naturally, using Markov models as sequence generators are only a approximation of the reality. The process at which real genes are generated is bound to be far more complex than any kind of mathematical model. However, using such models as approximations of the real processes provide a powerful tool for the identification of genes and enables us to reconstruct highly complex gene structures. Some limitations of the Genscan model, however, include: (1) Only protein-coding genes are considered, as the pattern for RNA genes are quite different and would need separate models to be detected [32]. (2) Only introns in between protein-coding exons are modeled, and not UTR introns for example. (3) Overlapping and alternatively spliced genes are not handled.

2.2.4.2 Reducing Computational Complexity

Gene prediction, as well as sequence analysis in general, involves dealing with large quantities of data, and an important question is how feasible the calculations of the HMM algorithms are in practice. We can address this by estimating the number of multiplications (or additions, if we use logarithms) required for each forward and backward variable. The emission distribution $b_i(Y_{t+1}^{t+d}|Y_1^t)$ is commonly calculated

---
Algorithm 4 The Genscan model

/* Initialize */

$l = 1$
$p_0 = 0$
Choose X_1 according to π
Choose state duration d_1 according to $f_{X_1}(\cdot)$
$p_1 = d_1$

/* Generate sequence */

while $p_l \leq T$ **do**
    Emit $Y_{p_{l-1}+1}^{p_l}$ according to $b_{X_l}(\cdot|\cdot)$
    Jump to state X_{l+1} according to $a_{X_l, X_{l+1}}$
    $l = l + 1$
    Choose state duration d_l according to $f_{X_l}(\cdot)$
    $p_l = p_{l-1} + d_l$
end while

/* Truncate the last state to get $p_L = T$ */

if $p_l > T$ **then**
    Emit $Y_{p_{l-1}+1}^{T}$ according to $b_{X_l}(\cdot|\cdot)$
end if

---

as the product of d single-base probabilities $\mathbf{P}(Y_t|Y_{t-1}, \dots, Y_{t-k})$, each of which is typically conditioned on a number k of previous bases. Therefore, if D is the maximum duration of a state and N the number of states, the order of computing a forward variable $\alpha_i(t)$ is $O(ND^2)$. If T is the length of the DNA sequence, there are NT such variables, leading to a total of $O(TN^2D^2)$ operations to compute the forward algorithm, and $O(TN)$ bytes to store the variable values. This means that the complexity of the problem is linear in the length of the sequence being analyzed, which is a desired property. However, there is a lot of structure in the topology of the model in Fig. 2.11 that we can take advantage of to get a significant reduction in computational complexity. First, we partition the state space into exon states E and intron and intergenic states I, where $S = E \cup I$, and let N_E and N_I denote the number of separate states in each class, such that $N = N_E + N_I$. Using the structure of the state space, we can actually reduce the memory requirement from $O(TN)$ to $O(T N_I)$ by storing $\alpha_i(t)$ only for the I-states.

Looking at the model, it is clear that an E-state *must* be followed by an I-state. Also, because of the alternation between E- and I-states, at any given time and state, either the previous state, or the state before that, was an I-state. Returning to the forward recursion in (2.69), we can separate the summation over the previous state j between the two-state classes E and I.

$$\alpha_i(t) = \sum_{j \in S} \sum_{d=1}^{D} \alpha_j(t-d) a_{ji} f_i(d) b_i(Y_{t-d+1}^t | Y_1^{t-d})$$

$$= \sum_{j \in I} \sum_{d=1}^{D} \alpha_j(t-d) a_{ji} f_i(d) b_i(Y_{t-d+1}^t | Y_1^{t-d}) \tag{2.79a}$$

$$+ \sum_{j \in E} \sum_{d=1}^{D} \alpha_j(t-d) a_{ji} f_i(d) b_i(Y_{t-d+1}^t | Y_1^{t-d}). \tag{2.79b}$$

The first term (2.79a) depends on forward variables for previous I-states only, and needs no further modification. To make the same come true for the second term (2.79b) we need to step back one state further, which then must be an I-state. The memory-reduced forward recursion can then be written as

$$\alpha_i(t) = \sum_{j \in I} \sum_{d=1}^{D} \alpha_j(t-d) a_{ji} f_i(d) b_i(Y_{t-d+1}^t | Y_1^{t-d}) \tag{2.80a}$$

$$+ \sum_{j \in E} \sum_{k \in I} \sum_{d=1}^{D} \sum_{e=1}^{D} \Big[a_{kj} \alpha_k(t-e-1) f_j(e) b_j(Y_{t-d-e+1}^{t-d} | Y_1^{t-d-e}) \tag{2.80b}$$

$$\cdot a_{ji} f_i(d) b_i(Y_{t-d+1}^t | Y_1^{t-d}) \Big]. \tag{2.80c}$$

In (2.80b), instead of referring to the forward variables of the E-states, we can move one step further back and use the forward variables of the preceding I-states, and include the contribution of the E-states explicitly. As a result we do not have to store the forward variables of the E-states.

Further simplifications can be made. First, recall that the I-states always have duration $d \equiv 1$. As a result, the summation over d can be dropped and $f_i(d)$ can be set to 1. Second, when leaving an exon $E_{i,j}$ we jump to I_j with probability 1. Thus, the transition probability $a_{ji} = 1$ for $j \in E$. Note also that the only way we can jump directly between I-states is via self-transitions, which means that the first sum over I-states only has one positive term. Third, we note that for any given pair of I-states (i, k) with an intervening E-state, the connecting E-state is unique, call it $E_{k,i}$. Thus, the summation over $j \in E$ in (2.80b) has only one positive term as well, the one involving $E_{k,i}$. Finally, the forward recursion becomes

$$\alpha_i(t) = b_i(Y_t | Y_1^{t-1}) \Big[a_{ii} \alpha_i(t-1)$$

$$+ \sum_{k \in I} \sum_{e=1}^{D} \alpha_k(t-e-1) a_{k,E_{k,i}} f_{E_{k,i}}(e) b_{E_{k,i}}(Y_{t-e}^{t-1} | Y_1^{t-e-1}) \Big]. \tag{2.81}$$

As a result, the required memory needed to store the forward variables becomes $O(T N_I)$ and the number of operations $O(T N_I^2 D^2)$. Since $N_I = 4$ and $N = 20$ we have achieved an 80% reduction in memory usage and a 96% reduction of the number of operations needed. Depending on the choice of exon emission distribution $b_{E_{i,j}}$ it may be possible to reduce the number of operations further, and significantly boost the performance of the algorithm. For instance, if it is possible to make a lookup table to compute exon probabilities in a small number of operations, then the complexity of the forward calculation may be reduced by a factor D to $O(T N_I^2 D)$.

If we use the Genscan reduction of the state space, an extra need for a silent begin and end state arises. Without an end state, the likelihood of the observed data would be calculated using

$$\mathbf{P}(Y_1^T) = \sum_{i=1}^{N} \alpha_i(T). \tag{2.82}$$

In the reduced model, however, we only store $\alpha_i(T)$ for the I-states, $i \in I$. Although we do not allow the sequence (or predictions of the sequence) to begin or end in the middle of an exon, we still want it to be possible to predict an exon with a boundary at Y_1 or Y_T, respectively, and the sum in (2.82) would have to run over all states, not just the I-states. We get around this by adding silent begin and end states to the model. For the begin state X_0 we set the initial distribution to be positive only for I-states. The initialization conditions are therefore

$$\alpha_i(0) = \pi_i, \quad i \in I. \tag{2.83}$$

Similarly, for the end state X_{L+1} we set transition probabilities from X_L to be positive only for transitions to I-states. The termination conditions become

$$\alpha_i(T + 1) = \mathbf{P}(Y_1^T, X_{L+1} = i), \quad i \in I. \tag{2.84}$$

As before, the expression for $\alpha_i(T + 1)$ is the same as for the other forward variables in (2.81), except it does not have the output term b_i. The likelihood of the observed data can then be calculated as usual

$$\mathbf{P}(Y_1^T) = \sum_{i \in I} \alpha_i(T + 1). \tag{2.85}$$

In the same manner, the backward and the Viterbi algorithms can be optimized to reduce memory and computational complexity. The backward algorithm simplifies to

$$\beta_i(t) = \beta_i(t + 1) a_{ii} b_i(Y_{t+1}|Y_1^t)$$

$$+ \sum_{k \in I} \sum_{e=1}^{D} \beta_k(t + e + 1) a_{i, E_{i,k}} f_{E_{i,k}}(e) b_{E_{i,k}}(Y_{t+1}^{t+d}|Y_1^t) b_j(Y_{t+e+1}|Y_1^{t+e}), \tag{2.86}$$

with initialization $\beta_i(T + 1) = 1, i \in S$.

The optimized Viterbi algorithm becomes

$$\delta_i(t) = b_i(Y_t|Y_1^{t-1})$$

$$\cdot \max\left\{\delta_i(t-1)a_{ii}, \max_{\substack{k \in I \\ 1 \le e \le D}}\left\{\delta_k(t-e-1)a_{k,E_{k,i}} f_{E_{k,i}}(e)b_{E_{k,i}}(Y_{t-e}^{t-1}|Y_1^{t-e-1})\right\}\right\} \quad (2.87)$$

with initialization $\delta_i(0) = \pi_i$, $i \in I$ and termination as in (2.87) but without the b_i term. The backtracking procedure has to be changed slightly for the optimized algorithms. We record the previous I-state that achieved the maximum, and if this max involved passing through an exon state we need to record the maximizing duration of that exon as well. Otherwise, we record that the max was achieved via a self-transition with a duration $d = 1$.

It is possible to speed things up further by using the fact that certain features are (almost) always present in some of the states. For example, every initial exon must start with a start codon ATG, and every terminal exon must end with one of three possible stop codons; TAA, TAG, or TGA. Similarly, almost all donor sites have consensus GT as the first two bases of the intron, and almost all acceptor sites have an AG consensus as the final two bases of the intron. If we are willing to limit ourselves to genes matching these rules only, we can restrict the summation over the state length in the forward algorithm to sum only over lengths that are consistent with these rules. The extent to which this reduces the computations will depend on the composition of the sequence being analyzed.

Repeat masking can also help in speeding up the algorithms. The key observation is that certain repeats (in particular long interspersed repeats such as the *Alu* repeat) never occur in coding exons. Therefore, it is possible to substantially reduce the number of potential exons to be considered (and summed over in the algorithm). The effect on the computational complexity will depend on the frequency of repeats and their structure in the sequence being analyzed.

2.2.4.3 Exon Probabilities

One of the attractive features of using a GHMM for gene prediction is that it provides a natural way of computing the *posterior* probability of a predicted exon, given the observed data. Say that we have predicted an exon E^* of type s_e between bases a to b, such that the length of the exon is $d = (b - a + 1)$. We would like to compute the probability that the prediction is correct, i.e., the probability that the exon is part of a real gene and is predicted in the correct frame.

$$\mathbf{P}(E^* \in s_e \text{ is correct}|Y_1^T) = \mathbf{P}\left(\bigcup_{l=1}^{L}(X_l = s_e, d_l = d, p_{l-1} = a - 1)\middle|Y_1^T\right)$$

$$= \frac{\mathbf{P}\left(Y_1^T, \bigcup_{l=1}^{L}(X_l = s_e, d_l = d, p_{l-1} = a - 1)\right)}{\mathbf{P}(Y_1^T)}. \tag{2.88}$$

The denominator is simply the probability calculated in (2.85). To simplify the numerator in (2.88), we recall that for any given pair of I-states surrounding an E-state, the connecting E-state is uniquely defined. The opposite holds true as well; every exon type is surrounded by a unique pair of I-states. Thus, if $X_l = s_e$ we can determine the previous and next I-states, call them i^- and i^+. The union in the numerator in (2.88) can then be split into two unions, one for the preceding I-state i^- and one for the subsequent exon and I-state i^+, and the desired probability can be calculated using intermediate values of the forward and the backward algorithms.

$$\mathbf{P}\left(Y_1^T, \bigcup_{l=1}^{L}(X_l = s_e, d_l = d, p_{l-1} = a - 1)\right)$$

$$= \mathbf{P}\left(Y_1^T, \bigcup_{l=1}^{L}(X_l = i^-, p_l = a - 1), \bigcup_{l=1}^{L}(X_l = s_e, X_{l+1} = i^+, d_l = d, p_l = b)\right)$$

$$= \mathbf{P}\left(Y_1^{a-1}, \bigcup_{l=1}^{L}(X_l = i^-, p_l = a - 1)\right)$$

$$\cdot \mathbf{P}\left(Y_a^b, \bigcup_{l=1}^{L}(X_l = s_e, X_{l+1} = i^+, d_l = d, p_l = b)\middle|Y_1^{a-1}, \bigcup_{l=1}^{L}(X_l = i^-, p_l = a - 1)\right)$$

$$\cdot \mathbf{P}\left(Y_{b+1}^T\middle|Y_1^b, \bigcup_{l=1}^{L}(X_l = i^+, p_l = b + 1)\right)$$

$$= \alpha_{i^-}(a - 1)\,a_{i^-,e}\,f_{s_e}(d)\,b_{s_e}(Y_a^b|Y_1^{a-1})\,\beta_{i^+}(b + 1). \tag{2.89}$$

This probability can be interpreted as the probability that there is an exon of particular type s_e running exactly from positions a to b in the sequence. The forward variable $\alpha_{i^-}(a - 1)$ represents the probability of all possible parses to the left of the exon that ends in the appropriate I-state i^-, while the backward variable $\beta_{i^+}(b + 1)$ captures the probability of all parses to the right of the exon, beginning in I-state i^+ (see Fig. 2.13). Thus, the probability is constituted by the sum over all possible pairings of parses to the left and the right of the exon in question, and not only by intrinsic properties of the exon model itself. While the exon scores in many other gene finders depend on local properties such as splice signals and codon composition, the exon probability in Genscan is affected by the entire sequence. For instance, the probability of an initial exon will be boosted if it is preceded by a strong promoter signal at an appropriate distance upstream of a. This procedure, generally referred to

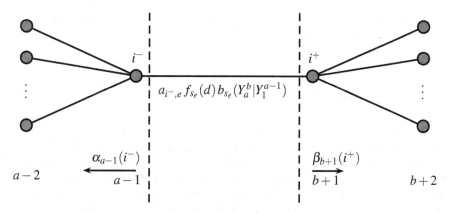

Fig. 2.13 Illustration of the forward–backward procedure for calculating the probability of a given predicted exon

as the *forward–backward algorithm*, is presented in [30] as a method for reestimating parameters in the training process, and is discussed in more general terms in Sect. 6.5.

We might prefer an exon probability that is less specific than the one in (2.89). For instance, we might want to know the probability that there is any kind of exon at all in a certain region, rather than having to specify the exon type. There is a second kind of probability that can help address this issue.

Note that the probability of being in (and at the end of) state i at time t is given by

$$\alpha_i(t)\beta_i(t) = \mathbf{P}(Y_1^T, \bigcup_{l=1}^{L}(X_l = i, p_l = t)). \tag{2.90}$$

This yields

$$\sum_{i\in I}\alpha_i(t)\beta_i(t) = \mathbf{P}(Y_1^T, \text{state } i \text{ at time } t \text{ is an } I\text{-state}). \tag{2.91}$$

Therefore, if we normalize (2.91) by the probability of the entire sequence $\mathbf{P}(Y_1^T)$, and subtract the result from 1, we get

$$\mathbf{P}(\text{the hidden state at time } t \text{ is some kind of exon} \mid Y_1^T) =$$

$$= 1 - \frac{\sum_{i\in I}\alpha_i(t)\beta_i(t)}{\mathbf{P}(Y_1^T)}. \tag{2.92}$$

This offers an alternative kind of probability to the exon probability in (2.89). It does not help much in determining what kind of exon it is or where its boundaries are, but may help indicating alternative candidate exon regions. Such regions could be missed by the Viterbi algorithm since the Viterbi only determines the single most likely sequence of exons, and there could be highly probable alternative splicings of a gene that goes undetected.

2.3 Interpolated Markov Models (IMMs)

Markov models have successfully been used as content sensors in DNA sequence analysis, both as discriminators between coding and noncoding sequences (see Chap. 5) as well as the detection of regular motifs such as eukaryotic promoters [25]. Usually higher order Markov models are required, due to long-range dependencies within a sequence. For instance, a model for coding regions should at least be of 2nd-order, because of the organization of nucleotides into codons. A 5th-order or higher would be even more preferred, since neighboring codons tend to be dependent as well. Basically, the higher the order the more sensitive the model is. The drawback, however, is that as the order increases, the required size of the training data grows exponentially. For instance, in a training model of order k there are 4^{k+1} probabilities to estimate. Thus, for a 5th-order model the training set has to be large enough to contain all 4096 possible hexamers, and sufficiently many times to provide reliable estimates, which is rarely the case in gene finding. The training data gets even more sparse, when used for the recognition of regulatory motifs such as promoters [25], or for automatic correction of sequence errors in low-quality data such as ESTs [39]. A common solution to the problem of sparse training data is to "smooth" the parameter estimates in some way in order to avoid zero probabilities. Smoothing strategies include using *pseudocounts*, *backing-off procedures*, or *interpolation techniques*. Using pseudocounts simply involves various ways of adding extra counts to the observed frequencies (see Sect. 6.2). Backing-off procedures involve setting the model to operate on a lower order when training data is insufficient.

In *interpolated Markov models* (IMMs) the order of the model is not fixed. Instead an interpolation of several Markov models of different orders is used. In pattern recognition these models are called *variable-order Markov models* [2], in data compression they go under the name of *variable-length Markov models* or *context trees* [31], and in speech recognition they are commonly referred to as *stochastic language models* [36]. The idea of IMMs is that although some oligomers occur too rarely to give reliable estimates, some may be very frequent, and would provide useful information to the prediction if included. Thus, instead of falling back to a lower order Markov model altogether, an IMM attempts to use the extra strength of the higher order, whenever there is data to support a longer context. As a result IMMs can capture both large and small dependencies in the sequence corresponding to the available statistics in the training set. Although an IMM is usually less powerful than a hidden Markov model, it has proved successful for many applications, where the problem of sparse training sets is frequent.

2.3.1 Preliminaries

The likelihood of a sequence Y_1^T can be decomposed as

$$\mathbf{P}(Y_1^T) = \prod_{t=1}^{T} \mathbf{P}(Y_t|Y_1^{t-1}). \tag{2.93}$$

However, using the entire previous sequence as context requires a huge training set, and is very expensive computationally. Thus a common approximation is to use an upper limit k of the context length.
The resulting model becomes a kth-order Markov model

$$\mathbf{P}(Y_1^T) \approx \prod_{t=1}^{T} \mathbf{P}(Y_t|Y_{t-k}^{t-1}). \tag{2.94}$$

Now, assume that we want to classify a sequence into one of N possible states, or classes $S = \{s_1, \ldots, s_N\}$. The conditional probabilities in (2.94) need to be estimated for each class, from a training set of known classification. The sequence is then classified into the state with the highest likelihood

$$s^* = \underset{s_i \in S}{\operatorname{argmax}} \ \mathbf{P}(Y_1^T|s_i). \tag{2.95}$$

The maximum likelihood (ML) estimates of the conditional probabilities in (2.94) are given by

$$\hat{P}(Y_t|Y_{t-k}^{t-1}) = \frac{f(Y_{t-k}^t)}{f(Y_{t-k}^{t-1})}, \tag{2.96}$$

where $f(Y_a^b)$ denotes the frequency count of the sequence Y_a^b. One problem, with the ML-estimates, however, is that when some k-mers are very infrequent, they may yield very unreliable estimates, or probability zero even. Even though some such k-mers may actually not belong to the specific class, and should yield a zero count, others may be missing due to sparse training data. The trick used in IMMs to overcome this problem, is to combine the Markov models of different orders. The next section describes two different interpolation schemes used to combine k-mers of different lengths in order to smooth the estimates of low-frequent oligomers.

2.3.2 Linear and Rational Interpolation

Interpolated Markov models can be seen as a generalization of fixed-order Markov models, where a combination of models of different orders is used. Instead of one

fixed order, the conditional probabilities in (2.94) are estimated using a combination
of the ML-estimates in (2.96). *Linear interpolation* [36] can be written as

$$\tilde{P}(Y_t|Y_{t-k}^{t-1}) = \rho_0 \frac{1}{L} + \rho_1 \hat{P}(Y_t) + \rho_2 \hat{P}(Y_t|Y_{t-1}) + \cdots + \rho_k \hat{P}(Y_t|Y_{t-k}^{t-1}), \quad (2.97)$$

where $\hat{P}(Y_t|Y_{t-k}^{t-1})$ is the ML estimate in (2.96). The coefficients ρ_i are positive
constants that sum to one, such that $\tilde{P}$ is still a probability, and the factor $1/L$ is a
kind of "pseudocount" that ensures that none of the conditional probabilities are set
to zero. The ML-estimates $\hat{P}$ of the different order models are based on counts of
oligomers in the training data. The coefficients ρ_i can be optimized with respect to
the likelihood using the EM-algorithm described in Sect. 6.4, where they are treated
as hidden variables in an HMM.

One problem with linear interpolation is that all the different orders in (2.97)
are treated equally, although some estimates may be less reliable than the others.
Rational interpolation [36] includes a weight function that scores the reliability of a
context Y_{t-i}^{t-1}, such that

$$\tilde{P}(Y_t|Y_{t-k}^{t-1}) = \frac{\displaystyle\sum_{i=0}^{k} \rho_i \cdot g(Y_{t-i}^{t-1}) \cdot \hat{P}(Y_t|Y_{t-i}^{t-1})}{\displaystyle\sum_{i=0}^{k} \rho_i \cdot g(Y_{t-i}^{t-1})} \quad (2.98)$$

where the denominator is needed for normalization. The weight function g can be
chosen different. For instance in [36] a sigmoid function is chosen, where the shape
depends on a constant bias C

$$g(Y_{t-i}^{t-1}) = \frac{f(Y_{t-i}^{t-1})}{f(Y_{t-i}^{t-1}) + C}. \quad (2.99)$$

For $C = 0$ the model reduces to linear interpolation, but for $C > 0$ we obtain the
rational interpolation model

$$\tilde{P}(Y_t|Y_{t-k}^{t-1}) = \frac{\displaystyle\sum_{i=0}^{k} \rho_i \frac{f(Y_{t-k}^t)}{f(Y_{t-k}^{t-1}) + C}}{\displaystyle\sum_{i=0}^{k} \rho_i \frac{f(Y_{t-k}^{t-1})}{f(Y_{t-k}^{t-1}) + C}}. \quad (2.100)$$

The bias has the most impact on small datasets, and the larger the training set,
the smaller the influence of C. One problem with rational interpolation is that the

EM-algorithm cannot be used to estimate the coefficients ρ_i. Instead a gradient descent method can be used to reach a local optimum [36].

2.3.3 GLIMMER: A Microbial Gene Finder

Microscopic organism that are too small to be observed by the naked eye are often referred to as *microbes*, or *microorganisms*. They do not constitute a specific classification, but can be found in almost all different taxa of life, including bacteria, animals, fungi, and plants, and include both prokaryotes and eukaryotes. Microbes are typically unicellular, and exist anywhere in the biosphere where there is liquid water, and they can survive extreme conditions such as heat, cold, acidity, salt, and darkness. Besides their importance in a wide variety of areas such as food production, water treatment (removing contaminants), and energy production (fermenting ethanol), microbes pose as important models and tools for biotechnology, biochemistry, genetics, and molecular biology. Examples of popular microbes are the budding yeast *Saccharomyces cerevisiae* and the bacterium *Escherichia coli*.

Splicing is rare in microbial genomes (as in prokaryotes in general, see Sect. 2.1.7). Thus, the issue of microbial gene finding is not to find actual coding sequences and determine the gene structures, but to identify the correct reading frame, and to separate overlapping genes. The main component in a microbial gene finder is, thus, the content sensor that scores coding potential and captures dependencies between nucleotides in open reading frames (ORFs). In addition, only some preprocessing to select potential ORFs, and some post-processing to handle overlapping gene candidates are necessary.

GLIMMER (Gene Locator and Interpolated Markov ModelER) is a microbial gene finder, particularly suited for bacteria, archaea, and viruses [9, 10, 35]. It uses an IMM to discriminate between coding and noncoding regions. The program consists of two sub-modules; one that builds the IMM from training data, and one that uses the resulting model to score new sequences.

2.3.3.1 Gene Prediction

GLIMMER scores a new sequence Y_1^T using a kth-order IMM

$$\mathbf{P}(Y_1^T) = \sum_{t=1}^{T} \text{IMM}_k(Y_{t-k}^t), \tag{2.101}$$

where IMM_k is calculated as

$$\text{IMM}_k(Y_{t-k}^t) = \rho_k(Y_{t-k}^{t-1}) \cdot \mathbf{P}(Y_t|Y_{t-k}^{t-1}) + (1 - \rho_k(Y_{t-k}^{t-1})) \cdot \text{IMM}_{k-1}(Y_{t-k}^t). \tag{2.102}$$

Table 2.3 Example in [45] of the gene prediction output of GLIMMER

orfID	start	end	frame	score
--------	------	-----	--	-----
> *Escherichia coli* O157:H7				
orf00001	11952	98	-3	2.84
orf00003	351	133	-1	5.25
orf00004	312	2816	+3	11.33
orf00005	2854	3750	+1	10.02
orf00007	3751	5037	+1	13.63

The training of the probabilities $\mathbf{P}(Y_t|Y_{t-k}^{t-1})$ and the coefficients ρ_k is described in the next section. The gene prediction procedure of GLIMMER goes as follows:

1. Identify all ORFs longer than a given threshold in the input sequence.
2. Score the ORFs in each of the six reading frames, using (2.101).
3. Select ORFs scoring higher than a given threshold for further analysis.
4. Examine selected ORFs for overlaps.
5. Report ORFs that passed the overlap analysis.

Instead of scoring the entire input sequence, open reading frames (ORFs) exceeding a given minimum length are selected and scored using (2.101). Gene dense genomes usually contain overlapping genes, and such occurrences are investigated further. If two ORFs of different reading frames overlap by more than some preset minimum, the overlapping region is scored separately in all six reading frames, and if the longer ORF scores higher than the shorter in this region, the shorter ORF is dropped from the analysis. GLIMMER outputs a set of predicted genes, along with notes on overlaps that may need further examination. An example of the output format of the final gene predictions is given in Table 2.3 (taken from [45]). The columns represent the ORF identifier, the start and stop coordinates of the genes, the reading frame, and the gene score.

A further extension of the IMMs used in GLIMMER are the *Interpolated context models* (ICMs). While an IMM can use as many bases in the context as the training data allows, an ICM can use *any* bases in the given context. That is, bases not necessarily adjacent to one another. The idea is to capture the high dependence on codon position when scoring a base. Often the ICM will choose a context identical to that the corresponding IMM would have chosen, but sometimes, as in the case with the third codon position, a slightly different model will be used.

2.3.3.2 Training the IMM

Seven submodels are trained by GLIMMER on a set of known sequences; one for each reading frame (three for each strand), and one for noncoding regions. Each submodel is trained by counting the occurrences of all oligomers of lengths $1, \ldots, k+1$ where k is a predefined maximum context depth (default is $k = 8$ in GLIMMER). The last base of the oligomer defines the frame, and the preceding bases represent the context of that base. The likelihoods $\mathbf{P}(Y_t | Y_{t-k}^{t-1})$ in (2.102) are estimated directly from these frequency counts using

$$\mathbf{P}(Y_t | Y_{t-k}^{t-1}) = \frac{f(Y_{t-k}^t)}{\sum_{y \in V} f(Y_{t-k}^{t-1}, y)}, \tag{2.103}$$

where (Y_{t-k}^{t-1}, y) denotes the concatenation of the context sequence and symbol $y \in \{A, C, G, T\}$.

The coefficients $\rho_k(Y_{t-k}^{t-1})$ are determined using two criteria: if the context Y_{t-k}^{t-1} occurs frequently enough, the actual frequency is used and the weight ρ_k is set to 1. The frequency threshold used is 400, and gives about 95 % confidence that the estimated probabilities are within ± 0.05 of their true value [35]. Otherwise, if a context occurs less than 400 times in the training set, the frequency of the context is compared to the IMM-value of the one base shorter context, to see if the longer context adds information to the prediction.

The comparison is performed using a χ^2-test

$$\rho_k(Y_{t-k}^{t-1}) = \begin{cases} 0 & \text{if } c < 0.5, \\ \dfrac{c}{400} \displaystyle\sum_{y \in V} f(Y_{t-k}^{t-1}, y) & \text{if } c \geq 0.5 \end{cases} \tag{2.104}$$

where c is the probability, taken from the χ^2-distribution, that the frequencies of the longer context differ from the IMM-values of the shorter context. That is, if we let

$$X^2 = \sum_{y \in \{A,C,G,T\}} \frac{\left[f(Y_{t-k}^{t-1}, y) - \text{IMM}_{k-1}(Y_{t-k+1}^{t-1}, y) \right]^2}{\text{IMM}_{k-1}(Y_{t-k+1}^{t-1}, y)}, \tag{2.105}$$

where X^2 is χ^2-distributed with 3 degrees of freedom, the probability c is given by

$$c = \mathbf{P}(\chi_3^2 \geq X^2). \tag{2.106}$$

In effect, the frequencies of the longer context are compared to the IMM-values of the one base shorter context, and if there is a significant difference between contexts the longer context serve as a better predictor and gets a higher value on the coefficient ρ. If there is little difference, meaning that the longer context adds no significant information, the longer context model gets a lower value on ρ.

2.3.3.3 GlimmerM

Eukaryotes such as the yeast *S. cerevisiae*, or the malaria parasite *Plasmodium falciparum*, are commonly analyzed using gene finders optimized for human. While these genomes have a gene density that is significantly lower than for microbes, they are still very gene rich, and a prokaryote gene finder may perform better than a human gene finder. GlimmerM is based on the GLIMMER method, but is optimized for gene densities around 20%, and has incorporated the *GeneSplicer* splice site detector [27]. Moreover, GlimmerM uses a combination of decision trees and IMM-based exon scoring. The decision trees, built by the OC1 system [24], estimate the probability that a given sequence is coding, and the resulting gene models are accepted if the IMM-score for the coding sequence is above a certain threshold.

2.4 Neural Networks

Artificial neural networks were first developed in an attempt to mimic the information processing and learning of the biological nervous system, such as the brain, in order to acquire some of their immense computational power. While the artificial neurons used in neural networks today remain quite far from their biological counterpart, their computational power has proved useful in a number of fields. Neural network models have traditionally been used in speech and image recognition, but has become more and more popular as components in DNA sequence analysis. In general, neural networks are suitable for classification problems with computationally complex patterns and many hypotheses to be evaluated in parallel.

Neural networks are essentially nonlinear mappings between a set of input variables and a set of output variables. An advantage of neural networks over other such mappings is that while many other techniques grow exponentially with the dimension of the input space, neural networks typically only grow linearly, or quadratically, with input dimension. We give a brief overview of the neural networks most commonly used in computational biology, the backpropagated feed-forward neural networks. For a more thorough treatment, see for instance [1].

2.4.1 Biological Neurons

Biological nervous systems, such as the brain, consist of myriads of *neurons*, which are specialized cells designed to process and transmit information. Learning, for instance, takes place when neurons communicate with each other. Each neuron can connect to several thousands of others, and multiple neurons can fire in parallel. Hence, the human brain, consisting of something like 10^{11} neurons, constitute a parallel processor with a capability that is vastly superior to the most advanced computer clusters that exist today.

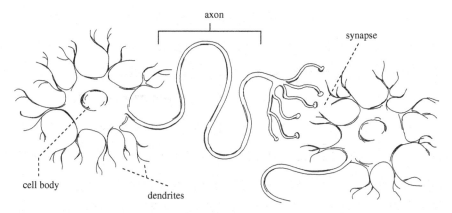

Fig. 2.14 A biological neuron consists of a cell body, a dendritic tree, and an axon. The space between the axon and the dendrites of the next neuron is called the synapse. The neuron receives signals on its dendrites, and transport them through the axon and into the synapse over to the next neuron

A neuron is typically composed of a cell body, a *dendritic tree*, and an *axon* (see Fig. 2.14). The neuron receives signals on the dendrites and releases (fires) signals through the axon. Connected neurons are separated by a small physical gap called a *synapse*. The information is carried through the system in the form of electrochemical pulses, or *action potentials*, that are passed on from neuron to neuron. A neuron can receive thousands of such pulses from different neurons, and each pulse may change the potential of the dendritic membrane, either by inhibiting or exciting the generation of further pulses. If the sum of these pulses exceeds a certain threshold the neuron "fires" by generating a new pulse that travels into the synapse and over to the next neuron.

2.4.2 Artificial Neurons and the Perceptron

An *artificial neuron* is, similarly to a biological neuron, composed by a cell body, dendrites, and an axon (see Fig. 2.15). The inputs, that are received through the dendrites, get integrated in some manner, and if the result exceeds a given threshold, the neuron transmits an output. Thus, an artificial neuron is simply a computational unit, that maps input values to one or more outputs. The computation is done in two steps: first the input values $\mathbf{x} = (x_1, \ldots, x_N)$ are integrated into a single value a, through some *integration function* $a = h(\mathbf{x})$, and then this value is transformed by some nonlinear function g, called the *activation function*, to produce an output value $y = g(a) = g(h(\mathbf{x}))$.

The simplest kind of artificial neurons, first proposed by McCulloch and Pitts [22], uses binary values (0 or 1) both for inputs and the output. The integration function is

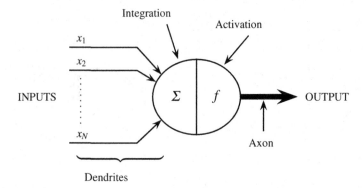

Fig. 2.15 An artificial neuron attempts to mimic a biological neuron, and consists of a cell body, dendrites, and an axon. The inputs are weighted and summed, before the activation function decides whether the neuron should fire or not

an unweighted sum of *excitatory* and *inhibitory* edges. If any of the inhibitory edges is 1, the neuron is inhibited and the output is 0.

Otherwise, if all inhibitory edges are 0, the integrated value is the sum of the excitatory edges

$$a = \sum_{i=1^N} x_i. \tag{2.107}$$

The activation function is the *Heaviside step function* (or *threshold function*)

$$\phi(a) = \begin{cases} 1 & \text{if } a > \theta, \\ 0 & \text{otherwise}, \end{cases} \tag{2.108}$$

where θ is called the *threshold*, or the *bias*. If the integrated value a exceeds the threshold, and ϕ takes value 1, the neuron fires. Otherwise it takes the deactivated value 0.

However, although very useful for the computation of logical functions in finite automatons, the McCulloch–Pitts neurons are rather limited. A generalization of the McCulloch–Pitts neuron, called the *perceptron*, was developed by Rosenblatt [33]. In its simplest form, it is basically the McCulloch–Pitts neuron with real-valued inputs and associated weights. The input values $x_1, \ldots, x_N, x_i \in \mathbb{R}$, are fed into the node through edges with associated weights $w_1, \ldots, w_N$. The integration function is the weighted sum

$$a = \sum_{i=1}^{N} w_i x_i, \tag{2.109}$$

Fig. 2.16 A single-layer
network diagram

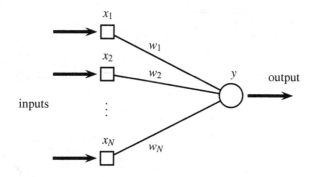

and the activation function is the same threshold function (2.108) as in the McCulloch–
Pitts neuron. A network only consisting of one neuron like this, is sometimes called a
single-layer network, because it consists of a single layer of weights (see Fig. 2.16).
In analogy with the biological neuron, the inputs x_i represent the level of activity of
the neurons connected to the current neuron, and the weights w_i signify the strengths
of these connections.

A further generalization of the perceptron is to allow for more general activation
functions. The activation function somehow determines how powerful the output of
the neuron should be. While biological neurons choose between "fire" or "not fire",
mathematically it is more convenient with a smoother (differentiable) activation
function.

A popular choice is the *logistic sigmoid* function

$$\phi(a) = \frac{1}{1 + e^{-\sigma(a+\theta)}}, \tag{2.110}$$

where θ is the bias that moves the curve away from zero, and σ a parameter that
affects the steepness of the curve. The bias θ can be viewed as the number of pulses
needed for the neuron to fire. Training a neural network involves estimating the
values of the edge weights and of the bias parameter. For convenience it is common
to invoke the bias into the network by adding an extra input variable $x_0 \equiv 1$, and an
associated edge $w_0 = -\theta$ (see Fig. 2.17). Using this, and assuming $\sigma = 1$, gives the
simpler form of the activation function

$$\phi(a) = \frac{1}{1 + e^{-a}}. \tag{2.111}$$

A small modification to (2.110) gives the 'tanh' activation function

$$\phi(a) = \frac{e^a - e^{-a}}{e^a + e^{-a}}, \tag{2.112}$$

Fig. 2.17 A single-layer
network diagram with an
added bias node

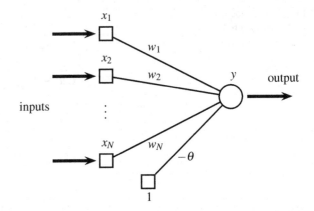

which is symmetric and therefore may achieve faster convergence of the training
algorithms in some cases.

2.4.3 Multilayer Neural Networks

The architecture of a neural network is usually either *feed-forward* or *recurrent*. A
feed-forward network is devoid of loops; it is a directed acyclic graph where the
information moves in only one direction, from the input nodes, possibly through one
or more hidden layers, and to the output nodes. The counterpart, recurrent networks,
contain cycles. We will only consider feed-forward networks here, since almost all
applications in computational biology use layered feed-forward network models.

A multilayer neural network is a further generalization of the single-layer network,
where the network function is composed by several successive functions. In a network
architecture, this can be seen as successive layers of nodes, or processing units, with
connections running from the nodes in one layer to the next. A node can be either
hidden or *visible*, where visible nodes are typically those connected to the outside
world, such as the input and the output nodes, and the hidden nodes occupy layers in
between. A layer that consists of only hidden nodes is called a *hidden layer*, and the
total number of layers define the *depth* of the network. A layered network does not
contain cycles, and usually each node in one layer is connected to each of the nodes
in the next. Figure 2.18 illustrates a two-layer feed-forward network.

Note that we choose not to include the input layer when counting the depth of
the network. This is because the input nodes are not really processing units, but
only holders of the input values. With this convention, the depth corresponds to the
layers of weights to be estimated from training data. Also, a multilayer network does
not have to be fully connected as in Fig. 2.18; a more economical model would be
preferred whenever possible.

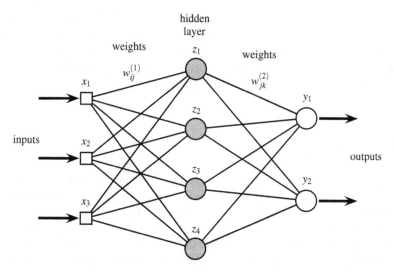

Fig. 2.18 A two-layer feed-forward network diagram

Consider a two-layer network with N input units $(x_1, \ldots, x_N)$, a hidden layer of M hidden units $(z_1, \ldots, z_M)$, and K output units $(y_1, \ldots, y_k)$.
If ψ is the activation function of the hidden units and ϕ the activation of the output units, the network can be represented mathematically as

$$y_k = \psi\left(\sum_{j=1}^{M} w_{jk}^{(2)} \cdot z_j\right), \quad k = 1, \ldots, K$$

$$= \psi\left(\sum_{j=1}^{M} w_{jk}^{(2)} \cdot \phi\left(\sum_{i=1}^{N} w_{ij}^{(1)} x_i\right)\right). \tag{2.113}$$

While it is possible to use different activation functions for different layers, it is common to use the same for all, such that $\psi = \phi$. A *multilayer perceptron* is a multilayer network, with either the threshold function (2.108) or the logistic sigmoid function (2.110) as activation function. The advantage with the sigmoid function is that it is differentiable, which enables the use of a very powerful training procedure called the *backpropagation algorithm* described in Sect. 6.7.

2.4.4 GRAIL: A Neural Network-Based Gene Finder

GRAIL [43, 44] is a neural network-based gene finder that scores potential exons by combining the scores of a number of content and signal sensors. Four types of exons are recognized: initial, internal, terminal, and single exons. These exon types represent open reading frames in combination with their specific boundaries: start codon to donor site (initial), acceptor to donor site (internal), acceptor site to stop codon (terminal), or start to stop codon (single). The gene prediction is performed in four separate steps:

1. Extract all possible exon candidates.
2. Remove improbable exons.
3. Score remaining exons.
4. Construct gene models.

The first step is a preprocessing step, where all possible exons in the sequence are extracted. A candidate exon consists of an open reading frame surrounded by the corresponding exon boundaries. This first step produces a huge number of candidates, typically several thousands just in a sequence of 10,000 bp [43]. Thus, in the second step a number of heuristic rules are applied to remove improbable exons. In the third step, all remaining exon candidates are scored by a feed-forward neural network, which has been trained by the backpropagation algorithm described in Sect. 6.7. The input to the network is a feature vector of various coding measures and splice site scores for each exon candidate. In the fourth and final step, the scored exon candidates are combined into frame-consistent gene models.

The GRAIL neural network consists of 13 input nodes, two hidden layers with seven, and three nodes, respectively, and one output node. A network diagram is shown in Fig. 2.19. The hidden layer of seven nodes, not shown in the figure, is part of the splice site scoring. A mathematical representation of the network can be written as

$$y = \phi \left(\sum_{k=1}^{3} w_k^3 \, \phi \left(\sum_{j=1}^{7} w_{kj}^2 \, \phi \left(\sum_{i=1}^{13} w_{ji}^1 \, x_i \right) \right) \right), \tag{2.114}$$

where ϕ is the logistic activation function

$$\phi(x) = \frac{1}{1 + e^{-x}}. \tag{2.115}$$

The weights w are trained using the backpropagation algorithm.
During training, the output is evaluated using a matching function M, that measures the overlap of the candidate exon with the true exon(s),

$$M(\text{candidate}) = \frac{\sum_i m_i}{\text{length}(\text{candidate})} \frac{\sum_i m_i}{\sum_j \text{length}(\text{exon}_j)}, \tag{2.116}$$

where $\sum_i m_i$ is the number of bases of the candidate exon that overlap true exons, and $\sum_j \text{length}(\text{exon}_j)$ is the total length of all exons that overlap the candidate. Thus, $0 \leq M \leq 1$ with

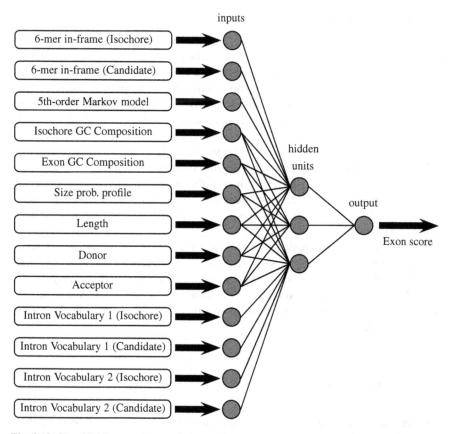

Fig. 2.19 The GRAIL neural network for scoring candidate exons. The network consists of 13 input nodes, two hidden layers of seven (not shown), and three nodes, respectively, and one output node delivering the final exon score. The figure is reproduced from [43], ©1996 IEEE

$$M = \begin{cases} 1 & \text{if prediction is correct} \\ 0 & \text{if no overlap with true exons.} \end{cases} \qquad (2.117)$$

The feature vector fed into the GRAIL neural network consists of 13 measures for each candidate exon, including various coding measures and splice site measures [43]. Coding potential is scored using both a frame-dependent 6-tuple preference model and a fifth-order inhomogeneous Markov model. These measures are not independent, but by applying supervised learning (labeled training examples), the weights are adapted for all features together. The splice site detector in GRAIL is in itself a neural network, that combines the scores from several measures. Neural networks applied to splice site detection are described in more detail in Sect. 5.4.4.

2.5 Decision Trees

A *decision tree* is a kind of tree diagram that can be used to choose between different decisions for an object, by connecting series of tests on different features of the object. Decision trees are a common ingredient in clinical research, in which various features of the patient lie as ground for diagnosis into one of two or more clinical categories. Traditional statistical methods struggle in such situations, where the set of possible features may be large, or the interactions between the features are complex, or the feature values do not follow a known distribution. Moreover, the outcome of the analysis may be difficult to interpret, for instance if diagnosis is presented in terms of probabilities. An advantage of decision trees is that it enables the reduction of rather complex datasets into simple and comprehensible data structures. In addition, being a nonparametric technique, decision trees avoid the problems of making assumptions about the distribution.

Decision trees can be applied to classification problems, in which objects need to be classified into different classes based on a set of features, or attributes, that characterize the object. In this context decision trees are also called *classification trees*. Here we give a brief overview of the decision tree theory applied to single species gene finding. For a more thorough treatment, confer for instance the books by Breiman et al. [6] or Quinlan [29].

2.5.1 Classification

Decision trees can be used to classify an object based on a set of features that characterize the object. A decision tree consists of internal nodes and leaf nodes. The leaf nodes contain the class labels, and each internal node performs a test on one specific features. A new object is classified by passing it down from the root of the tree, through a series of tests on its features, finally ending up in one of the leaf nodes. In each node the corresponding feature is tested, and depending on the answer the object is passed down into one of its child nodes. The process is recursed until the object reaches a leaf node and receives its classification. In other words, given a set of features, a decision tree represents a series of rules that are used for classification of the corresponding object. The features can be of any type, binary, categorical, or numerical, while the class labels must be qualitative.

Using an existing decision tree for classification is easy. The trick to decision tree analysis is the actual construction of the tree, called *decision tree learning*, using a training set of objects with corresponding feature values and known class labels. Given a large set of possible features, decision tree learning techniques have been developed to choose both which features that are relevant, and in which order they are to be tested. Example 2.5 illustrates a simple dataset, borrowed from [28], containing only categorical feature values.

Table 2.4 A simple decision tree training set

Object	Features				Class
	Outlook	Temperature	Humidity	Windy	
1	Sunny	Hot	High	False	N
2	Sunny	Hot	High	True	N
3	Overcast	Hot	High	False	P
4	Rain	Mild	High	False	P
5	Rain	Cool	Normal	False	P
6	Rain	Cool	Normal	True	N
7	Overcast	Cool	Normal	True	P
8	Sunny	Mild	High	False	N
9	Sunny	Cool	Normal	False	P
10	Rain	Mild	Normal	False	P
11	Sunny	Mild	Normal	True	P
12	Overcast	Mild	High	True	P
13	Overcast	Hot	Normal	False	P
14	Rain	Mild	High	True	N

With kind permission from Springer Science + Business media: [28, p. 87, Table 1]

Example 2.5 A simple decision tree training set
The following example is borrowed from [28]. In this training set the observed objects are *Saturday mornings*. Suppose we use a number of different weather features to determine whether we will undertake a certain activity or not. The classification of the objects is thus either *P* or *N* for positive or negative instances, respectively, where a positive instance means that the activity will take place. The weather features and the corresponding values used are

Feature	Values
Outlook	Sunny, overcast, rain
Temperature	Cool, mild, hot
Humidity	Normal, high
Windy	True, false

The dataset is presented in Table 2.4. Given this training set we would like to build a decision tree that, based on the feature values of a new Saturday morning, can be used to determine whether the activity in question will happen or not. In the next section, we describe how to build such a decision tree from data. □

2.5.2 Decision Tree Learning

Depending on which order the features are tested, there are many ways to build a complete decision tree from the same training set. By the principle of *Occam's Razor*, the shortest hypothesis should always be preferable. Or, in terms of decision trees, the tree that is optimal for a given dataset is the smallest one. However, creating an algorithm that, for a general set of features, always finds the smallest tree is an NP-complete problem, basically meaning that it cannot be solved in reasonable time. Therefore, numerous algorithms have been created that search for close to optimal trees, among the most noted ones being ID3 [28], C4.5 [29], and CART [6]. These algorithms typically use a greedy recursive procedure which, while creating reasonable trees, cannot guarantee to find the optimal solution. Such algorithms typically consist of the following basic steps:

1. Determine the feature that best splits the data.
2. For each *pure* subset (all of the same class), create a leaf node with that class. For each impure subset, return to 1.
3. Stop when no more splits are possible and all paths end with a leaf node.

We call a set of objects *pure* if all objects belong to the same class, and *impure* otherwise. For instance, for a given feature, we can group the objects according to their feature values. If that grouping corresponds completely with the grouping according to class label, it represents a *pure split* of the dataset.

Which feature that best splits the data is determined using some kind of *measure of impurity*. A popular measure, for instance used by the ID3 algorithm, is the *Shannon entropy*, or simply *entropy*. Suppose that we have a training set D of n objects each characterized by a set of features $A_1, \ldots, A_p$, and each with a known class label $c_i \in C, i = 1, \ldots, n$, where C is the set of all classes. The entropy of such a set can be written as

$$H(D) = -\sum_{c \in C} p_c \log_2 p_c, \tag{2.118}$$

where the sum runs over all possible classes, and where p_c is the probability of belonging to class $c \in C$. The entropy basically measures the uncertainty, level of randomness, or *information content* of the dataset. The more uniform the distribution is, the higher the entropy. The base 2 of the logarithm transforms the value into "bits" commonly used in information theory. The entropy assigns measure zero to pure sets and reaches its maximum when all classes have equal probabilities. Alternative impurity measures include the *Gini index* and the *twoing rule*.

$$\text{Gini} = 1 - \sum_c p_c^2, \tag{2.119}$$

$$\text{Twoing} = \frac{|T_L||T_R|}{n^2} \left(\sum_{c \in C} \left| \frac{L_c}{|T_L|} - \frac{R_c}{|T_R|} \right| \right)^2, \tag{2.120}$$

where, for a split at node T containing n objects, $|T_L|$ and $|T_R|$ are the numbers of objects to the left and to the right of the split, respectively, and L_c and R_c are the numbers of objects having class label c to the left and the right of the split, respectively. The Gini index chooses the split attempting to separate as large a class from the rest as possible, while the twoing rule attempts to split the data as central as possible. Which splitting rule that works best depends on the application (cf. [5, 6]).

The feature that best splits the training data is the one that causes the largest decrease in impurity. The goal is to create descendant subsets that are purer than its parents. This decrease in impurity is calculated using a measure called the *information gain*: for a set D of n objects the information gain of splitting over a specific feature A is given by

$$IG(D, A) = H(D) - \sum_{v \in A} \frac{|D_v|}{|D|} H(D_v), \qquad (2.121)$$

where the sum runs over all possible feature values of A, D_v is the set of objects in D that take value v for feature A, and $|D_v|$ and $|D|$ denote the numbers of objects in each set (i.e., $|D| = n$). The second part of (2.121) in fact corresponds to an entity known as the *conditional entropy* $H(D|A)$ of D, given the attribute values of A.

Now we can calculate the information gain of splitting the dataset in each of the features. Then the feature with the highest information gain is chosen to be tested first and the test is placed in the root of the tree. Branches are created for each possible value of the feature, the dataset is split into subsets according to their values on the chosen feature, and the procedure is repeated in the child nodes.

Example 2.6 A simple decision tree training set (cont.)
We illustrate how the decision tree for the data in Table 2.4 is built using entropy and information gain. First, in order to calculate the entropy $H(D)$ in (2.118) of the entire dataset, we estimate the class probabilities by the relative frequencies for class labels P and N:

$$p_P = 9/14, \quad p_N = 5/14.$$

Thus, the entropy becomes

$$H(D) = -(9/14) \log_2(9/14) - (5/14) \log_2(5/14) \approx 0.940.$$

Next, if we were to split the data according to attribute 'Outlook', we would split the dataset into groups according to the feature values 'sunny', 'overcast', or 'rain'.

Outlook	Class	Outlook	Class	Outlook	Class
Sunny	N	Overcast	P	Rain	P
Sunny	N	Overcast	P	Rain	P
Sunny	N	Overcast	P	Rain	P
Sunny	P	Overcast	P	Rain	N
Sunny	P			Rain	N

The entropies of the subsets become

Sunny: $H(D_v) = -(3/5)\log_2(3/5) - (2/5)\log_2(2/5) \approx 0.971$
Overcast: $H(D_v) = -(4/4)\log_2(4/4) = 0$
Rain: $H(D_v) = -(3/5)\log_2(3/5) - (2/5)\log_2(2/5) \approx 0.971$

The resulting information gain for 'Outlook' thus becomes

$$IG(D, \text{Outlook}) = 0.940 - \left(\frac{5}{14} \cdot 0.971 + \frac{4}{14} \cdot 0 + \frac{5}{14} \cdot 0.971\right) \approx 0.247.$$

Similarly, we get for the other features

$$IG(D, \text{Temperature}) \approx 0.029$$
$$IG(D, \text{Humidity}) \approx 0.152$$
$$IG(D, \text{Windy}) \approx 0.048$$

We see that 'Outlook' achieves the highest information gain and we therefore place it in the root node. We draw three branches from this node, one for each of the values of 'Outlook', and continue. Next we note that the 'overcast' group is pure (all objects have label P), and we therefore insert a leaf node with class label P. The other two subsets are impure and need to be split further. The information gain is now calculated over the corresponding subsets of objects. For instance, the subset 'sunny' now contains $n = 5$ objects, and the information gain is calculated for the features 'Temperature', 'Humidity', and 'Windy' for this subset,

$$IG(D_{\text{sunny}}, \text{Temperature}) \approx 0.571$$
$$IG(D_{\text{sunny}}, \text{Humidity}) \approx 0.971$$
$$IG(D_{\text{sunny}}, \text{Windy}) \approx 0.020$$

Humidity achieves the highest information gain for this subset, and is placed in the corresponding node. The procedure continues until all subsets are pure and can be finished off with leaf nodes. The resulting tree is shown in Fig. 2.20. Note that the feature 'Temperature' is never used. The 'Temperature' feature is very impure, meaning that it has very weak (if any) association with the classification, and the tree reaches its leaf nodes without having to take that feature into consideration.   □

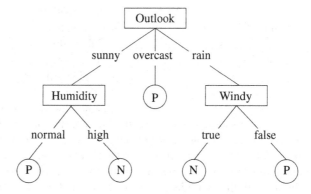

Fig. 2.20 The resulting tree of the data in Table 2.4. With kind permission from Springer Science + Business media: [28, p. 87, Fig. 2]

The resulting decision tree classifies the objects in the training set perfectly. The risk is, however, that the tree is too specific to the training set, and will not be able to correctly classify new objects presented to it. This problem is known as *overfitting*, and is commonly solved by some kind of *pruning* procedure. Pruning basically means that parts of the tree will be cut off by turning internal nodes into leaf nodes. This makes the tree less specific to the training set, but more flexible to new data. Confer for instance [6, 29] for more details.

We have treated only categorical or binary feature values so far, but the feature values are allowed to be numerical as well. The node tests in the decision tree would then typically involve inequalities such as $x_i \geq 4.2$ versus $x_i < 4.2$, or possibly separation into several subintervals. There are many different methods for dealing with numerical values, but most of them involve discretizing the data in some manner in order to treat them as categorical values. A rather different treatment is introduced by the OC1 algorithm [24], used by the MORGAN gene finder presented next. OC1 does not split the data for one specific feature, but uses linear combinations of the feature values to determine the best decision tree.

2.5.3 MORGAN: A Decision Tree-Based Gene Finder

MORGAN (Multi-frame Optimal Rule-based Gene ANalyzer) [34] is a gene finder that combines decision trees with dynamic programming and signal sensor algorithms. The dynamic programming algorithm is used to search through all possible parses of the sequence, while the decision tree algorithm and the signal sensors provide scores of the different parts of the potential gene. The decision trees are built using the OC1 system [24], which uses something called *oblique tests* in the decision tree nodes. In order to estimate probabilities of a potential exon or intron, the OC1 also includes a random component which means that it can produce different trees for the same data each time it is run.

Before the MORGAN system can be trained, the training set, consisting of raw genomic DNA sequences with known exons and introns, is transformed into the form of objects, class labels, and features. This is done by first identifying all potential start, stop, donor, and acceptor sites, scoring above a certain threshold, in the training sequences. Next, candidate exons are identified by combining the corresponding boundary sites (start-donor for initial exons, acceptor–donor for internal exons, and acceptor-stop for terminal exons), and requiring an open reading frame (ORF) between the sites. Similarly, potential introns are identified by pairing up donor and acceptor sites, with an additional length constraint (between 20 and 16,000 bp), but without the ORF requirement. For each of the three types of exons and the intron, a decision tree is constructed. Since the true exons and introns are known in the training set, the identified candidate exons and introns receive a label that is either 'true' or 'false'. Thus, the objects are the potential exons and introns, and the class labels are 'true' or 'false' revealing which objects are real or not.

The features used by MORGAN to characterize the objects include boundary site scores, an in-frame hexamer statistic, and a position asymmetry statistic. The signal sensors used to score the boundary sites are a first-order Markov model for the start sites based on the *Kozak sequence* (see Sect. 5.3.2), and second-order Markov models for the splice sites. Since no consensus sequence is known for the sequence surrounding the stop sites, the stop codons are simply identified directly. These type of submodels are discussed further in Chap. 5.

The in-frame hexamer statistic for a subsequence between positions i and j in the sequence is given by

$$IF_6(i, j) = \begin{cases} \sum_{k=0,3,6,\dots,j-6} \log(f_k/F_k) \\ \sum_{k=1,4,7,\dots,j-6} \log(f_k/F_k) \\ \sum_{k=2,5,8,\dots,j-6} \log(f_k/F_k) \end{cases} \tag{2.122}$$

where f_k is the frequency of the hexamer starting in position k in coding sequences, and F_k is the frequency of the hexamer among all hexamers in the training set, in all reading frames [41]. The position asymmetry statistic, presented in [12], counts the frequency of each nucleotide in each of the three codon positions.

The OC1 system [24], used to build the decision trees, is specifically designed to handle numerical feature values. OC1 does not split the data according to their feature values, but uses *linear discriminant* kind of tests, where, instead of using interval tests such as $x_i \geq 4.2$, each internal node contains a linear combination of one or more features,

$$a_1 x_1 + a_2 x_2 + \cdots a_p x_p \geq a_{p+1}. \tag{2.123}$$

Since this linear combination represents a hyperplane that is nonparallel to the axes in feature space, this is called an *oblique split*.

After the decision tree is built, OC1 prunes the tree using a method called *complexity pruning* [6]. Basically, a complexity measure is calculated for each internal node based on the number of misclassifications that would result on the training set if that node were turned into a leaf, combined with the size of the subtree rooted at that node. The node with the largest complexity measure is then turned into a leaf. The series of increasingly smaller trees are then tested on a separate part of the training set, and the tree with the highest accuracy on this set is kept as the output of the system.

2.6 Conditional Random Fields

In gene prediction we want to connect the observed sequence data to a sequence of labels corresponding to the underlying gene model. A successful approach to this has been to employ hidden Markov models (HMMs), described earlier in this chapter. One disadvantage with HMMs, however, is that in order to make computations feasible, two rather strong independence assumptions have to be made: (i) given the current state (i.e., current sequence label), the next state is conditionally independent of everything else, and (ii) the observed output from each state only depends on the underlying state. With these assumptions, the HMM machinery comes together very nicely, but often the observed sequence include complex interdependencies that when ignored may significantly hurt classification performance. Conditional random fields (CRFs) [19] were developed mainly to fill this gap. CRFs offer an alternative to HMMs, where, instead of making simplifying assumptions, the model is extended to include interdependence features. The cost of this added flexibility, however, is increased computational complexity and a less straightforward interpretation of the parameters. This section gives a brief encounter of CRFs, in the context of computational gene prediction. More general and detailed descriptions can be found for instance in [19, 42].

2.6.1 Preliminaries

We recall from Sect. 2.1.1 that a *random process* is a collection of random variables that is indexed by some ordered set T. Such a collection can typically be used to model the evolution of a system or the development of a physical process over time, where the system or process switches randomly between *states*, or *phases*. If the index set T is ordered it is often referred to as "time", and the indexed collection of random variables can be lined up as in a chain of events. Hidden Markov models (HMMs), described in Sect. 2.1, are a special kind of random processes, that consist of two interrelated process: a Markov process that is hidden from the observer, corresponding to the state labels we want to predict (e.g., exons, introns, intergene, etc.), and an observed process corresponding to the observed output we wish to annotate (e.g., the DNA sequence).

A *random field* is a generalization of random processes where the process evolves in a multidimensional space, and the time index is replaced by a corresponding multidimensional coordinate vector. Random fields are useful for instance to model spatial data such as the pixels in image analysis, where both the position and the value (*attribute*) of the process are of interest. As for random processes, there are many kinds of random fields, but a family of models relevant to this section are the *Markov random fields*, also known as *Markov networks*. A Markov random field is a collection of random variables having a similar Markov property as for Markov chains, that can readily be described by an *undirected graph*. Basically, the Markov property for random fields state that given the neighbors in the graph, a random variable is conditionally independent of everything else. Markov random fields are similar to *Bayesian networks*, described in Sect. 5.4.7, in how the dependency structure is represented. The difference is that Bayesian networks are directed and acyclic, while Markov random fields are undirected and possibly cyclic. A *conditional random field* (CRF) is an extension of Markov random fields in the same manner as an HMM is an extension of a Markov chain. That is, a CRF is a Markov random field in which each random variable can be conditioned upon a set of global observations.

2.6.2 Generative Versus Discriminative Models

Before we move on we need to introduce some new notation. In the HMM framework described earlier in this chapter, the hidden label sequence is denoted $\mathbf{X}$ and the observed sequence $\mathbf{Y}$. In the CRF community, however, this notation is usually switched. Therefore, to avoid confusion, throughout this section the observed sequence, also called the *input* sequence, is denoted $\mathbf{O}$, and the hidden *output* sequence is denoted $\mathbf{H}$.

Generative models is a family of models where the joint probability of the hidden and the observed sequence can be factorized as

$$\mathbf{P(O, H)} = \mathbf{P(O)P(H|O)}. \tag{2.124}$$

A generative model thus allows us to draw samples from it, in order to "generate" synthetic examples of the observed sequence given the hidden. However, due to high dimensionality and complex dependencies the distribution of the observed sequence may be difficult to render, which is why numerous independence assumptions often need to be made to make the computations tractable. *Discriminative models*, on the other hand, is a family of conditional distributions $\mathbf{P(H|O)}$ where the hidden sequence to be classified is modeled directly. The distribution of the observed sequence is ignored and thereby the need for independency assumptions on the observed sequence is avoided. By supplying a model for the marginal distribution of the observed sequence, the conditional distribution of the discriminative model could be used to compute the joint distribution as in (2.124), but since the conditional distribution is all we need for classification, this is usually not done.

In this manner, there are *generative-discriminative* model pairs, where one model can be converted into the other using *Bayes' rule* (see Sect. 5.4.7). One such pair is the *naive Bayes classifiers* and the *logistic regression*. Assume that we want to determine a single classification label H, based on a vector of observations or *features* $\mathbf{O} = (O_1, O_2, \ldots, O_n)$. The naive Bayes classifier is based on the joint probability of the classification label and the observations, which can be factorized as

$$\mathbf{P}(H, \mathbf{O}) = \mathbf{P}(H) \prod_{i=1}^{n} \mathbf{P}(O_i | H). \tag{2.125}$$

The logistic regression classifier is instead based on the conditional probability and assumes that the logarithm of the conditional distribution, $\log \mathbf{P}(H | \mathbf{O})$, is a linear function of $\mathbf{O}$, such that

$$\mathbf{P}(H | \mathbf{O}) = \frac{1}{Z(\mathbf{O})} \exp \left(\theta_H + \sum_{i=1}^{n} \theta_{H,i} O_i \right) \tag{2.126}$$

where $Z(\mathbf{O})$ is a normalization factor and θ_H is a *bias weight* corresponding to the initial $\log \mathbf{P}(H)$ component in the naive Bayes formula in (2.125). To write this in more compact form we can define *feature functions* that are indicator functions for a single class only. That is, we let $f_{H',j}(H, \mathbf{O}) = 1_{\{H'=H\}} O_j$ represent the feature weights and $f_{H'}(H, \mathbf{O}) = 1_{\{H'=H\}}$ the bias weights. By instead using a common index k for all different feature functions f_k and their corresponding weights θ_k, the logistic regression model can be written as

$$\mathbf{P}(H | \mathbf{O}) = \frac{1}{Z(\mathbf{O})} \exp \left(\sum_{k=1}^{K} \theta_k f_k(H, \mathbf{O}) \right). \tag{2.127}$$

By training the naive Bayes classifier in (2.125) to maximize the conditional likelihood, we achieve the logistic regression classifier, and if the logistic regression classifier is trained to maximize the joint distribution we achieve the naive Bayes. In a similar manner, HMMs and CRFs are a generative-discriminative pair, and for suitable choices of feature functions in the CRFs we can convert one model into the other.

An important note is that while the two models in a generative-discriminative pair exactly mirror one another in theory, this is rarely true in practice. In order for this to hold we need access to the true distributions, but in practice we are usually left to work with estimations and approximations resulting from only having samples of the true distributions. Therefore, it matters which model we choose, generative or discriminative, and the choice for a given application may not be obvious as both approaches have their pros and cons. If we focus merely on the classification task, discriminative models can be highly superior, both in terms of computational complexity and in terms of the level of dependencies they can include. They impose conditional independence assumptions on the hidden sequence pretty much in the same manner as in generative models, and they describe how the hidden sequence may depend on the observed, while interdependencies within the observed sequence need not be explicitly stated. This way discriminative models can include very complex dependencies and overlapping features which may improve the classification accuracy. However, generative models are usually more flexible, in particular when it comes to training, and are more easily interpreted. Also, generative models are better at handling missing, latent, or partially labeled data, and can sometimes perform better than a discriminative model as a result. Therefore, which approach to use has to be guided by the application in question [3, 21].

2.6.3 Graphical Models and Markov Random Fields

In many statistical applications we have prior knowledge about the ordering of a set of variables, either of the temporal ordering of events or in terms of dependency structures. Such knowledge can often be illustrated in a graphical model $G = (V, E)$, where V are the vertices and E the connecting edges. The vertices correspond to the random variables and the edges represent the dependency structure between these variables. Graphical models can be divided into two main classes: *directed acyclic graphs* (DAGs) and *undirected graphs*. Two important models for our purposes are *Bayesian networks* which are a kind of DAGs, described in Sect. 5.4.7, and *Markov random fields*, which are undirected graphs that will be discussed a little further in this section. For a more comprehensive treatment on graphical models and random fields, see for instance [23].

We say that random variables A and B are *conditionally independent* given a third random variable C if and only if

$$\mathbf{P}(A, B|C) = \mathbf{P}(A|C)\mathbf{P}(B|C), \quad A, B, C \in V. \tag{2.128}$$

Conditional independence is a powerful concept as it can be used to factorize complex multivariate distributions into products of factors acting on smaller subsets of the random variables. Any joint distribution of a set of random variables can be represented by a DAG, where the edges correspond to conditional dependencies between the variables, and the absence of an edge implies conditional independence between the variables of the corresponding vertices.

Now, let $\mathbf{X} = (X_v)_{v \in V}$ be a collection of random variables. Recall from (2.1) that the probability of any such set and for any ordering can be decomposed into a product of conditional probabilities

$$\mathbf{P}(\mathbf{X}) = \mathbf{P}(X_1) \prod_{v=2}^{V} \mathbf{P}(X_v|X_1, \ldots, X_{v-1}). \tag{2.129}$$

For a graph $G = (V, E)$, if there exists an ordering $v_1, \ldots, v_d$ of the vertices (i.e., of the random variables) that is consistent with the graph, meaning that a directed edge $v_i \rightarrow v_j \in E$ implies the ordering $i < j$, then G is called *directed acyclic graph* (DAG). We define the *parents* $\pi(v)$ of a vertex $v \in V$ as the set of vertices having a directed edge to v. A *directed model* is then a family of distributions that factorize as

$$P(X) = \prod_{v \in V} P(X_v | X_{\pi(v)}) \tag{2.130}$$

where X_v is the random variable at vertex v and $X_{\pi(v)}$ is the set of random variables of the parent vertices of v. Because of the recursiveness in this decomposition, the resulting graph is *acyclic*, meaning that it does not contain any loops, resulting in a DAG (see Fig. 2.21a for an illustration). A common family of directed acyclic models are Bayesian networks, described in Sect. 5.4.7, and hidden Markov models and neural networks described earlier in this chapter can both be considered special cases of Bayesian networks.

In Markov random fields, on the other hand, the underlying graph is *undirected* and may be cyclic, representing a correlation between the random variables rather than a causality. An *undirected graph* is a graph where the edges have no direction. That is, for two vertices $i, j \in V$ the edges $\langle i, j \rangle$ and $\langle j, i \rangle$ are equivalent. Since there is no direction of the edges, there is no ordering of the random variables, meaning that the distribution can no longer be factorized according to a set of parents as in (2.130). Instead, an undirected graph can represent a family of distributions that each factorize according to a set of *factors*. A factor can be any strictly positive, real-valued function, and do not necessarily correspond to a conditional probability, which is why we also need a normalization factor to achieve a proper probability distribution. Formally, given a set of random variables X and a collection of A subsets $\{X_a\}_{a=1}^{A}$, an undirected graphical model is the set of distributions that can be written as

$$P(X) = \frac{1}{Z} \prod_{a=1}^{A} \Psi_a(X_a) \tag{2.131}$$

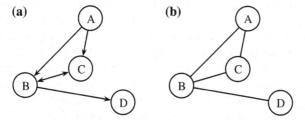

Fig. 2.21 An illustration of a graphical representation of four random variables, A, B, C, D. **a** A directed acyclic graph where the joint distribution factorizes as $P(A, B, C, D) = P(A)P(B|A, C)P(C|A, B)P(D|B)$. **b** The corresponding undirected cyclic graph. Each node with its former parents form a complete subgraph of the graph

for any choice of positive factors $\Psi_a(\mathbf{X}_a) > 0$ for all $\mathbf{X}_a$. The constant Z, also known as the *partition function*, is a normalization factor

$$Z = \sum_{\mathbf{X}} \prod_{a=1}^{A} \Psi_a(\mathbf{X}_a), \qquad (2.132)$$

where the sum runs over all possible assignments to the set $\mathbf{X}$. The factors Ψ_a are also called *local functions*, because they act on local subsets of the graph vertices, or *compatibility functions*, because they represent how compatible the values in a subset $\mathbf{X}_a$ are with each other.

There is a clear connection between directed and undirected graphs. To see this, assume that the distribution of $\mathbf{X}$ factorizes with respect to an undirected acyclic graph. Instead of talking about the parents of a vertex as in directed graphs, we now talk about the *neighbors* $n(v)$ of $v \in V$, meaning all vertices connected to v by an edge. In a directed graph a random variable X_v is conditionally independent of all predecessors in the graph, given its parents $\pi(v)$. In undirected graphs the corresponding conditional independence structure is represented by simple graph separation. It may be tempting to think that a DAG can be converted into an undirected graph simply by dropping the direction of the edges as in Fig. 2.21, but this does not hold in general. A v-shape in a DAG with edges $A \rightarrow B \leftarrow C$ would in the undirected graph result in a structure $A - B - C$ where A and C are conditionally independent given B, which clearly is not true. Instead we need to add an edge between A and C in the undirected graph to indicate their connection. This way of linking "unmarried" nodes is called *moralization*. Unfortunately, we loose some information of the DAG in the process, and we therefore cannot move in the other direction, creating a DAG from an undirected graph.

A special type of conditional independence structure is given by a Markov property formulation, similar to that of Markov chains, and that can be defined at three different levels: the *global Markov property*, the *local Markov property*, and the *pairwise Markov property*. The *global Markov property* of an undirected graph states that any two subset of random variables are conditionally independent given a separating subset. That is, for three subsets of vertices $A, B, C \subset V$ we say that $\mathbf{X}_A$ is conditionally independent of $\mathbf{X}_B$ given $\mathbf{X}_C$ if and only if the vertices in C separates those in A from those in B. In essence, this means that if we remove all the vertices in C from the graph, the sets A and B are no longer connected. The *local Markov property* states that a random variable X_v is conditionally independent of all other variables in the graph, given its neighbors, and the *pairwise Markov property* states that two random variables not connected by an edge, are conditionally independent given everything else. The global property implies the local, which in turn implies the pairwise property. However, if we add the assumption that the distribution of the random variables is *positive*, meaning that $\mathbf{P}(X_v) > 0$ for all $v \in V$, we achieve equivalence between the three Markov properties. A *random field* is a generalization of random processes in which a collection of random variables are indexed by a multidimensional space. In a *Markov random field* the index space is an undirected graph $G = (V, E)$ that fulfills the local Markov property. That is, for each vertex $v \in V$, given its neighbors $n(v)$ the corresponding random variable X_v is conditionally independent of everything else. That is,

$$\mathbf{P}(X_v | \mathbf{X} \backslash X_v) = \mathbf{P}(X_v | \mathbf{X}_{n(v)}), \ v \in V \tag{2.133}$$

where $\mathbf{X} \backslash X_v$ denotes all variables in $\mathbf{X}$ except X_v.

The conditional independences of an arbitrary distribution can be difficult to sort out, and a convenient subclass of Markov random fields are those that use the *maximal cliques* of the graph as the factorization subsets. A *clique* is a subgraph of G that is fully connected, meaning that there is an edge between every pair of vertices in the subgraph. Furthermore, a *maximal clique* is a clique that cannot be extended further without breaking the full connectedness property. The set of factors operating on the maximal cliques of G are called *potential functions*. A joint distribution, factorized by its maximal cliques, is then proportional to the product of the potential functions. The *Hammersley–Clifford theorem* [14, 23] gives that any positive distribution that satisfies the local Markov property can be factorized according to its maximal cliques. Such a Markov random field is sometimes called a *Gibbs random field*, which is popular in statistical physics because it can be represented by a *Gibbs distribution* for appropriate potential functions. A *Gibbs distribution* is a measure that factorizes over the maximal cliques $\mathscr{C}$ of the undirected graph G, and where the distribution takes the log-linear form

$$\mathbf{P}(\mathbf{X}) = \frac{1}{Z} \exp(-H(\mathbf{X})). \tag{2.134}$$

where $H(\mathbf{X}) > 0$ is called the *energy function* of configuration $\mathbf{X}$. The meaning of an energy function can be somewhat abstract, but it relates to the energy used to describe the organization of atoms in thermodynamical systems. For instance, the more ordered the atoms in a metal are, the lower the energy (see Sect. 3.2.8 for more on this). In the Gibbs distribution, the factors in (2.131) thus take the form

$$\Psi_c(\mathbf{X}_c) = \exp(-H(\mathbf{X}_c)) \tag{2.135}$$

where H is the energy of the subset of random variables in clique $c \in \mathscr{C}$, and the energy function $H(\mathbf{X})$ sums over the maximal cliques $\mathscr{C}$. We can now give a more formal statement of the Hammersley–Clifford theorem:

Theorem 2.5 (The Hammersley–Clifford Theorem) *A positive distribution is a Markov random field if and only if it is a Gibbs random field.*

An important note is that although the maximal clique factorization corresponds to the conditional independence structure of the graphical model, the potential functions in themselves do not necessarily have a probabilistic interpretation. They merely represent constraints on the underlying random variables, which in turn effect the global probability distribution, but that do not directly translate into probabilistic terms.

2.6.4 Conditional Random Fields (CRFs)

A *conditional random field* (CRF) is a Markov random field where each random variable in the field may also be conditioned upon a set of global observations [19]. A CRF can be seen as an extension of logistic regression where the hidden variables are conditioned on the observed sequence. CRFs are also closely related to the hidden Markov models (HMMs) described earlier in this chapter. In fact, for a suitable choice of clique potentials, HMMs and CRFs form a generative-discriminative model pair in the same way as naive Bayes and logistic regression discussed above. The main difference between HMMs and CRFs is that while HMMs model the joint distribution $P(\mathbf{H}, \mathbf{O})$ of an observed input sequence $\mathbf{O}$ and a hidden output $\mathbf{H}$ and (recall that we changed the HMM notation from $(\mathbf{X}, \mathbf{Y})$ to $(\mathbf{H}, \mathbf{O})$), CRFs focus on the conditional distribution $P(\mathbf{H}|\mathbf{O})$ of the hidden sequence, given the observed. Thus, in the conditional setting, the observed distribution $P(\mathbf{O})$ does not need to be modeled explicitly, leading to a simpler model which can allow the inclusion of complex dependencies within the observed sequence. Another advantage is that the potential functions can depend on the data, for instance by incorporating global features into the local potential functions, something that is very hard to do in generative models. The main disadvantage of CRFs is that they need to be trained on labeled data and the training process is typically very computer intense. General graphs typically become intractable fast, while graphs with a chain or tree structure may still be manageable.

The choice of potential functions for CRFs is closely related to the maximum entropy method described in Sect. 5.4.6. In order to include interdependencies within the observed sequence as well as other local and global knowledge of the data, we define a set of input *features*. A CRF is then a Markov random field where the clique potentials are conditioned on this feature set, denote it $\mathbf{K}$,

$$P(\mathbf{H}|\mathbf{O}, \mathbf{K}) = \frac{1}{Z(\mathbf{O}, \mathbf{K})} \prod_{c \in \mathscr{C}} \Psi_c(\mathbf{H}_c|\mathbf{O}, \mathbf{K}). \tag{2.136}$$

Finding the maximum entropy distribution that satisfies K features f_k is an optimization problem under constraints, and if we choose log-linear clique potentials as in logistic regression we get for clique $c \in \mathscr{C}$

$$\Psi_c(\mathbf{H}_c|\mathbf{O}) = \exp\left(\sum_{k=1}^{K} \lambda_{ck} f_{ck}(\mathbf{H}_c, \mathbf{O}) \right) \tag{2.137}$$

where λ_{ck} is the Lagrangian multiplier associated with feature f_{ck}. Note also that we can make the features clique specific. The resulting CRF distribution becomes

$$P(\mathbf{H}|\mathbf{O}) = \frac{1}{Z(\mathbf{O})} \exp\left(\sum_{c \in \mathscr{C}} \sum_{k=1}^{K} \lambda_{ck} f_{ck}(\mathbf{H}_c, \mathbf{O}) \right). \tag{2.138}$$

Linear-Chain CRFs

For sequence models the common choice is a *linear-chain CRF*, which models the correlation between adjacent hidden variables in a linear sequence similarly to HMMs. In a linear graph each vertex only has two neighbors, and the maximal cliques simply constitute each pair of adjacent vertices connected by an edge. Therefore, we can define the clique potentials on the edges $e = \langle i - 1, i \rangle$ instead on vertex subsets

$$\Psi_e(H_e) = \exp\left(\sum_{k=1}^{K} \lambda_k f_k(H_{i-1}, H_i, \mathbf{O}) + \sum_{k=1}^{K} \mu_k g_k(H_i, \mathbf{O}) \right) \qquad (2.139)$$

where f_k are feature functions of the local transitions and the global observed sequence, and g_k feature functions of the sequence label at position i and the observed sequence. The features f_k and g_k are, thus, closely connected to the transitions and emissions in an HMM. In fact, by choosing features exactly corresponding to the logarithm of the transition and emission probabilities in an HMM, the conditional distribution $\mathbf{P}(\mathbf{H}|\mathbf{O})$ rendered from the joint distribution $\mathbf{P}(\mathbf{H}, \mathbf{O})$ in an HMM, is a CRF. That is, by rewriting the joint distribution as

$$\mathbf{P}(\mathbf{O}, \mathbf{H}) = \frac{1}{Z} \prod_{t=1}^{T} \exp\left(\sum_{i,j \in S} \theta_{ij} \mathbf{1}_{\{H_t=j\}} \mathbf{1}_{\{H_{t-1}=i\}} + \sum_{i \in S} \sum_{v \in V} \mu_{iv} \mathbf{1}_{\{H_t=i\}} \mathbf{1}_{\{O_t=v\}} \right)$$
$$(2.140)$$

and defining the parameters as

$$\begin{aligned}
\theta_{ij} &= \log \mathbf{P}(H_t = j | H_{t-1} = j) \\
\mu_{iv} &= \log \mathbf{P}(O_t = v | H_t = i) \\
Z &= 1
\end{aligned} \qquad (2.141)$$

we achieve a direct correspondence between the HMM and the related CRF. To see this, by using a generic notation f_k for features and θ_k for the corresponding parameters, we transfer the the formula in (2.140) into

$$\mathbf{P}(\mathbf{O}, \mathbf{H}) = \frac{1}{Z} \prod_{t=1}^{T} \exp\left(\sum_{k=1}^{K} \theta_k f_k(H_{t-1}, H_t, O_t) \right). \qquad (2.142)$$

The conditional distribution achieved by using (2.142) in

$$\mathbf{P}(\mathbf{H}|\mathbf{O}) = \frac{\mathbf{P}(\mathbf{O}, \mathbf{H})}{\sum_{\mathbf{O}'} \mathbf{P}(\mathbf{O}', \mathbf{H})} \qquad (2.143)$$

is then a special type of linear-chain CRFs that only include features for the transitions and emissions modeled by a standard HMM. However, general linear-chain CRFs are not limited to the use of indicator functions, but can use any real-valued set of functions in place of the feature functions f_k in (2.142). For instance, since CRFs do not model the observed input sequence, we can let the feature functions depend on the entire observation sequence without having to alter the dependency structure in the graphical model.

2.6.5 Conrad: CRF-Based Gene Prediction

Generalized (GHMMs), described in Sect. 2.2, have proved very powerful in gene prediction, as they are flexible, easy to train, and easily interpreted probabilistically. The disadvantages include the difficulties to include external information, such as various homology sources, long-ranging sequence features, and unknown dependencies both within and between the external sources. Since CRFs avoid the problems of modeling the observed input data, they can easily incorporate various sources of information, regardless of unknown dependencies and long-range effects.

Conrad [8] is a gene prediction software based on semi-Markov CRFs (SMCRFs), which has inherited the generalized (semi-Markov) features of GHMMs and combined them with discriminative features from the CRF framework. As before, we have a hidden state sequence $\mathbf{H}$ of labels to be predicted, and an observed sequence $\mathbf{O}$ corresponding to the given DNA sequence to be labeled. Again, a CRF expresses the conditional probability $\mathbf{P}(\mathbf{H}|\mathbf{O})$ as opposed to GHMMs which model the joint probability $\mathbf{P}(\mathbf{H}, \mathbf{O})$ of the hidden and the observed data. The conditional probability is as before expressed in log-linear form

$$\mathbf{P}(\mathbf{H}|\mathbf{O}) = \frac{1}{Z_\lambda(\mathbf{O})} \exp\left(\sum_j \lambda_j F_j(\mathbf{H}, \mathbf{O})\right) \tag{2.144}$$

where λ_j is the feature weight, F_j a feature function, which in itself is a sum of features (see below), and $Z_\lambda(\mathbf{O})$ the normalizing factor.

The hidden sequence is assumed to be a linearly structured vector of labels such as "exons", "introns", and "intergenes", with one label per nucleotide in the observed sequence. Or conversely, the observed sequence can be segmented into p intervals $\{I_i\}_{i=1}^p = \{(t_i, u_i, v_i)\}_{i=1}^p$ of equally labeled segments (e.g., corresponding to an entire exon), with start at nucleotide t_i, end at u_i, and the same label v_i all through the segment. The segmentation p naturally varies and is determined as part of the prediction. As in GHMMs, Conrad assumes that each interval (t_i, u_i, v_i) only depends on the adjacent neighboring intervals I_{i-1} and I_{i+1}. The feature function F_j is therefore written as a sum of *localized* feature functions

$$F_j(\mathbf{H}, \mathbf{O}) = \sum_{i=1}^{p} f_j(t_i, u_i, v_i, v_{i-1}, \mathbf{O}). \tag{2.145}$$

The partitioning of the observed sequence is similar to the generalized (semi-Markov) feature of GHMMs and is what makes the CRF semi-Markov. The prediction of hidden labels produced by the SMCRF for a given observed sequence is the segmentation $\mathbf{H}$ that maximizes then the conditional probability $\mathbf{P}(\mathbf{H}|\mathbf{O})$.

Feature Selection

The major issues when applying SMCRFs to gene prediction is the construction of suitable feature functions f_j, and the training of their corresponding weights λ_j. The advantage over GHMMs, as mentioned earlier, is that these features are not required to be independent or to have a probabilistic interpretation. Conrad is constructed to use both *generative features*, inherited from GHMMs, and *discriminative features*, with the result that Conrad can behave either as a pure GHMM or as a SMCRF or anywhere in between. The generative features in Conrad are:

- *Reference features*: modeling the internal sequence composition of the different model states, using a third-order Markov model. These features do not include the segmentation boundaries such as start and stop codons, or splice sites.
- *Length features*: modeling the state length distributions of exons, introns, and intergenic regions. The intergene lengths are modeled using an exponential distribution (the continuous counterpart of the geometric distribution), and exon and intron lengths are modeled by a mixture of two gamma distributions.
- *Transition feature*: modeling the transition probabilities between states.
- *Boundary features*: modeling state boundary signals such as start and stop codons and splice sites.
- *Phylogenetic features*: modeling species homology through state-specific multiple alignments.

By using only reference, length, transition, and boundary features with all weights set to $\lambda_j = 1$, Conrad is equivalent to the conditional probability computed by the corresponding GHMM by taking

$$\mathbf{P}_{GHMM}(\mathbf{H}|\mathbf{O}) = \frac{\mathbf{P}_{GHMM}(\mathbf{H}, \mathbf{O})}{\mathbf{P}_{GHMM}(\mathbf{O})}. \tag{2.146}$$

In the GHMM, we let a_{ij} denote the transition probability between states $i, j \in S$, π_i the initial probability of $i \in S$, and $q_j(O_{t_i}, O_{u_i})$ the emission probability for the segment $O_{t_i}, \ldots, O_{u_i}$, now including the duration probability as well (emission and duration were separated in Sect. 2.2). The joint probability then takes the form

$$\mathbf{P}_{GHMM}(\mathbf{H}|\mathbf{O}) = \pi_{v_1} \prod_{i=2}^{p} a_{v_{i-1}, v_i} q_{v_i}(O_{t_i}, O_{u_i}) \tag{2.147}$$

and the features in Conrad translates to

$$f_{GHMM}(v_{i-1}, v_i, t_i, u_i, \mathbf{O}) = \begin{cases} \log(q_{v_i}(O_{t_i}, O_{u_i})) + \log(\pi_{v_i}) & \text{if } t_i = 1 \\ \log(q_{v_i}(O_{t_i}, O_{u_i})) + \log(a_{v_{i-1}, v_i}) & \text{if } t_i > 1. \end{cases}$$

(2.148)

This version of Conrad (called ConradG-1) is similar to Genscan [7] described in Sect. 2.2.4. ConradG-2, which includes phylogenetic features for two-species comparisons is similar to Twinscan [17] described in Sect. 4.1.2.

Discriminative features are features lacking a probabilistic interpretation. The use of discriminative features enables the ability to incorporate long-range effects and unknown dependencies, or any other type of information that may be difficult to model probabilistically. Conrad incorporates a few discriminative features that represent information commonly used when annotations are curated manually, but that is difficult to include in a probabilistic setting. The discriminative features are:

- *Gap features*: modeling gaps in the multiple alignments that are not captured by the phylogenetic features.
- *Footprint features*: modeling the positions at which the different species in the multiple alignment are aligned.
- *EST features*: modeling the connection between the EST alignments and the state fragmentation of the hidden label sequence.

For instance, the gap feature for a specific exon E takes the form

$$f_{GAP,E}(v_{i-1}, v_i, t_i, u_i, \mathbf{O}) = \sum_{k=t_i}^{u_i} \begin{cases} 1 \text{ if } v_i = E \text{ and gap of length 1 or 2 (mod 3) at } k \\ 0 \text{ otherwise,} \end{cases}$$

(2.149)

thus counting the number of gaps in the alignment that would cause a frameshift in the coding sequence. The features are similar for introns and intergenes. Also, the footprint and EST features work the same way, by summing similar indicator functions while scanning through the state segment.

Parameter Training

The feature weights λ_j are trained from labeled example sequences. The common approach to train the weights in CRFs is to use *conditional maximum likelihood* (CML) described in Sect. 6.8. That is, for a single pair of training sequences $(\mathbf{H}^0, \mathbf{O}^0)$, the CML estimator is given by

$$\hat{\lambda}_{CML} = \underset{\lambda}{\text{argmax}} \, (\log \mathbf{P}(\mathbf{H}^0 | \mathbf{O}^0)).$$

(2.150)

The maximum is typically found using a gradient-based technique (see Sect. 6.6), where the specific choice of algorithm depends on the formulation of the CRF. For SMCRFs the common approach is to use dynamic programming algorithms similar to the forward and the backward algorithms in HMMs.

Another approach, introduced by the Conrad group, is to use something called *maximum expected accuracy* (MEA). CML optimizes the accuracy of the prediction indirectly by maximizing the likelihood of the hidden sequences given in the training set. Instead, one would like to optimize the accuracy directly, but this becomes intractable since changing the weights causes changes in the segmentation, which in turn changes the accuracy in a discontinuous way. Instead the objective function is defined as the expected accuracy over the entire distribution of segmentations defined by the SMCRF. However, in order to compute this, we first need to need a similarity metric. We define a similarity function S between the training set $(\mathbf{H}^0, \mathbf{O}^0)$ and a certain label sequence $\mathbf{H}$ as

$$S(\mathbf{H}, \mathbf{H}^0, \mathbf{O}^0) = \sum_{t=1}^{T} s(H_{t-1}, H_t, H_{t-1}^0, H_t^0, \mathbf{O}^0, t) \qquad (2.151)$$

where s are some kind of similarity functions over dinucleotides that can be set as suited. For gene prediction, the function S is divided into two parts, corresponding to splice sites and internal nucleotides. The nucleotide similarity score is simply counting the number of correctly labeled nucleotides in each state, while the splice site similarity scores consider both the labeling and the placement of the splice boundary.

The objective function used to optimize the weights is then defined as the expectation of the similarity function

$$A_{MEA}(\lambda) = E_\lambda[S(\mathbf{H}, \mathbf{H}^0, \mathbf{O}^0)] = \sum_y P_\lambda(\mathbf{H}|\mathbf{O}^0) S(\mathbf{H}, \mathbf{H}^0, \mathbf{O}^0) \qquad (2.152)$$

and MEA estimator is given by

$$\hat{\lambda}_{MEA} = \underset{\lambda}{\text{argmax}}\ A_{MEA}(\lambda). \qquad (2.153)$$

This maximum is again achieved by using gradient-based methods. However, since the objective function is not concave in λ, there is no guarantee that the global maximum is reached. To achieve the best results, the initial weights are set by using the CML estimates.

References

1. Baldi, P., Brunak, S.: Bioinformatics: The Machine Learning Approach. MIT Press, Cambridge (2001)
2. Begleiter, R., El-Yaniv, R., Yona, G.: On prediction using variable order Markov models. J. Artif. Intell. **22**, 385–421 (2004)
3. Bishop, C.M., Lasserre, J.: Generative or discriminative? Getting the best of both worlds. Bayesian Stat. **8**, 3–24 (2007)

4. Blattner, F.R., Plunkett, G., Bloch, C.A., Perna, N.T., Burland, V., Riley, M., Collado-vides, J., Glasner, J.D., Rode, C.K., Mayhew, G.F., Gregor, J., Davis, N.W., Kirkpatrick, H.A., Goeden, M.A., Rose, D.J., Mau, B., Shao, Y.: The complete genome sequence of *Escherichia coli* K-12. Science **277**, 1453–1469 (1997)
5. Breiman, L.: Some properties of splitting criteria. Mach. Learn. **24**, 41–47 (1996)
6. Breiman, L., Friedman, J., Stone, C.J., Olshen, R.A.: Classification and Regression Trees. Chapman & Hall, New York (1984)
7. Burge, C., Karlin, S.: Prediction of complete gene structures in human genomic DNA. J. Mol. Biol. **268**, 78–94 (1997)
8. DeCaprio, D., Vinson, J.P., Pearson, M.D., Montgomery, P., Doherty, M., Galagan, J.E.: Conrad: gene prediction using conditional random fields. Genome Res. **17**, 1389–1398 (2007)
9. Delcher, A.L., Harmon, D., Kasif, S., White, O., Salzberg, S.L.: Improved microbial gene identification with GLIMMER. Nucleic Acids Res. **27**, 4636–4641 (1999)
10. Delcher, A.L., Bratke, K.A., Powers, E.C., Salzberg, S.L.: Identifying bacterial genes and endosymbiont DNA with Glimmer. Bioinformatics **23**, 673–679 (2007)
11. Durbin, R., Eddy, S., Krogh, A., Mitchison, G.: Biological sequence analysis. Probabilistic Models of Proteins and Nucleic Acids. Cambridge University Press, Cambridge (1998)
12. Fickett, J.W., Tung, C.-S.: Assessment of protein coding measures. Nucleic Acids Res. **20**, 6441–6450 (1992)
13. Gusfield, D.: Algorithms on Strings, Trees and Sequences: Computer Science and Computational Biology. Cambridge University Press, Cambridge (1997)
14. Hammersley, J., Clifford, P.: Markov fields on finite graphs and lattices.http://www.statslab. cam.ac.uk/~grg/books/hammfest/hamm-cliff.pdf
15. Jukes, T.H., Osawa, S.: The genetic code in mitochondria and chloroplasts. Experientia **46**, 1117–1126 (1990)
16. Karlin, S., Taylor, H.M.: A First Course in Stochastic Processes, 2nd edn. Academic Press, New York (1975)
17. Korf, I., Flicek, P., Duan, D., Brent, M.R.: Integrating genomic homology into gene structure prediction. Bioinformatics **17**, S140–S148 (2001)
18. Koski, T.: Hidden Markov Models for Bioinformatics. Springer, Berlin (2001)
19. Lafferty, J., McCallum, A., Pereira, F.: Conditional random fields: probabilistic models for segmenting and labeling sequence data. In: Proceedings of International Conference Machine Learning, pp. 282–289 (2001)
20. Larsen, T., Krogh, A.: EasyGene—a prokaryotic gene finder that ranks ORFs by statistical significance. BMC Bioinform. **4**, 21–35 (2003)
21. Ng, A.Y., Jordan, M.I.: On discriminative versus generative classifiers: a comparison of logistic regression and naive Bayes. In: NIPS (2001)
22. McCulloch, W.S., Pitts, W.: A logical calculus of the ideas immanent in nervous activity. Bull. Math. Biol. **52**, 99–115 (1943)
23. Murphy, K.P.: Machine Learning: A Probabilistic Perspective. MIT Press, Cambridge (2012)
24. Murthy, S.K., Kasif, S., Salzberg, S.L.: A system for induction of oblique decision trees. J. Artif. Intell. Res. **2**, 1–32 (1994)
25. Ohler, U., Harbeck, S., Niemann, H., Nöth, E., Reese, M.G.: Interpolated Markov chains for eukaryotic promoter recognition. Bioinformatics **15**, 362–369 (1999)
26. Perna, N.T., Plunkett, G., Burland, V., Mau, B., Glasner, J.D., Rose, D.J., Mayhew, G.F., Evans, P.S., Gregor, J., Kirkpatrick, H.A., Pósfai, G., Hackett, J., Klink, S., Boutin, A., Shao, Y., Miller, L., Grotbeck, E.J., Davis, N.W., Lim, A., Dimalanta, E.T., Potamousis, K.D., Apodaca, J., Anantharaman, T.S., Lin, J., Yen, G., Schwartz, D.C., Welch, R.A., Blattner, F.R.: Genome sequence of enterohaemorrhagic *Escherichia coli* O157:H7. Nature **409**, 529–533 (2001)
27. Pertea, M., Lin, X., Salzberg, S.L.: GeneSplicer: a new computational method for splice site prediction. Nucleic Acids Res. **29**, 1185–1190 (2001)
28. Quinlan, J.R.: Induction of decision trees. Mach. Learn. **1**, 81–106 (1986)
29. Quinlan, J.R.: C4.5: Programs for machine learning. Morgan Kaufmann Publishers, San Mateo (1993)

30. Rabiner, L.R.: A tutorial on hidden Markov models and selected applications in speech recognition. Proc. IEEE **77**, 257–286 (1989)
31. Rissanen, J.: A universal data compression system. IEEE Trans. Inf. Theory **29**, 656–664 (1983)
32. Rivas, E., Eddy, S.R.: Noncoding RNA gene detection using comparative sequence analysis. BMC Bioinform. **2**, 8 (2001)
33. Rosenblatt, F.: The perceptron: a probabilistic model for information storage and organization in the brain. Psychol. Rev. **65**, 386–408 (1958)
34. Salzberg, S.L., Delcher, A.L., Fasman, K.H., Henderson, J.: A decision tree system for finding genes in DNA. J. Comput. Biol. **5**, 667–680 (1998)
35. Salzberg, S.L., Delcher, A.L., Kasif, S., White, O.: Microbial gene identification using interpolated Markov models. Nucleic Acids Res. **26**, 544–548 (1998)
36. Schukat-Talamazzini, E.G., Gallwitz, F., Harbeck, S., Warnke, V.: Rational interpolation of maximum likelihood predictors in stochastic language modeling. In: Proceedings of Eurospeech'97, pp. 2731–2734. Rhodes, Greece (1997)
37. Sharp, P.M., Cowe, E.: Synonymous codon usage in Sacharomyces cerevisiae. Yeast **7**, 657–678 (1991)
38. Shmatkov, A.M., Melikyan, A.A., Chernousko, F.L., Borodovsky, M.: Finding prokaryotic genes by the 'frame-by-frame' algorithm: targeting gene starts and overlapping genes. Bioinformatics **15**, 874–886 (1999)
39. Shmilovici, A., Ben-Gal, I.: Using a VOM model for reconstructing potential coding regions in EST sequences. Comput. Stat. **22**, 49–69 (2007)
40. Skovgaard, M., Jensen, L.J., Brunak, S., Ussery, D., Krogh, A.: On the total number of genes and their length distribution in complete microbial genomes. Trends Genet. **17**, 425–428 (2001)
41. Snyder, E.E., Stormo, G.D.: Identification of protein coding regions in genomic DNA. J. Mol. Biol. **248**, 1–18 (1995)
42. Sutton, C., McCallum, A.: An introduction to conditional random fields. Found. Trends Mach. Learn. **4**, 267–373 (2011)
43. Xu, Y., Mural, R.J., Einstein, J.R., Shah, M.B., Uberbacher, E.C.: GRAIL: a multi-agent neural network system for gene identification. Proc. IEEE **84**, 1544–1552 (1996)
44. Xu, Y., Uberbacher, E.C.: Computational gene prediction using neural networks and similarity search. In: Salzberg, S.L., Searls, D.B., Kasif, S. (eds.) Computational Methods in Molecular Biology, pp. 109–128. Elsevier Science B.V., Amsterdam (1998)
45. http://www.cbcb.umd.edu/glimmer/

Chapter 3
Sequence Alignment

A fundamental task in biological sequence analysis is to reveal the evolutionary relationships between biological sequences, such as protein or DNA sequences. Moreover, by comparing novel sequences to already characterized ones, the hope is that regions of high sequence similarity can be used to infer both the structure and the function of novel genes. The underlying idea is that homologous sequences, originating from the same ancestral sequence, have transformed into their current states through a series of changes, or point mutations, to the sequence. Therefore, high levels of sequence similarity can be used to infer homology between sequences. It is important to note, however, that sequence similarity does not necessarily imply homology. If the compared regions are too short, or if the sequence is repetitive or of low complexity, the sequences may appear similar just by chance. Therefore, the challenge is to quantify the notions of sequence similarity to separate spurious hits with those revealing functionally important elements in the sequence. *Sequence alignments* have turned out to be a suitable format for the comparison of biological sequences. In this chapter, we introduce a number of scoring schemes and algorithms for pairwise and multiple alignments.

3.1 Pairwise Sequence Alignment

Sequence alignment can be seen as a form of approximate string matching [38]. Methods for string matching have a long history in a wide range of areas besides molecular biology, including error control of noisy radio channels, automatic string-editing/correction of keyboard inputs, string comparisons/pattern matching of computer files, and speech recognition [59]. Common to all these areas is the need for methods that can handle various types of sequence differences.

The sequences are compared in sequence alignment by arranging them in rows on top of each other such that matched residues are arranged in successive columns. By inserting spaces at various positions and in varying numbers, the number of matching residues can be optimized. The resulting alignment is an assembly of operations, such as *matches*, *mismatches*, *insertions* and *deletions* (see Table 3.1).

© Springer-Verlag London 2015
M. Axelson-Fisk, *Comparative Gene Finding*, Computational Biology 20,
DOI 10.1007/978-1-4471-6693-1_3

Table 3.1 Illustration of a sequence alignment

```
        50       .    :    .    :    .    :    .    :    .    :
Human:  247 GGTGAGGTCGAGGACCCTGCA  CGGAGCTGTATGGAGGGCA    AGAGC
            |:   ||  ||||:  |||| --:||  ||| |::|    |||---||||
Mouse:  368 GAGTCGGGGGAGGGGGCTGCTGTTGGCTCTGGACAGCTTGCATTGAGAGG


        100      .    :    .    :    .    :    .    :    .    :
Human:  292 TTC         CTACAGAAAAGTCCCAGCAAGGAGCCACACTTCACTG
            |||----------|| |   |::| |: ||||::|:||:-||  |||:| |
Mouse:  418 TTCTGGCTACGCTCTCCCTTAGGGACTGAGCAGAGGGCT CAGGTCGCGG
```

Matches correspond to alignments of identical residues, while mismatches, or *substitutions*, correspond to alignments of different residues. Insertion and deletions are signified by a lack of a corresponding match in the other sequence; residues in one sequence are matched to empty spaces, or *gaps*, in the other sequence. Sequence alignments void of gaps are often referred to as *ungapped* alignments. While insertion and deletions represent different events biologically, in a string matching context, insertions and deletions constitute inverse operations of one another, and are therefore commonly referred to as *indels* [59]. Another important operation is that of *transpositions*, which involve moving a sequence segment from one location to another, resulting in a swap in the sequence order when comparing two sequences. Transpositions are the most difficult operations to handle in sequence comparisons, and are typically not treated in biological sequence alignment.

The simplest, and the most direct way of detecting sequence similarity is to arrange the sequences in a *dot plot*, described in Sect. 3.1.1. If nothing is known about the evolutionary relationship between the sequences compared, a dot plot provides a graphical illustration of the level of similarity, and the location of conserved elements in the sequences. However, in terms of revealing the best alignment in some sense, the dot plot method is limited. For this, we need a scoring scheme that is able to quantify the different possible alignments.

A sequence alignment measures the evolutionary distance, or the degree of (dis)similarity, between two related sequences in some sense. Alignment algorithms attempt to determine the optimal alignment between sequences by modeling the mutational process that, starting from an (unknown) common ancestor, has given rise to the observed sequences. There are many possible ways to align two sequences, and in order to select the best one, we need means to quantify their relative quality. The idea is to assign a score to each alignment, and then choose the one with the optimal score. The scoring schemes used for alignments typically include a substitution matrix and a gap penalty function. The substitution matrix is used to score matches and mismatches, and the gap penalty function scores insertion and deletion events. The resulting alignment score, which is the sum of the scores of the individual events, gives a measure of the quality of the current alignment. Different substitution models and gap penalty models are described in Sects. 3.1.2–3.1.4.

Since there are many ways to align two sequences, we need means to efficiently sift through the possible alignments in search for the best one. The *dynamic programming* algorithm is a popular method that has been applied to both *global* and *local alignments*. A global alignment involves the matching of two sequences in their entirety, while local alignments only search for subsequences of high similarity. Global alignments are useful when comparing sequences that have not diverged substantially, or when the sequences constitute a single element, such as a protein domain. If the sequences are highly diverged or have become rearranged during evolution, a local alignment might be more suitable. An important note, however, is that the optimal alignment might not be the most biologically meaningful. A good general strategy is therefore to review several suboptimal (near optimal) local alignments, before choosing the "best." The *Needleman–Wunsch* algorithm for global alignments using dynamic programming is described in Sect. 3.1.5, and the *Smith–Waterman* algorithm for local alignments is described in Sect. 3.1.6.

3.1.1 Dot Plot Matrix

Dot plots are probably one of the oldest ways to compare sequences in molecular biology [30], and provide a simple and direct means for identifying regions of similarity between evolutionary related sequences. Dot plots are represented by a two-dimensional array, where the sequences to be compared are placed on the axes, and where a dot is placed in each cell corresponding to matching residues in the respective sequence positions (see Fig. 3.1). In the resulting plot, regions of similarity will appear as diagonal stretches of dots. If the two sequences are identical, the main diagonal will be filled with dots, insertions, and deletions between the sequences will appear as lateral displacements of the diagonals, and duplications appear as parallel diagonal lines in the plot.

However, while dot plots provide a lot of information directly and are easy to interpret, their usefulness is rather limited, in particular when comparing nucleotide sequences. Even for unrelated sequences, the dot plot of nucleotide sequences will contain at least 25 % dots, and even more if the base composition is skewed, making it difficult to distinguish true homology from noise. This problem has been addressed by applying

Fig. 3.1 A dot plot of two DNA sequences

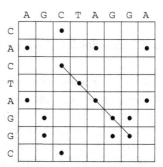

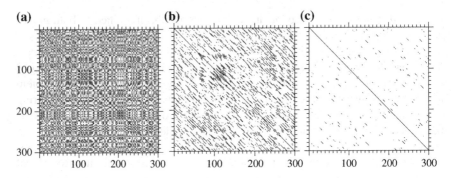

Fig. 3.2 A dot plot of a human hemoglobin subsequence against itself using window length $n = 10$. **a** Without filter. **b** Minimum number of matches $m = 5$. **c** Minimum number of matches $m = 7$

various filtering methods to the plots [63, 78, 87]. The purpose of filtering is to remove spurious matches from the plot and only present the dots displaying true homology. There are a wide variety of filters suggested, but generally they all involve the grouping of residues in some manner. The method suggested by Maizel and Lenk [63] slides a window of n residues over the pair of sequences, and place a dot when the window contains at least m matches. Other methods choose to highlight the entire window when the window match is above a certain threshold. Figure 3.2 illustrates the dot plot of a human hemoglobin against itself, first without a filter, then with a filter using a window length of $n = 10$, and the minimum number of matches required in a window set to $m = 5$ and $m = 7$, respectively.

Dot plots of protein sequences are often less messy, largely due to the larger alphabet of amino acids, but noise reduction is still very useful. However, an additional issue when comparing protein sequences is that we might not only want to highlight exact matches, but also take into account chemical and structural similarities between amino acids. Staden [86] introduced a method that included amino acid weights in the alignment algorithm. The weights were based on a relatedness odds matrix, or a *substitution matrix*, produced by Dayhoff [16]. This work by Dayhoff and colleagues later evolved into the now commonly used PAM matrix [18]. PAM and the similar BLOSUM matrices [42] are described in the next section.

3.1.2 Nucleotide Substitution Models

An important problem in biological sequence analysis is to determine the evolutionary distance between sequences. Dealing with actual time scales is often impossible, however, in particular since the substitution rate tends to vary over time both between and within sequences. Therefore, a more convenient measure of evolutionary distance is based on the number of substitutions that have occurred between the common ancestor and the current sequences. A very direct measure would, thus, be the proportional number of differences (mismatches) in an ungapped alignment of the two sequences

$$d = k/n \qquad\qquad (3.1)$$

where n is the sequence length, and k the number of mismatches. In terms of scoring schemes, the underlying substitution matrix would thus take the form

$$
\begin{array}{c|cccc}
 & \text{A C G T} \\
\hline
\text{A} & 1\ 0\ 0\ 0 \\
\text{C} & 0\ 1\ 0\ 0 \\
\text{G} & 0\ 0\ 1\ 0 \\
\text{T} & 0\ 0\ 0\ 1 \\
\end{array}
$$

If we ignore gaps for now, the alignment

```
ATCG--G
AC-GTCA
```

would then receive the alignment score

$$S = s(A, A) + s(T, C) + s(G, G) + s(G, A) = +1 + 0 + 1 + 0 = 2.$$

One problem with this approach, however, is that as the evolutionary distance increases, the number of *observed* substitutions is often less than the *actual* number of substitutions. The reason is that as time passes, the probability of having a second substitution in an already changed position increases, making a simple count insufficient for long evolutionary distances. To solve this and other issues, a number of different substitution models have been proposed.

Most substitution models assume that sequence positions are independent of each other and of previous events in the evolution history, and that substitution rates and base composition remain the same across all sites and over time. Moreover, since we generally have no information about the common ancestral sequence, substitution models are usually assumed to be *time-reversible* (see Sect. 2.1.1). That is, the probability of changing one sequence into another is the same as the probability of the process going in the opposite direction. As a result, when comparing two sequences Y and Z, instead of modeling two separate mutational processes evolving from some common ancestor, we can model the process of changing sequence Y into sequence Z directly (see Fig. 3.3). This way we can ignore the fact that the ancestor sequence is unknown.

A common model of substitution is to use a homogeneous, continuous-time, time-reversible, stationary Markov chain, such as those described in Sect. 2.1.1. Let $\{X(t) : t \geq 0\}$ denote such a substitution process, where each instance, or state, $X(t)$ of the process represents a (new) version of the initial sequence $X(0)$. Since we assume that sequence positions evolve independently, it is enough to consider the substitution process of one specific, but arbitrary, position k in the sequence, in

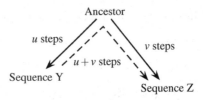

Fig. 3.3 If the substitution process is time-reversible, we can swap the direction of one of the processes originating in a common ancestor, and achieve a single process going from Y to Z in $u + v$ steps directly

order to draw conclusions on the substitution process as a whole. Let $X_k(t)$ denote the residue in position k after t time units. $X_k(t)$ takes values in some state space S, which is the set of nucleotides for DNA sequences and the set of amino acid residues for protein sequences. We let $\pi_i, i \in S$ denote the initial probabilities of this process

$$\pi_i = \mathbf{P}(X_k(0) = i), \quad i \in S \tag{3.2}$$

and denote the transition matrix $\mathbf{A}(t) = \left(a_{ij}(t)\right)_{i,j \in S}$ where

$$a_{ij}(t) = \mathbf{P}(X_k(t) = j | X_k(0) = i). \tag{3.3}$$

Recall from Sect. 2.1.1 that, in contrast to discrete-time Markov chains, the transition matrix is not enough to characterize the development of the process. The transition probability $a_{ij}(t)$ only gives the probability that state i has changed into state j at time t, but not how many changes that have occurred in between. For this, we need a transition rate matrix $\mathbf{Q} = \left(\mu_{ij}\right)_{i,j \in S}$, giving the "instantaneous" substitution rates between states i and j. Recall also that the connection between the transition matrix and the transition rate matrix is approximately given by

$$a_{ij}(t) \approx \begin{cases} \mu_{ij}t & \text{if } i \neq j \\ 1 + \mu_{ii}t & \text{if } i = j. \end{cases} \tag{3.4}$$

where $\mu_{ii} = -\sum_{j \neq i} \mu_{ij}$, such that each row of $\mathbf{Q}$ sums to zero. Assuming that the evolutionary processes of two homologous sequences are identically distributed and stationary, we can use the transition rates to calculate an interesting property,

$$K = 2t \sum_{i \in S} \pi_i \mu_i \tag{3.5}$$

where $\mu_i = -\mu_{ii}$. K can be interpreted as the *mean number of substitutions per site* [89]. If the time t to the common ancestor is known, the expression in (3.5) can be used to estimate the substitution rate, and vice versa.

The Jukes-Cantor Model

The simplest model for nucleotide substitution, is the Jukes-Cantor model [49], where each site is assumed to evolve according to a Poisson process, independently of the rest, and where all substitution rates are set to be equal. The substitution rate matrix can be written as

$$\mathbf{Q} = \begin{pmatrix} \cdot & \alpha & \alpha & \alpha \\ \alpha & \cdot & \alpha & \alpha \\ \alpha & \alpha & \cdot & \alpha \\ \alpha & \alpha & \alpha & \cdot \end{pmatrix} \tag{3.6}$$

Recall that the rows sum to 0, such that the diagonal elements are given by taking the negative sum of the others, -3α in this case.

The evolutionary distance between two sequences Y and Z can be defined as the estimated number of changes per site. The *Jukes-Cantor distance* (or *correction*) is given by

$$d_{JC}(Y, Z) = -\frac{3}{4} \log \left(1 - \frac{4}{3} D \right) \tag{3.7}$$

where D is the proportional number of differences between Y and Z in an ungapped alignment. The Jukes-Cantor distance has the desired property that it increases linearly with the number of accumulated mutations, while the increase in the proportional mismatch count in (3.1) slows down after a while, finally reaching an asymptote of 0.75. However, the variance of the Jukes-Cantor distance, derived by Kimura and Ohta [56],

$$Var(d_{JC}) = \frac{H(1 - H)}{n\left(1 - \frac{3}{4} H \right)^2}, \tag{3.8}$$

goes to infinity as the distance increases, indicating that the measures get increasingly less reliable as mutations accumulate.

The Kimura Model

The Jukes-Cantor model is somewhat unrealistic as it assumes equal substitution rates for all types of substitutions. In reality, however, purines (A and G) are more likely to change into a purine than into a pyrimidine (C and T), and vice versa. Substitutions between the same type of nucleotides (purines to purines, or pyrimidines to pyrimidines), also known as *translations*, can happen readily, while substitutions across types (purine to pyrimidine or vice versa), also known as *transversions*, are much less frequent. Kimura [53] suggested a two-parameter model using uniform base frequencies and a Poisson process for the substitutions, but with different rates for transitions and transversions.

Fig. 3.4 In the Jukes-Cantor model, all changes are assumed to occur with equal probabilities, while the Kimura 2-parameter model has different rates for transitions and transversions

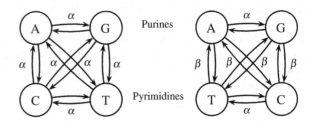

If we let α denote the transition rates, and β the transversion rates, the substitution rate matrix, with rows and columns ordered as {A,C,G,T}, is given by

$$\mathbf{Q} = \begin{pmatrix} \cdot & \beta & \alpha & \beta \\ \beta & \cdot & \beta & \alpha \\ \alpha & \beta & \cdot & \beta \\ \beta & \alpha & \beta & \cdot \end{pmatrix} \tag{3.9}$$

Figure 3.4 illustrates the difference between the Jukes-Cantor and the Kimura 2-parameter model. If p and q denote the proportional numbers of transitional and transversal differences, respectively, the Kimura distance is given by

$$d_{K2p}(\mathbf{a}, \mathbf{b}) = -\frac{1}{2}\log(1 - 2p - q) - \frac{1}{4}\log(1 - 2q). \tag{3.10}$$

In the special case of $\alpha = \beta$, we get $P = Q/2$ and (3.10) reduces to the Jukes-Cantor distance. A more general approach, presented in [54], allowed for different substitution rates for all substitutions. Again, ordering the rows and columns as {A,C,G,T}, the substitution matrix can be written as

$$\mathbf{Q} = \begin{pmatrix} \cdot & \gamma & \alpha & \beta \\ \gamma & \cdot & \beta & \alpha \\ \alpha & \beta & \cdot & \gamma \\ \beta & \alpha & \gamma & \cdot \end{pmatrix} \tag{3.11}$$

The special case of $\gamma = \beta$ gives the 2-parameter model above.

The Felsenstein Model

After the Kimura model, several models were proposed, with an increasing number of parameters. One problem with both the Kimura and the Jukes-Cantor models is that the base composition is assumed to be uniform. However, real sequences often contain an $A + T$ or $G + C$ bias, and coding sequences tend to have different frequencies for all four bases. Felsenstein [24] suggested that the substitution rate of a nucleotide only depends on its base frequency, and proposed a method that uses base frequencies $\pi = (\pi_A, \pi_C, \pi_G, \pi_T)$ estimated from the actual sequences. The Felsenstein transition rate matrix can be written as

$$
Q = \begin{pmatrix}
\cdot & \alpha\pi_A & \alpha\pi_A & \alpha\pi_A \\
\alpha\pi_C & \cdot & \alpha\pi_C & \alpha\pi_C \\
\alpha\pi_G & \alpha\pi_G & \cdot & \alpha\pi_G \\
\alpha\pi_T & \alpha\pi_T & \alpha\pi_T & \cdot
\end{pmatrix}
\tag{3.12}
$$

The parameter α can no longer be interpreted as the mutation rate, however.

The Tamura and Nei Model

Various developments and generalizations of the Felsenstein model have been implemented. Hasegawa et al. [39] invoked the transition/transversion bias introduced by Kimura into the Felsenstein model. Tamura and Nei took this one step further by using one parameter for transversions but two for transitions, one for transitions between purines and one for pyrimidines [88]. Their substitution rate matrix is given by

$$
Q = \begin{pmatrix}
\cdot & \beta\pi_C & \alpha_1\pi_G & \beta\pi_T \\
\beta\pi_A & \cdot & \beta\pi_G & \alpha_2\pi_T \\
\alpha_1\pi_A & \beta\pi_C & \cdot & \beta\pi_T \\
\beta\pi_A & \alpha_2\pi_C & \beta\pi_G & \cdot
\end{pmatrix}
\tag{3.13}
$$

General Time-Reversible (GTR) Model

The most general model suggested is probably the General Time Reversible (GTR) model, first proposed by Tavaré [89]. This model can include as many as 12 parameters, but reduces to 9 under the assumption of time reversibility. Using the parametrization given in [99], the substitution rate matrix can be written

$$
Q = \begin{pmatrix}
\cdot & \alpha\pi_C & \beta\pi_G & \gamma\pi_T \\
\alpha\pi_A & \cdot & \rho\pi_G & \sigma\pi_T \\
\beta\pi_A & \rho\pi_C & \cdot & \tau\pi_T \\
\gamma\pi_A & \sigma\pi_C & \tau\pi_G & \cdot
\end{pmatrix}
\tag{3.14}
$$

Thus, the substitution rates differ between each distinct pair of nucleotides, which makes it more general than the previous models. It requires, however, that the substitution rates are the same in both directions within a pair, which makes it time reversible. This is not a serious limitation, though. With nonreversible substitution models, we would be faced with something of a paradox: the distance between two sequences X and Y would differ depending on if we measured from X to Y or from Y to X.

The main difference between the substitution models presented are the number of parameters included. These parameters need to be estimated from a training set, typically using maximum likelihood estimation (see Sect. 6.3). Which model to use depends highly upon the size of the training set. Of all the models described above, the GTR is the most sensitive, but requires a rather large training set. Moreover, since nucleotide sequence alignments only involve a 4-letter alphabet, while the amino acid alphabet consists of 20 residues, amino acid alignments provide a more sensitive measure of homology. Therefore, if the sequences in question are assumed to be protein coding, it might be a good idea to translate them into their corresponding amino acid sequences, and use a amino acid substitution matrix for scoring. We might not know, however, the reading frame of a sequence, which is why many alignment methods translate the sequences in all frames and compare all possible combinations.

3.1.3 Amino Acid Substitution Models

There are several ways to score the pairings of amino acids in an alignment. Just as in nucleotide substitution, the simplest scoring scheme for pairs of amino acids is to simply count the number of matches in the alignment. However, while this approach works reasonably well for closely related sequences but, the performance is poor for larger evolutionary distances. What is more, such an approach completely ignores the extra level of information contained in an amino acid sequence. There is a wide variety of structural and chemical properties that affects how likely one amino acid is to change into another. The development of amino acid substitution models has gone from focusing on the matches in the alignment (Unitary Matrix), to take the underlying nucleotide sequence into account (Genetic Code Matrix), to letting structural and chemical properties affect the scoring (Dayhoff-type matrices) [25]. The latter approach is almost universally used today. In general, properties to consider in order to construct biologically relevant substitution matrices include:

- Matches should receive a higher score than mismatches.
- Conservative substitutions should receive a higher score than nonconservative.
- Different evolutionary distances require different scoring schemes.

Dayhoff et al. [18] pioneered the work of estimating substitution rates from real observations, which gave rise to the popular *PAM* matrices. The underlying dataset consisted of protein sequences that were close enough to guarantee unambiguous alignments. This work was closely followed by Henikoff and Henikoff [42] in the equally popular *BLOSUM* series. Several other models have been derived since, including the GONNET matrix [31], which is based on an exhaustive matching of the entire Swiss-Prot database [12], and the JTT matrix [48], which is an update of the PAM matrix. Here, we give a brief overview of the construction of the PAM, BLOSUM, and GONNET matrices.

All substitution matrices are essentially *log-odds* matrices [3]. A log-odds score is the logarithm of the likelihood ratio of two models; the evolutionary substitution model, assuming that the sequences are related, and the random model, assuming independence between sequences. The evolutionary model returns the joint probability q_{ij} of observing residues i and j together in an alignment, while the random model returns the probability $p_i\, p_j$ of observing two independent residues. Assuming independence between sequence positions, the likelihood ratio of an alignment of two sequences $Y = Y_1, \ldots, Y_T$ and $Z = Z_1, \ldots, Z_T$ becomes

$$LR(\text{alignment}) = \prod_{t=1}^{T} \frac{\mathbf{P}(Y_t, Z_t)}{\mathbf{P}(Y_t)\mathbf{P}(Z_t)} = \prod_{t=1}^{T} \frac{q_{Y_t, Z_t}}{p_{Y_t}\, p_{Z_t}}. \tag{3.15}$$

Note how this formulation is related to that of using the transition matrix of the underlying evolutionary model

$$LR(\text{alignment}) = \prod_{t=1}^{T} \frac{\mathbf{P}(Z_t | Y_t)\mathbf{P}(Y_t)}{p_{Y_t}\, p_{Z_t}} = \prod_{t=1}^{T} \frac{a_{Y_t, Z_t}}{p_{Z_t}}, \tag{3.16}$$

where a_{ij} is the transition probability from residue i to j. The log-odds ratio is more convenient, however, since it transforms products to sums,

$$S = \sum_{t=1}^{T} \log \frac{\mathbf{P}(Z_t | Y_t)\mathbf{P}(Y_t)}{p_{Y_t}\, p_{Z_t}} = \sum_{t=1}^{T} \log \frac{a_{Y_t, Z_t}}{p_{Z_t}}. \tag{3.17}$$

The entries of a substitution matrix, thus, takes the form

$$S_{ij} = \log \frac{a_{ij}}{p_j}. \tag{3.18}$$

The PAM Matrix

The work of constructing a statistically rigorous substitution model was originated by Dayhoff et al. [18]. Their PAM series is based on observed *percent accepted (point) mutations* in a large number of high-quality alignments. The notion 'accepted' comes from the fact that in order to observe an evolutionary change of amino acids, a mutation must not only occur; it must also be kept, or "accepted," by the species. It was observed by Dayhoff and colleagues that changes between structurally and chemically similar amino acids were accepted to a higher extent than changes between less similar residues.

Before going through the steps of computing a PAM matrix it is important to note that there are two different types of matrices at work here; the *PAM transition matrix* estimating the transition probabilities of the underlying evolutionary process, and the *PAM substitution matrix* consisting of log-odds ratios used to score the alignment of amino acid pairs. The PAM transition matrix, corresponding to the evolutionary distance of 1 PAM, is computed such that applying it to a protein sequence renders the next "state" of the sequence to differ from the current state in on average 1 % of the positions. The distance unit of 1 PAM thus corresponds to the time interval needed for evolution to create that amount of change in a sequence. If we had a large number of aligned sequences, differing in on average 1 % of the positions, the construction of the PAM transition matrix would be very straightforward:

1. Count the observed number of accepted point mutations f_{ij} between each pair i and j of amino acids.
2. Estimate the transition probabilities by the observed relative frequencies

$$a_{ij} = \frac{f_{ij}}{\sum_k f_{ik}}.$$ (3.19)

However, in reality, we usually face two main problems. First, the aligned sequences are not strictly at a 1 % difference. Rather, the PAM matrices were constructed from sequence families of as much as 15 % differences. To deal with this, we apply a scaling factor on the observed frequencies. Second, since both the frequency of occurrence and the substitution rate vary between residues, the dataset will typically not contain all substitutions. At least not in enough numbers to provide a reliable estimate of the substitution rate. Therefore, a factor called the *relative mutability* is applied to the observed frequencies. The PAM transition matrix is constructed as follows:

1. Group the protein sequences into families of at least 85 % similarity.
2. Construct a phylogenetic tree of each group, inferring intermediate ancestor sequences of each evolutionary split.
3. Count the observed number of accepted point mutations f_{ij} between each pair i and j of amino acids.
4. Calculate the *relative mutability* of each amino acid

$$m_i = \frac{\sum_{j \neq i} f_{ij}}{f_i}$$ (3.20)

   where f_i is the observed frequency of amino acid i.
5. Compute the estimated *mutation probability matrix*

$$M_{ij} = \begin{cases} \dfrac{m_j f_{ij}}{\sum_i f_{ij}} & \text{if } i \neq j, \\ 1 - m_i & \text{if } i = j. \end{cases}$$ (3.21)

6. Scale the mutation probabilities to correspond to on average 1 % change when applied to a protein sequence

$$a_{ij} = \begin{cases} \lambda M_{ij} & \text{if } i \neq j, \\ 1 - \displaystyle\sum_{k \neq i} \lambda M_{ik} & \text{if } i = j, \end{cases}$$ (3.22)

where λ is a scaling factor set to yield an evolutionary distance of 1 PAM. That is,

$$\lambda = \frac{0.01}{\sum_{i=1}^{20} \sum_{j \neq i} f_i M_{ij}}. \tag{3.23}$$

Similarly to the nucleotide substitution models, we assume that sequence residues evolve independently of one another. Therefore, we can compute the scaling factor λ by considering the substitution process of a single, but arbitrary alignment position k. We let $X_k(t)$ denote the residue in that position at time t, taking values in a state space consisting of the 20 amino acids. The probability of observing a change after 1 PAM time unit is given by

$$\mathbf{P}(X_k(1) \neq X_k(0)) = \sum_{i=1}^{20} \sum_{j \neq i} \mathbf{P}(X_k(1) = j | X_k(0) = i) \mathbf{P}(X_k(0) = i)$$

$$= \lambda \sum_{i=1}^{20} \sum_{j \neq i} M_{ij} f_i. \tag{3.24}$$

Thus, to induce a 1 % change in 1 PAM time unit, we set

$$\mathbf{P}(X_k(1) \neq X_k(0)) = 0.01 \tag{3.25}$$

and get the solution in (3.23). The resulting matrix $\mathbf{A} = (a_{ij})_{i,j}$ is the PAM transition matrix of the underlying substitution process. The computation of the PAM *substitution matrix* is straightforward:

1. Calculate the *relatedness odds matrix*

$$R_{ij} = \frac{a_{ij}}{p_i}, \tag{3.26}$$

   where p_i is the relative frequency of amino acid i.
2. The final PAM1 log-odds matrix is then given by

$$\text{PAM}_{ij} = \log R_{ij}. \tag{3.27}$$

The PAM transition matrix $\mathbf{A} = (a_{ij})$ "defines" an evolutionary distance unit of 1 PAM. That is, used as a transition matrix applied to a target protein sequence, the next "state" will be a new sequence with on average 1 % differences from the current sequence. The distance unit of 1 PAM thus corresponds to the time interval needed for evolution to create that amount of change in a sequence. This distance varies both within and between species, as well as over time, such that there is no one-to-one translation between PAM distance and actual time. There is an approximative relationship, however, between the percent difference between compared sequences and their evolutionary distance in PAM units (Fig. 3.5). The average percent difference

Table 3.2 The correspondence between percent differences in evolutionary related sequences and evolutionary distance in PAM units

% difference	1	5	10	15	20	25	30	35	40	45	50	55	60	65	70	75	80	85
PAM distance	1	5	11	17	23	30	38	47	56	67	80	94	112	133	159	195	246	328

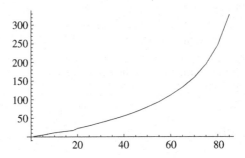

Fig. 3.5 A plot of the percent difference against PAM distance

of amino acids for each matrix presented in Table 3.2 corresponds to

$$100\left(1 - \sum_i p_i a_{ii}\right).$$ (3.28)

The PAM model assumes that the amino acid composition remains constant over time, and that the substitution process is time-homogeneous. As a result, the substitution matrix for longer evolutionary times can be achieved directly by applying the PAM1 matrix successively to the target sequence. The PAMn matrix can therefore be achieved directly by matrix multiplication

$$\text{PAM}_{ij}(n) = \log R_{ij}(n).$$ (3.29)

where $R_{ij}(n)$ corresponds to the n-step relatedness odds matrix $\mathbf{R}^n$, which, analogously to (2.15), is given by

$$R_{ij}(n) = \sum_{k=1}^{20} R_{ik} R_{kj}(n-1).$$ (3.30)

Equivalently, by scaling the elements in (3.22) by a different scaling factor λ, we can achieve higher order PAMs.

That is, the PAMn matrix has scaling factor

$$\lambda = \frac{n/100}{\displaystyle\sum_{i=1}^{20}\sum_{j\neq i} f_i M_{ij}}$$ (3.31)

Table 3.3 The PAM250 log-odds matrix using log-base 10 and multiplying the numbers by 10 for readability

	A	R	N	D	C	Q	E	G	H	I	L	K	M	F	P	S	T	W	Y	V
A	2	-2	0	0	-2	0	0	1	-1	-1	-2	-1	-1	-3	1	1	1	-6	-3	0
R	-2	6	0	-1	-4	1	-1	-3	2	-2	-3	3	0	-4	0	0	-1	2	-4	-2
N	0	0	2	2	-4	1	1	0	2	-2	-3	1	-2	-3	0	1	0	-4	-2	-2
D	0	-1	2	4	-5	2	3	1	1	-2	-4	0	-3	-6	-1	0	0	-7	-4	-2
C	-2	-4	-4	-5	12	-5	-5	-3	-3	-2	-6	-5	-5	-4	-3	0	-2	-8	0	-2
Q	0	1	1	2	-5	4	2	-1	3	-2	-2	1	-1	-5	0	-1	-1	-5	-4	-2
E	0	-1	1	3	-5	2	4	0	1	-2	-3	0	-2	-5	-1	0	0	-7	-4	-2
G	1	-3	0	1	-3	-1	0	5	-2	-3	-4	-2	-3	-5	0	1	0	-7	-5	-1
H	-1	2	2	1	-3	3	1	-2	6	-2	-2	0	-2	-2	0	-1	-1	-3	0	-2
I	-1	-2	-2	-2	-2	-2	-2	-3	-2	5	2	-2	2	1	-2	-1	0	-5	-1	4
L	-2	-3	-3	-4	-6	-2	-3	-4	-2	2	6	-3	4	2	-3	-3	-2	-2	-1	2
K	-1	3	1	0	-5	1	0	-2	0	-2	-3	5	0	-5	-1	0	0	-3	-4	-2
M	-1	0	-2	-3	-5	-1	-2	-3	-2	2	4	0	6	0	-2	-2	-1	-4	-2	2
F	-3	-4	-3	-6	-4	-5	-5	-5	-2	1	2	-5	0	9	-5	-3	-3	0	7	-1
P	1	0	0	-1	-3	0	-1	0	0	-2	-3	-1	-2	-5	6	1	0	-6	-5	-1
S	1	0	1	0	0	-1	0	1	-1	-1	-3	0	-2	-3	1	2	1	-2	-3	-1
T	1	-1	0	0	-2	-1	0	0	-1	0	-2	0	-1	-3	0	1	3	-5	-3	0
W	-6	2	-4	-7	-8	-5	-7	-7	-3	-5	-2	-3	-4	0	-6	-2	-5	17	0	-6
Y	-3	-4	-2	-4	0	-4	-4	-5	0	-1	-1	-4	-2	7	-5	-3	-3	0	10	-2
V	0	-2	-2	-2	-2	-2	-2	-1	-2	4	2	-2	2	-1	-1	-1	0	-6	-2	4

for $n = 1, 2, 3, \ldots$. The PAMn matrix corresponds to an evolutionary time interval of n PAMs, and the PAM0 matrix is simply the identity matrix, representing the situation where no residues have changed yet. In the limit, as n increases, the matrix values approaches the amino acid composition. The common representation of the PAM250 matrix (sometimes referred to as the mutation data matrix MDM78 in [17]) is shown in Table 3.3. The PAM matrices are used extensively in various comparative algorithms, and although it constitutes a fairly good approximation of the evolutionary process for the most part, it has some limitations. The PAM model is based on several assumptions:

- Amino acid composition is constant over time.
- The substitution process is time-homogeneous such that the PAM1 can be scaled to arbitrary evolutionary distances.
- The substitution rate is constant throughout the sequence.
- The transition probabilities at any site only depends on the current amino acid.
- Compared sequences have the same average amino acid composition.

In reality, these assumptions are often clearly violated. Both amino acid compositions and substitution rates vary over time, within the same sequence, and between species, and the mutability of a site is clearly position dependent. Moreover, by comparing closely related sequences, we ensure the occurrence of at most one substitution per site, but instead we yield the problem that many substitutions occur too infrequently to provide reliable estimates. When the mutation rates are extrapolated to longer evolutionary distances, any such errors will propagate through the process and become magnified in higher order matrices. A possible solution would be to use counts from more distantly related sequences, such as in the GONNET matrices [31] presented below. One problem with using longer distances, however, is that it becomes much harder to resolve the "true" alignments on which the matrix is based.

The BLOSUM Matrix

The PAM model assumes constant substitution rates throughout the sequences. This is clearly not true in reality, as certain positions are more important and thereby more conserved than others. The BLOSUM (BLOcks SUbstitution Matrix) series [42] attempts to utilize the observation that distantly related sequences tend to have highly conserved regions, or *blocks*, intervened by less-conserved stretches of sequence. While the PAM matrix is constructed from closely related sequences and then extrapolated to varying distances, the BLOSUM matrices are constructed empirically from multiple alignments of sequences at various evolutionary distances. The multiple alignments are derived from the BLOCKS database [41, 43], and consist of ungapped blocks of the most highly conserved regions in the proteins. The BLOSUM80 matrix shown in Table 3.4 is constructed as follows.

Table 3.4 The BLOSUM80 matrix

	A	R	N	D	C	Q	E	G	H	I	L	K	M	F	P	S	T	W	Y	V
A	5	-2	-2	-2	-1	-1	-1	0	-2	-2	-2	-1	-1	-3	-1	1	0	-3	-2	0
R	-2	6	-1	-2	-4	1	-1	-3	0	-3	-3	2	-2	-4	-2	-1	-1	-4	-3	-3
N	-2	-1	6	1	-3	0	-1	-1	0	-4	-4	0	-3	-4	-3	0	0	-4	-3	-4
D	-2	-2	1	6	-4	-1	1	-2	-2	-4	-5	-1	-4	-4	-2	-1	-1	-6	-4	-4
C	-1	-4	-3	-4	9	-4	-5	-4	-4	-2	-2	-4	-2	-3	-4	-2	-1	-3	-3	-1
Q	-1	1	0	-1	-4	6	2	-2	1	-3	-3	1	0	-4	-2	0	-1	-3	-2	-3
E	-1	-1	-1	1	-5	2	6	-3	0	-4	-4	1	-2	-4	-2	0	-1	-4	-3	-3
G	0	-3	-1	-2	-4	-2	-3	6	-3	-5	-4	-2	-4	-4	-3	-1	-2	-4	-4	-4
H	-2	0	0	-2	-4	1	0	-3	8	-4	-3	-1	-2	-2	-3	-1	-2	-3	2	-4
I	-2	-3	-4	-4	-2	-3	-4	-5	-4	5	1	-3	1	-1	-4	-3	-1	-3	-2	3
L	-2	-3	-4	-5	-2	-3	-4	-4	-3	1	4	-3	2	0	-3	-3	-2	-2	-2	1
K	-1	2	0	-1	-4	1	1	-2	-1	-3	-3	5	-2	-4	-1	-1	-1	-4	-3	-3
M	-1	-2	-3	-4	-2	0	-2	-4	-2	1	2	-2	6	0	-3	-2	-1	-2	-2	1
F	-3	-4	-4	-4	-3	-4	-4	-4	-2	-1	0	-4	0	6	-4	-3	-2	0	3	-1
P	-1	-2	-3	-2	-4	-2	-2	-3	-3	-4	-3	-1	-3	-4	8	-1	-2	-5	-4	-3
S	1	-1	0	-1	-2	0	0	-1	-1	-3	-3	-1	-2	-3	-1	5	1	-4	-2	-2
T	0	-1	0	-1	-1	-1	-1	-2	-2	-1	-2	-1	-1	-2	-2	1	5	-4	-2	0
W	-3	-4	-4	-6	-3	-3	-4	-4	-3	-3	-2	-4	-2	0	-5	-4	-4	11	2	-3
Y	-2	-3	-3	-4	-3	-2	-3	-4	2	-2	-2	-3	-2	3	-4	-2	-2	2	7	-2
V	0	-3	-4	-4	-1	-3	-3	-4	-4	3	1	-3	1	-1	-3	-2	0	-3	-2	4

1. Retrieve protein blocks from the BLOCKS database.
2. In each block, cluster the sequences according to the given matrix level, and assign sequence weights so that the contribution of each cluster in a block is equal to one. For the BLOSUM80 matrix, sequences of at least 80% identity are clustered.
3. In each block column, count the number of matches and mismatches, resulting in a table of frequencies f_{ij} of observed pairs of amino acids i and j.
4. Estimate the probability of occurrence for each pair

$$q_{ij} = \frac{f_{ij}}{\sum_{i=1}^{20} \sum_{j=1}^{i} f_{ij}}. \tag{3.32}$$

5. Calculate the expected probability of occurrence (under the random model)

$$e_{ij} = \begin{cases} p_i\, p_j & \text{if } i = j, \\ 2\, p_i\, p_j & \text{if } i \neq j, \end{cases} \tag{3.33}$$

where p_i is the probability of amino acid i occurring in a pair

$$p_i = q_{ii} + \sum_{i \neq j} q_{ij}/2. \tag{3.34}$$

6. The log-odds ratio becomes

$$S_{ij} = \log_2 \frac{q_{ij}}{e_{ij}}. \tag{3.35}$$

7. The final BLOSUM matrix entry is achieved by multiplying the log-odds score by 2 and rounding off to the nearest integer.

The BLOCKS database consists of ungapped multiple alignments of highly conserved protein regions, such as in Fig. 3.6. In each column in each block, all pairwise counts are tallied, both matches and mismatches (recall that PAM only records substitutions). For instance, for the two blocks in Fig. 3.6 the count matrix becomes

	A	B	C	D
A	11			
B		22		
C	7	4	4	
D	3	12	5	12

Different levels of the BLOSUM matrix are created by clustering the blocks according to the specific percentage sequence similarity. In order to reduce sequence bias, the sequences within a cluster are weighted such that the total contribution of each

Fig. 3.6 Blocks are
constructed from ungapped
highly conserved multiple
alignments

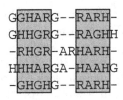

cluster is equal to that of a single sequence. For instance, for a BLOSUM80 matrix, the cluster threshold is set to 80 %. This means that if sequences *A* and *B* are at least 80 % identical, they belong to the same cluster. Furthermore, if sequence *C* is at least 80 % identical to either *A* or *B*, it is included in the cluster as well. That is, all pairs in a cluster need not have a similarity above the given threshold.

Which log-base to use in the log-odds ratio is arbitrary, since its main purpose is to transform multiplications to sums, but BLOSUM uses log-base 2 to enable the interpretation of scores in "bits" representing the information content. By multiplying the log-odds ratio by 2 and rounding to the nearest integer, the final BLOSUM scores are presented in half-bit units.

The GONNET Matrix

While the PAM matrix is derived from closely related sequences, and then extrapolated to longer evolutionary distances, Gonnet et al. [31] derived a more modern set of substitution matrices by performing an exhaustive match of all protein sequences available in the Swiss-Prot (now UniProt [92]) protein database [12]. In this match, proteins of varying evolutionary distances (between 6.4 to 100 PAMs) were pairwise aligned using the Needleman–Wunsch algorithm for global alignments (see Sect. 3.1.5), using classical substitution matrices and gap penalties. The alignments were then refined iteratively by recalculating the substitution scores and gap penalties from the new alignments, and realigning the sequences using the new scores. To speed things up, sequence pairs of potentially significant similarities were identified by organizing the Swiss-Prot database in a *Patricia tree* [69]. A Patricia tree is a special case of a *prefix tree*, which is a data structure used to enable efficient string searches in, for instance, various types of dictionaries. The GONNET matrix is constructed as follows:

1. Arrange all protein sequences in a Patricia tree.
2. Pairwise align all sequences within a preset distance in the tree, using the Needleman–Wunsch algorithm.
3. Recalculate the substitution matrix and the gap penalties based on the alignments.
4. Iterate the alignments and the recalculation of the scores until no further improvements can be made.

In order to identify all potentially significant homologies, the target score in the first round of alignments (step 2) was set rather liberal. A comparison between substitution matrices achieved from closely related sequences and from more distantly related sequences were found to differ, underscoring the problem with the extrapolation to

Table 3.5 The GONNET250 matrix

	A	R	N	D	C	Q	E	G	H	I	L	K	M	F	P	S	T	W	Y	V
A	2	-1	0	0	1	0	0	1	-1	-1	-1	0	-1	-2	0	1	1	-4	-2	0
R	-1	5	0	0	-2	2	0	-1	1	-2	-2	3	-2	-3	-1	0	0	-2	-2	-2
N	0	0	4	2	-2	1	1	0	1	-3	-3	1	-2	-3	-1	1	1	-4	-1	-2
D	0	0	2	5	-3	1	3	0	0	-4	-4	1	-3	-5	-1	1	0	-5	-3	-3
C	1	-2	-2	-3	12	-2	-3	-2	-1	-1	-2	-3	-1	-1	-3	0	-1	-1	-1	0
Q	0	2	1	1	-2	3	2	-1	1	-2	-2	2	-1	-3	0	0	0	-3	-2	-2
E	0	0	1	3	-3	2	4	-1	0	-3	-3	1	-2	-4	-1	0	0	-4	-3	-2
G	1	-1	0	0	-2	-1	-1	7	-1	-5	-4	-1	-4	-5	-2	0	-1	-4	-4	-3
H	-1	1	1	0	-1	1	0	-1	6	-2	-2	1	-1	-3	-1	0	0	-2	-2	-2
I	-1	-2	-3	-4	-1	-2	-3	-5	-2	4	3	-2	3	1	-3	-2	-1	-2	-1	3
L	-1	-2	-3	-4	-2	-2	-3	-4	-2	3	4	-2	3	2	-2	-2	-1	-1	0	2
K	0	3	1	1	-3	2	1	-1	1	-2	-2	3	-1	-3	-1	0	0	-4	-2	-2
M	-1	-2	-2	-3	-1	-1	-2	-4	-1	3	3	-1	4	2	0	-1	-1	-1	0	2
F	-2	-3	-3	-5	-1	-3	-4	-5	0	1	2	-3	2	7	-4	-3	-2	4	5	0
P	0	-1	-1	-1	-3	0	-1	-2	-1	-3	-2	-1	-2	-4	8	0	0	-5	-3	-2
S	1	0	1	1	0	0	0	0	0	-2	-2	0	-1	-3	0	2	2	-3	-2	-1
T	1	0	1	0	-1	0	0	-1	0	-1	-1	0	-1	-2	0	2	3	-4	-2	0
W	-4	-2	-4	-5	-1	-3	-4	-4	-1	-2	-1	-4	-1	4	-5	-3	-4	14	4	-3
Y	-2	-2	-1	-3	-1	-2	-3	-4	2	-1	0	-2	0	5	-3	-2	-2	4	8	-1
V	0	-2	-2	-3	0	-2	-2	-3	-2	3	2	-2	2	0	0	-1	0	-3	-1	3

longer distances done in the PAM matrix model. The GONNET250 matrix is given in Table 3.5. The matrix entries represent values for the log-odds ratio

$$10 \log \frac{q_{ij}}{p_i \, p_j} \tag{3.36}$$

normalized to an evolutionary distance of 250 PAMs and rounded to the nearest integer.

3.1.4 Gap Models

No matter how different two sequences are, by inserting appropriate amounts of gaps, the number of mismatches can be decreased to zero. Using gaps in such an unconstrained manner, however, will be a poor imitation of the biological reality. Instead, we assume that insertions and deletions are rare events, and therefore we penalize the use of gaps in an alignment.

Various models for gaps have been suggested, the simplest being a *linear* gap model, where each insertion or deletion is assumed to involve a single residue that is independent of everything else. A linear gap penalty function γ of a gap of length k takes the form

$$\gamma(k) = -\delta k, \tag{3.37}$$

where k is the length of the gap and δ the penalty for each individual gap residue. The problem with this model is that a gap of length k gets the same penalty as k gaps of length 1. The biological reality, however, is that a single insertion or deletion event tends to involve several residues at once. A more realistic gap model would, thus, penalize gap residues less and less the longer the gap gets. A *concave* function is a function that satisfies the condition

$$\gamma(k + 1) - \gamma(k) \le \gamma(k) - \gamma(k - 1), \quad \text{for all } k. \tag{3.38}$$

The implementation of a concave gap function is computationally expensive, however, requiring on the order of $O(k^3)$ in running time (number of operations), and $O(k^2)$ in memory usage. Waterman [96] improves slightly on this by presenting a gap function that requires $O(k^2 \log k)$ in running time,

$$\gamma(k) = -\delta - \varepsilon \log(k). \tag{3.39}$$

This model still becomes intractable for long sequences, however. Therefore, a common linear approximation of the concave gap function is to use an *affine* gap model. The affine gap model is composed of two parameters; a gap *opening* penalty δ for the first gap residue, and a gap *extension* penalty ε for any succeeding gap residues

$$\gamma(k) = -\delta - \varepsilon(k - 1). \tag{3.40}$$

The gap extension penalty is usually smaller than the gap opening penalty $\varepsilon < \delta$, resulting in a lower cost for long insertions than in a linear gap model. The gain in efficiency using affine gaps over more general gap models is that the alignment algorithm only needs to keep track of the previous pairing in the alignment, to see if a current gap is part of a longer indel or not.

Example 3.1 Linear and affine gap penalties
A model using linear gaps would score the following two alignments equally,

```
    TCAGGCTGGCCATG              TCAGGCTGGCCATG
    TCA--C---C-ATG              TCA------CCATG
```

while the affine gap penalty model, with a lower gap extension penalty $\varepsilon < \delta$, would penalize the rightmost alignment less.   □

3.1.5 The Needleman–Wunsch Algorithm

The number of possible alignments grows rapidly with the lengths of the sequences to be aligned, and the computationally challenging task is to sift through them all in

Fig. 3.7 The calculation of $M(t, u)$ builds on one of three previously calculated cell scores in the matrix

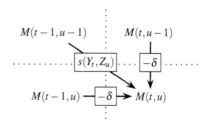

search for the optimal alignment. The Needleman–Wunsch algorithm [72] returns an optimal global alignment using dynamic programming (see Sect. 2.1.3). The extension to local alignments is presented in the Smith–Waterman algorithm in the next section.

The two sequences to be compared, $Y = (Y_1, \ldots, Y_T)$ and $Z = (Z_1, \ldots, Z_U)$, are organized in a matrix, much like a dot plot matrix (see Sect. 3.1.1), with sequence Y on the horizontal axis and Z on the vertical axis (see Fig. 3.8). Each alignment of Y and Z can then be represented by a pathway through the matrix, running from the top-left corner to the bottom right. The only permissible operations while moving along the path involve one step down, one step right, or one step diagonally downright (see Fig. 3.7). A diagonal step represents a match or a mismatch in the alignment (i.e., a pairing of two residues), a vertical step represents a gap in the Y sequence, and a horizontal step represents a gap in Z.

The scoring model consists of a substitution score and a gap penalty, where the substitution score typically comes from some kind of substitution matrix. We begin to illustrate the algorithm using a linear gap model, and then extend it to affine gaps in the next section. The calculations of the alignment algorithm are organized in an $(T + 1) \times (U + 1)$ dynamic programming matrix M, indexed as $t = 0, 1, 2, \ldots, T$, and $u = 0, 1, 2, \ldots, U$ where each cell $M(t, u)$ represents the score of the optimal alignment of the subsequences $Y_1, \ldots, Y_t$, and $Z_1, \ldots, Z_u$ up to residue pair (t, u). Recall from Sect. 2.1.3 that dynamic programming algorithms consist of three parts:

1. The recurrence relation.
2. The tabular computation.
3. The traceback.

The recurrence relation in the Needleman–Wunsch algorithm is given by

$$M(t, u) = \max \begin{cases} M(t-1, u-1) + s(Y_t, Z_u) & \text{match/mismatch} \\ M(t-1, u) - \delta & \text{gap in } Z \\ M(t, u-1) - \delta & \text{gap in } Y \end{cases} \tag{3.41}$$

where $s(Y_t, Z_u)$ is the substitution score of residues Y_t and Z_u, and δ the linear gap penalty. Thus, the score $M(t, u)$ is built upon one of three possible previous positions (see Fig. 3.7).

The tabular computation begins by initiating row and column 0 using the following scores

$$M(0, 0) = 0$$
$$M(t, 0) = -t\delta$$
$$M(0, u) = -u\delta. \tag{3.42}$$

The tabular computation proceeds by calculating the recurrence relation for cell $(1, 1)$, and moving from cell to cell one row at a time. In each cell, we keep pointers to the one of the previous three cells that was used to compose the current score. When the matrix is completely filled, the last cell (T, U) contains the score of the optimal global alignment of the two sequences under the model. In the traceback, the alignment corresponding to the score $M(T, U)$ is reconstructed by following the pointers back from cell (T, U) to the very first cell $(0, 0)$.

Example 3.2 The Needleman–Wunsch algorithm
Assume we were to align the two sequences

    ANVDR
    VANDR

using a gap penalty of $\delta = 8$ and the BLOSUM80 substitution matrix in Table 3.4. The tabular computation is illustrated in Fig. 3.8, where we see that the score of the optimal alignment becomes 7.

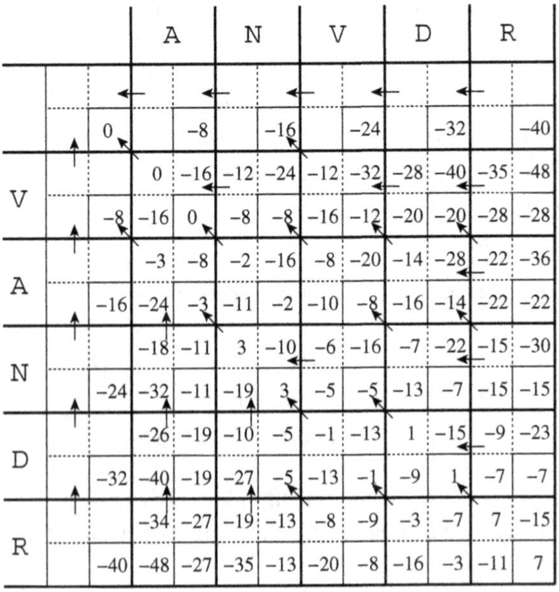

Fig. 3.8 Alignment of sequences ANVDR and VCNDR using gap penalty $\delta = 8$ and the BLOSUM80 matrix. The four subcells in each cell correspond to the scores for a diagonal move (*top-left*), a vertical move (*top-right*), a horizontal move (*bottom-left*), and the optimal cell score (*bottom-right*)

The traceback through the matrix, following the stored pointers, reveals that the optimal alignment achieving this score is

$$
\begin{array}{cccccc}
- & A & N & V & D & R \\
V & A & N & - & D & R
\end{array}
$$

□

3.1.5.1 Needleman–Wunsch Using Affine Gaps

In nature, insertions and deletions usually occur as block events. As a consequence, the probability of a gap residue should be increased if the previous position in the alignment contains a gap. Therefore, as discussed in Sect. 3.1.4, a more sensible gap penalty than the linear is to use an affine gap model. The Needleman–Wunsch algorithm can be adjusted to allow for a general gap function $\gamma(g)$. Following [95], the recurrence relation in (3.41) can be modified to

$$
M(t, u) = \max \begin{cases} M(t - 1, u - 1) + s(Y_t, Z_u) & \\ M(t - k, u) + \gamma(k) & k = 1, \ldots, t \\ M(t, u - k) + \gamma(k) & k = 1, \ldots, u. \end{cases} \tag{3.43}
$$

Figure 3.9 illustrates how position (t, u) in the alignment can be preceded by a gap of length $k = 1, \ldots, u$ in the first sequence (representing an insertion in the second), or a gap of length $k = 1, \ldots, t$ in the second sequence (representing an insertion in the first).

The drawback with this model is that an alignment of sequences $Y_1, \ldots, Y_T$ and $Z_1, \ldots, Z_U$, with $U \geq T$, requires $O(TU^2)$ in running time and $O(TU)$ in memory usage, since in each matrix cell (t, u) we need to loop through all previous positions in column t and row u. Gotoh presented a method that reduces the running time to $O(TU)$ for a model using affine gaps [32]. However, in this method we need to introduce several dynamic programming matrices. Instead of using (3.43), we

Fig. 3.9 With a general gap function, position (t, u) can be preceded with up to t gaps in the sequence Z, or u gaps in Y

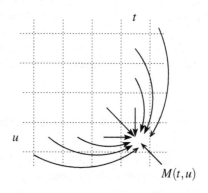

introduce a gap opening penalty δ and a gap extension penalty ε and we define the dynamic programming matrices as

$$M(t, u) = \max \begin{cases} M(t - 1, u - 1) + s(Y_t, Z_u), \\ V(t, u), \\ W(t, u). \end{cases} \tag{3.44}$$

$$V(t, u) = \max \begin{cases} M(t - 1, u) - \delta, \\ V(t - 1, u) - \varepsilon. \end{cases} \tag{3.45}$$

$$W(t, u) = \max \begin{cases} M(t, u - 1) - \delta, \\ W(t, u - 1) - \varepsilon. \end{cases} \tag{3.46}$$

$V(t, u)$ is the score of the optimal alignment of $Y_1, \ldots, Y_t$ and $Z_1, \ldots, Z_u$ when Y_t is aligned to a gap after Z_u. Similarly, $W(t, u)$ is the optimal score when Z_u is aligned to a gap after Y_t. In other words, either we align Y_t and Z_u directly and add a match/mismatch score $s(Y_t, Z_u)$ to $M(t, u)$, or Y_t is part of an insertion in Y in which case we use the $V(t, u)$ score, or Z_u is part of an insertion in Z (or a deletion in Y) in which case we use the $W(t, u)$ score.

$V(t, u)$: $W(t, u)$:

A D N V D Y_t D C Y_t - -
V C Z_u- - -   V C N A Z_u

The induction is completed in TU steps, where in each step the cell (t, u) is calculated in each of the three matrices. The matrices are initialized by setting $M(0, 0) = 0$ and

$$M(t, 0) = V(t, 0) = -\delta - (t - 1)\varepsilon, \quad t \geq 1,$$
$$M(0, u) = W(0, u) = -\delta - (u - 1)\varepsilon, \quad u \geq 1. \tag{3.47}$$

Only M needs to initialize both a row and column 0. The cell values $V(0, u)$ in column u and $W(t, 0)$ in row t will never be used.

Although the affine gap model is an improvement to the linear model, it is a very crude approximation of a concave function, since, after the first gap opening residue the affine model behaves like the linear.

3.1.6 The Smith–Waterman Algorithm

Through a simple modification of the Needleman–Wunsch algorithm, Smith and Waterman provided a dynamic programming algorithm for local alignments [83]. The reasoning for local alignments is simple that even if two sequences are too different overall to produce a meaningful global alignment, they may still share segments of high similarity. In particular, functional sequences, such as protein coding exons, tend to diverge much slower than the more random sequences in between. This is also the main motivation for comparative gene finding, which will be discussed more in detail in the next chapter. The basis for such comparative analyses lies in the ability to produce meaningful *local* alignments. The modification of the Needleman–Wunsch to the Smith–Waterman algorithm simply involves adding one term to the maximization in the recurrence relation

$$
M(t, u) = \max \begin{cases} M(t-1, u-1) + s(Y_t, Z_u), \\ M(t-1, u) - \delta, \\ M(t, u-1) - \delta, \\ 0. \end{cases} \tag{3.48}
$$

The term 0 can be seen as a way of "resetting" the alignment in regions where the overall similarity has become too low. The tabular computation is performed as before, with the only difference that the matrix is initialized with zeros (since negative values are no longer possible), $M(t, 0) = M(0, u) = 0$ for $t, u \geq 0$. After the tabular computation is terminated, the traceback starts in the cell (t, u) that holds the highest score $M(t, u)$. That is, the traceback can start anywhere in the matrix and not necessarily in the bottom-right cell (T, U). The traceback then proceeds up and left as before, following the stored pointers, until reaching a cell with score 0 (see Fig. 3.10). In order for this to work, the expected score of a random match must be negative. Otherwise, long random matches will get a high score just based on their length. Also, at least some residue pairs (a, b) must have positive scores $s(a, b) > 0$, otherwise all cells will be set to 0 and the algorithm will not find any alignments at all.

Example 3.3 The Smith–Waterman algorithm

Borrowing a popular example from [19], assume that we want to align the amino acid sequences

        HEAGAWGHEE
        PAWHEAE

using the Smith–Waterman algorithm with linear gap penalty $\delta = 8$ and the BLO-SUM80 matrix in Table 3.4. The highest score in the Smith–Waterman matrix (see Fig. 3.10) is 26, so that is the cell where the traceback starts. The optimal local alignment with score 26 is given by

                A W G H E
                A W - H E   □

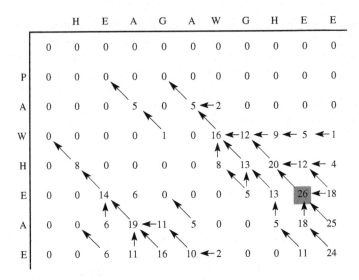

Fig. 3.10 The Smith–Waterman matrix

3.1.7 Pair Hidden Markov Models (PHMMs)

There are many ways to score an alignment, but one of the advantages of using HMMs over more heuristic scoring schemes is that it places the alignment in a probabilistic framework, and gives the problem of selecting an optimal solution a probabilistic interpretation. HMMs applied to pairwise sequence alignments are called *pair* HMMs (PHMMs). In theory, the PHMM could be extended to higher dimensions, but for practical purposes the model is mostly used for pairwise alignments. For multiple alignments, something called *profile* HMMs (see Sect. 3.2.9) are often used. The PHMM has also been applied to gene finding in the comparative gene finder DoubleScan [65], which is described in a little more detail in Sect. 4.3.1.

Preliminaries

Recall from Sect. 2.1 that an HMM is composed of two interrelated random processes: a hidden process of states and an observed process. The hidden process is a Markov chain that jumps between states in a state space, and is hidden from the observer, and the observed process generate output as a function of the hidden process. The basic mechanism of a PHMM is that of a standard HMM with the main difference being that instead of generating a single output in each step, it emits an *aligned pair* of symbols.

A PHMM can be illustrated in several ways (see [1] for a review), but a representation that has become a bit of a standard is the one given in Durbin et al. [19]. In this setting, the PHMM state space consists of three main states (see Fig. 3.11): $S = \{M, I, D\}$, where M stands for 'match', I for 'insertion', and D for 'deletion'.

Fig. 3.11 A PHMM state space consists of a match state M signifying a match or a mismatch, an insertion state I signifying an insertion in the first sequence, and a deletion state D signifying a deletion in the first sequence

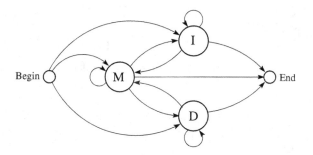

To account for the fact that we are dealing with finite sequences, we also include a silent begin and end state. When in the match state a pair of residues from some given alphabet (e.g., DNA or protein residues) is generated. The insertion and deletion states attempt to mimic the evolutionary process by assuming that the first input sequence has mutated into the second, thus an insertion is represented by a residue in the first sequence and a gap in the second, while a deletion results in a gap in the first and a residue in the second sequence. Note that an output from the insertion state may very well represent a deletion in the second sequence, rather than an insertion in the first. The assignment of first and second sequences is arbitrary, and there is no way of knowing if an indel event is actually an insertion in one sequence or a deletion in the other.

When using a PHMM to model sequence alignments, the input is two potentially homologous sequences Y and Z, and the output is the most probable alignment of the two, given the model. Just as with standard HMMs, each possible alignment can be represented by a path through a dynamic programming matrix. Let Y_1^T and Z_1^U denote the pair of sequences to be aligned, where $Y_a^b = Y_a, \ldots, Y_b$ denotes a sequence segment running from index a to index b. We can use the same notation as for standard HMMs, only modifying it to include two observed sequences rather than one. As for standard HMMs, we let X_1^L denote the hidden state sequence taking values in the state space $S = \{M, I, D\}$. We denote the initial probabilities

$$\pi_i = \mathbf{P}(X_0 = i), \quad i \in S, \tag{3.49}$$

and the transition probabilities

$$a_{ij} = \mathbf{P}(X_l = j | X_{l-1} = i), \quad i, j \in S. \tag{3.50}$$

Because of the occurrence of gaps in the produced alignment, the indices of the hidden and the observed processes will differ and need to be separated just as in GHMMs. Moreover, similarly to GHMMs, we can associate durations and partial sums with the observed sequences, to account for whether an output is made or not in that sequence. Let the durations corresponding to the states X_1^L be denoted $(d_1, \ldots, d_L)$ in the Y sequence, and $(e_1, \ldots, e_L)$ in the Z sequence, respectively.

Now we have that the pair of durations in each state follows one of three possible patterns

$$(d_l, e_l) = \begin{cases} (1, 1) & \text{if } X_l = M, \\ (1, 0) & \text{if } X_l = I, \\ (0, 1) & \text{if } X_l = D. \end{cases} \tag{3.51}$$

Thus, we do not need an explicit duration distribution, since the durations are given by the state. We use partial sums to keep track of the number of emitted symbols in each sequence,

$$p_l = \sum_{k=1}^{l} d_k \quad \text{and} \quad q_l = \sum_{k=1}^{l} e_k, \tag{3.52}$$

with $p_L = T$, $q_L = U$, and $p_0 = q_0 = 0$. Using the same terminology as in GHMMs, and using the convention that Y_a^b signifies a gap if $a > b$, the emission probability in state X_l is denoted as

$$b_{X_l}(Y_{p_{l-1}+1}^{p_l}, Z_{q_{l-1}+1}^{q_l} | Y_1^{p_{l-1}}, Z_1^{q_{l-1}}) = P(Y_{p_{l-1}+1}^{p_l}, Z_{q_{l-1}+1}^{q_l} | X_l, Y_1^{p_{l-1}}, Z_1^{q_{l-1}}). \tag{3.53}$$

Although very different in notation, this emission probability still corresponds to the same kind of substitution and gap models used in the Needleman–Wunsch algorithm. That is, for $p_l = t$ and $q_l = u$, and using a linear gap model, we can write

$$b_{X_l}(Y_{p_{l-1}+1}^{p_l}, Z_{q_{l-1}+1}^{q_l} | Y_1^{p_{l-1}}, Z_1^{q_{l-1}}) = \begin{cases} s(Y_t, Z_u) & \text{if } X_l = M, \\ -\delta & \text{if } X_l = I \text{ or } D, \end{cases} \tag{3.54}$$

where $s(a, b)$ is the substitution score for residues a and b, and δ is the linear gap penalty. Finally, the joint probability of the hidden and the observed data can be written as

$$P(Y_1^T, Z_1^U, X_1^L, d_1^L, e_1^L) = \prod_{l=1}^{L} a_{X_{l-1}, X_l} b_{X_l}(Y_{p_{l-1}+1}^{p_l}, Z_{q_{l-1}+1}^{q_l} | Y_1^{p_{l-1}}, Z_1^{q_{l-1}}), \tag{3.55}$$

where X_0 is the silent begin state with

$$a_{X_0, X_1} = \pi_{X_1}. \tag{3.56}$$

The Forward, Backward, and Viterbi Algorithms

The extension of the HMM algorithms to PHMMs is just as straightforward as for the joint probability in (3.55) above. Recall that Y_a^b corresponds to a gap if $a > b$. The recursive relation for the forward algorithm becomes

$$\alpha_i(t, u) = \mathbf{P}\left(Y_1^t, Z_1^u, \{\text{some hidden state } i \text{ ends at } (t, u)\}\right)$$

$$= \mathbf{P}\left(Y_1^t, Z_1^u, \bigcup_{l=1}^{L}(X_l = i, p_l = t, q_l = u)\right)$$

$$= \sum_{j \in S} \sum_{(d,e)} \alpha_j(t - d, u - e) a_{ji} b_i(Y_{t-d+1}^t, Z_{u-e+1}^u | Y_1^{t-d}, Z_1^{u-e})$$

$$= \alpha_M(t - 1, u - 1) a_{Mi} b_M(Y_t, Z_u | Y_1^{t-1}, Z_1^{u-1})$$
$$+ \; \alpha_I(t - 1, u) a_{Ii} b_I(Y_t, \emptyset | Y_1^{t-1}, Z_1^{u-1})$$
$$+ \; \alpha_D(t, u - 1) a_{Di} b_D(\emptyset, Z_u | Y_1^{t-1}, Z_1^{u-1}). \tag{3.57}$$

The sum over the durations run over only three values $(d, e) \in \{(1, 1), (1, 0), (0, 1)\}$, which coincide with the states, resulting in a summation over only three terms. We initialize the silent begin state X_0 with

$$\alpha_i(0, 0) = \pi_i,$$
$$\alpha_i(t, 0) = 0 \text{ if } t > 0, \tag{3.58}$$
$$\alpha_i(0, u) = 0 \text{ if } u > 0,$$

and terminate in the silent end state X_{L+1} with

$$\alpha_i(T + 1, U + 1) = \sum_{j \in S} \alpha_j(T, U) a_{ji}, \quad i \in S. \tag{3.59}$$

The probability of the pair of observed sequences, given the model, are then given by

$$\mathbf{P}(Y_1^T, Z_1^U) = \sum_{i \in S} \alpha_i(T + 1, U + 1). \tag{3.60}$$

Similarly, the backward algorithm becomes

$$\beta_i(t, u) = \mathbf{P}\left(Y_{t+1}^T, Z_{u+1}^U | Y_1^t, Z_1^u, \bigcup_{l=1}^{L}(X_l = i, p_l = t, q_l = u)\right)$$

$$= \beta_M(t + 1, u + 1) a_{iM} b_M(Y_{t+1}, Z_{u+1} | Y_1^t, Z_1^u)$$
$$+ \; \beta_I(t + 1, u) a_{iI} b_I(Y_{t+1}, \emptyset | Y_1^t, Z_1^u)$$
$$+ \; \beta_D(t, u + 1) a_{iD} b_I(\emptyset, Z_{u+1} | Y_1^t, Z_1^u). \tag{3.61}$$

We initialize with

$$\beta_i(T + 1, U + 1) = 1, \quad i \in S, \tag{3.62}$$

and terminate upon computation of $\beta_i(0, 0)$.

As before, the Viterbi algorithm is essentially the same as the forward, but with sums replaced by maxima. That is,

$$\delta_i(t, u) = \max_{X_1^{l-1}, d_1^{l-1}, e_1^{l-1}} \mathbf{P}\left(Y_1^t, Z_1^u, X_1^{l-1}, X_l = i, p_l = t, q_l = u\right) \quad (3.63)$$

$$= \max_{j,d,e} \delta_j(t - d, u - e) a_{ji} b_i(Y_{t-d+1}^t, Z_{u-e+1}^u | Y_1^{t-d}, Z_1^{u-1})$$

$$= \max \begin{cases} \delta_M(t - 1, u - 1) a_{Mi} b_M(Y_t, Z_u | Y_1^{t-1}, Z_1^{u-1}) \\ \delta_I(t - 1, u) a_{Ii} b_I(Y_t, - | Y_1^{t-1}, Z_1^{u-1}) \\ \delta_D(t, u - 1) a_{Mi} b_M(-, Z_u | Y_1^{t-1}, Z_1^{u-1}). \end{cases} \quad (3.64)$$

The initialization and termination is analogous to the forward, and the backtracking proceeds just as for the GHMM in Sect. 2.2.3. Generally, for the traceback, we would record three values in each cell of the dynamic programming matrix during the Viterbi computation; the previous state and the durations of the two observed sequences. But since the durations are given by the state, we only need to record the previous state.

Just as in the Needleman–Wunsch algorithm, we are not restricted to use a linear gap model. Using affine gaps, the Viterbi would take the same form as in (3.43), and could be split into three matrices as in (3.44)–(3.46) to reduce running time. The calculations of the forward and backward algorithms would get a little more complicated, however, but would still be doable.

3.1.8 Database Similarity Searches

In order to guarantee that the alignment algorithm always finds an optimal alignment, it has to search through all possible alignments. In effect, the algorithm has to consider all pairings of the residues in first sequence, to the residues in the second. Although dynamic programming methods efficiently utilize the recursive structure of the alignment problem, they still become intractable when dealing with large datasets. For instance, if we had a query sequence of 100 residues and wanted to search for homologies in a database consisting of 10,000 sequences, each being 100 residues long, the dynamic programming algorithm would require a running time proportional to $10,000 \times 100^2 = 10^8$. This is already infeasible, and the existing databases are much larger than this and with considerably longer sequences. For instance, Genbank contains over 76 million individual sequences constituting over 80 billion nucleotides [9], and the database is ever increasing.

Heuristic local alignment algorithms attempt to approximate the dynamic programming approach, while at the same time increase the computational speed by only search in small portions of the dynamic programming matrix. Wilbur and Lipman [97] pioneered this approach by noting that most sequences in a database will not match the query sequence, and should not be considered for alignment at all. Their approach was to rapidly search through the database by using heuristics to exclude irrelevant sequences. The basis of the method were to identify a list of short, but highly similar *words* of a fixed length. These words would then function as "seeds" for further analysis. The technique of Wilbur and Lipman remains the basis of two of the most widely used bioinformatical tools today; FASTA [60] and BLAST [6]. The idea is to first look for seeds in the query sequence, match them to the target sequence, extend the region around the seeds, and then only perform a dynamic programming alignment around the best extended regions. The gain in computational speed is huge, but at the cost of sensitivity; the optimal alignment is no longer guaranteed to be found.

3.1.8.1 FASTA

We expect evolutionary related sequences to contain at least short segments of complete identity. Heuristic database searches focus on these short segments and attempt to identify the most significant diagonals in a dot plot or dynamic programming matrix of two strings. Lipman and Pearson first developed a method called FASTP [60], which was later improved in the method FASTA [77]. FASTP was designed to only compare protein sequence, while FASTA includes nucleotide sequence comparisons as well as being more sensitive than the original algorithm. FASTA, alongside with BLAST described in the next section, is one of the most widely used tools for database searches.

We call a novel, uncharacterized sequence the *query* sequence, while the sequences in a database are called *target* sequences. The FASTA procedure of matching a query sequence to a given target sequence goes as follows (Fig. 3.12):

1. Begin by identifying alignment seeds of length *ktup* in the query sequence. The parameter *ktup* is typically 1–2 for proteins and 4–6 for DNA sequences. The seeds are stored in a lookup (or hash) table along with their positions in the sequence. The target sequence is then matched against the lookup table to identify all identical *ktup*-length matches, called *hotspots*, between the query and the target sequence. A hotspot is defined by a pair of coordinates (i, j), where i is the position of the hotspot in the query sequence, and j the corresponding position in the target sequence. The running time in this step will be linear in the size of the database (the sum of all sequence lengths).
2. Hotspots on the same diagonal are linked together into *diagonal runs* and scored by counting the hotspots and penalizing intervening mismatches. The 10 highest scoring runs are selected.
3. The selected diagonal runs are re-evaluated using a substitution matrix, which is typically the BLOSUM50 matrix for protein sequences and the identity matrix for DNA sequences. Diagonal runs scoring below a certain threshold are discarded and the highest scoring subalignment is denoted $init_1$.
4. Until now, insertions or deletions have not been allowed. In this step, we attempt to combine the "good" subalignments into a single longer approximate alignment allowing for gaps by applying a kind of "joining" penalty, similar to gap penalties.

This can be done by constructing a weighted graph, where the vertices are the subalignments and the weight of each vertex is the score assigned in the previous step. We extend an edge from subalignment a to subalignment b by moving horizontally or vertically in the alignment matrix. Each edge is given a negative weight depending on the number of gaps required, and the score of the entire path is the sum of the vertice weights and the edge penalties. If the highest scoring path, denoted $init_n$, scores above a certain threshold we continue to the next step.

5. In addition to $init_n$, FASTA calculates an alternative optimal score, denoted opt, using a *banded* Smith–Waterman algorithm centered around the diagonal run $init_1$. The band width required depends on the choice of *ktup* and the idea is that the best alignment path between the hotspots in the $init_1$ subalignment is likely to reside within a narrow band around the $init_1$ diagonal.

6. In the last step, all database sequences are ranked according to their $init_n$ or opt score, and the highest ranking candidates are realigned using a full dynamic programming algorithm.

7. FASTA displays the highest ranking alignments along with a histogram over the initial scores of the database sequences and a calculation of the statistical significance of the opt score of each alignment.

To improve speed, a lookup table of the database is usually precomputed and only updated when new entries are inserted. Because of the dynamic programming approach in the last step of the FASTA algorithm, the final score of a candidate match is comparable to any exact algorithm score. However, FASTA only determines the highest scoring alignment, not all high-scoring alignments between the sequences. Thus, repeated instances or multiple protein domains will be missed. The choice of *ktup* determines the sensitivity versus computational complexity. Choosing $ktup = 1$ achieves a sensitivity close to that of a local dynamic programming algorithm, while higher values increases the speed immensely, at the risk of missing true matches. Although it is possible to show instances where the optimal alignment is missed by FASTA, in most cases the reported alignments are comparable to the optimal, at the while the algorithm is much faster than any exact algorithm.

As an additional note, the *FASTA file format*, defined by Pearson and Lipman [77], is now one of the standard input formats for many other sequence analysis tools.

3.1.8.2 BLAST

BLAST [6], which stands for *Basic Local Alignment Search Tool*, is one of the most widely used computer tools in biology. The motivation for developing the algorithm was to increase the speed of heuristic searches even further, by looking for even fewer and better alignment seeds. This is accomplished by integrating a substitution matrix already in the first step. As a result, BLAST is empirically about 10 to 50 times faster than the Smith–Waterman algorithm.

Similarly to FASTA, BLAST is a word-based heuristic algorithm that identifies short segments of high similarity, which are then extended into longer alignments.

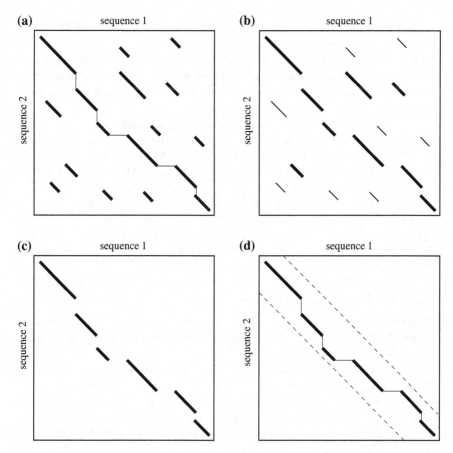

Fig. 3.12 Illustration of the FASTA algorithm. **a** Hotspots are identified and linked together in diagonal runs. **b** Selected diagonal runs are rescored and the highest scoring are filtered out. **c** Subalignments are joined. **d** The highest ranking subalignments are realigned using dynamic programming

The short segments, or *words*, are longer in BLAST than in FASTA, with typical lengths of $w = 3$ to 5 for protein and $w = 11$ to 12 for DNA sequences. BLAST uses an approximation of the Smith–Waterman algorithm, sometimes referred to as the *maximal segment pairs algorithm*. The algorithm identifies ungapped local alignments of a query sequence and a target sequence, and only include the first and the last elements in the Smith–Waterman recurrence relation in (3.48).

$$M(t, u) = \max \begin{cases} M(t - 1, u - 1) + s(Y_t, Z_u), \\ 0. \end{cases} \qquad (3.65)$$

Given a query sequence and a target sequence, the procedure of BLAST is as follows:

1. List all words of length w in the query sequence.
2. Compare the words to all possible w-lengthed words (20^w possibilities for amino acids and 4^w for nucleotides), and score the alignment using a substitution matrix. The scoring scheme used is typically a PAM120 matrix for protein sequences and match score 5 and mismatch score-4 for DNA sequences.
3. Organize the words that score above a threshold T in a search tree.
4. Scan the target sequence for identical matches to the words in the search tree. These matches, which are called hotspots in FASTA, are called *hits* in BLAST. Note the difference between BLAST and FASTA here; FASTA lists *all* identical matches as hotspots, while BLAST focuses only on the high-scoring ones.
5. Extend each hit to see if it is contained in a longer, ungapped local alignment, called a *high-scoring segment pair* (HSP). Continue the extension both to the right and to the left of the hit until the score drops more than a certain amount below the maximum score seen yet. List all HSPs that score above a (different) threshold S.
6. Display the HSPs ranked according to a measure of statistical significance. This measure is described further in the next section.

Note that the search tree in step 3 may be empty for a given query sequence and a given threshold T. The trade-off between speed and sensitivity is regulated by the setting of T.

BLAST is typically more sensitive than FASTA for protein alignments. The reason is the modifications to the heuristic algorithm, where conservative substitutions are allowed in the HSPs as long as the HSP score is above a given threshold. Since this modification has less effect for DNA sequences because of the smaller alphabet to be matched, FASTA tends to be more sensitive than BLAST for DNA sequences.

Gapped BLAST

The extension step of BLAST (step 5 above), where a hit is extended into a longer segment, accounts for 90 % of the computational time. Therefore, considerable improvements to the computational complexity have been gained by the so-called *two-hit method* [7], where an extension of a hit is made only if *two* hits, within a certain distance from each other, are found on the same diagonal of the alignment matrix. Moreover, the new version of BLAST, called *Gapped BLAST*, allows for gaps in the extension step. The Gapped BLAST algorithm is an improvement both in speed and sensitivity compared to the original algorithm, with a computational time about three times faster. With steps 1 to 4 being the same as in the original algorithm, the gapped BLAST method proceeds as follows:

5. In the dynamic programming matrix of the sequences, search for pairs of nonoverlapping hits on the same diagonal, at a distance of at most A steps. Concatenate such pairs.
6. Look for the highest scoring window of a given length l (typically $l = 11$) in the two-hit segments. Extend the alignment around the middle pair of residues

in the window, using the Smith–Waterman algorithm. The alignment is extended in both directions, allowing for gaps, until the score drops more than a certain amount below the maximum score seen yet.

In order to maintain the same sensitivity as in the original algorithm, the threshold value T in step 3 above has been lowered in Gapped BLAST. As a result, more single hits are found, but only a small fraction constitute a two-hit pair. The resulting gapped Smith–Waterman alignment will be optimal if two conditions are met: (1) the alignment is extended until score 0 is reached. Using a higher score as stopping rule saves computing time, but at a small risk of missing the optimal alignment. (2) The middle pair of residues chosen as starting point for the extension must be a part of the optimal alignment achieved in a full Smith–Waterman alignment.

PSI-BLAST

Position-Specific Iterative BLAST, or PSI-BLAST [7], is yet another extension of the BLAST algorithm. Here the highest scoring hits of a regular BLAST query are used to construct a profile matrix (see Sect. 3.2.9), which is then used to search the database again. The profile is then iteratively refined until the list of highest scoring hits no longer changes. Each iteration of PSI-BLAST takes about the same time as running Gapped BLAST, but is much more sensitive for distantly related sequences. The algorithm has a drawback, however. If the query sequence contains a strongly conserved domain, the profile will tend to be biased toward that domain, and may miss out on picking up relevant, but weaker, homologies in other regions of the sequence.

3.1.9 The Significance of Alignment Scores

It would be desirable to have some sort of measure of significance of an alignment score, to be able to judge if the match indicates true homology or have arisen by chance. Unfortunately, the statistical theory for optimal alignments is very complex. Typically, one would like to compare the optimal score of a pair of aligned sequences to the corresponding score distribution of the alignments of random, unrelated sequences. This can be done by employing various kinds of Monte Carlo methods, where large samples of representative pairs of random sequences are aligned and compared to the original alignment. However, constructing such random sequences is far from straightforward, and a number of issues need to be considered [28], since different sampling methods can yield very different score distributions. Regardless of method, in the case of *global* optimal alignments any sample of random sequences can only give indications regarding the mean and standard deviation of the alignment score distribution, while the behavior of the distribution at large remains mainly unknown. For local alignments scores, on the other hand, much more is known, and particularly for local ungapped alignments a rigorous statistical theory for the score distribution has been provided [50].

The optimal alignments of a sample of random sequences provide us with a sample of the score distribution for unrelated sequences. An important note is that the corresponding scores each constitute the *maximum* of all possible scores for the respective pair of sequences. While the well-known *central limit theorem* states that the sum of a large sample of independent, identically distributed random variables tends toward a normal distribution, the *extreme value theorem* gives us the limiting distributions of minima and maxima of large samples. Karlin and Altschul [50] showed that the score distribution of ungapped Smith–Waterman alignments tend to a distribution called the *extreme value type I distribution*, or the *Gumbel distribution* [36]. The Gumbel distribution models the extremes of ordered samples. That is, if we generate a large number of samples from some distribution and only keep the maximum value in each sample, the sample of maxima will approximately follow a Gumbel distribution. The Gumbel distribution for the maximum M of a sample is given by

$$\mathbf{P}(M \le x) = \exp\{e^{-\frac{x-\mu}{\sigma}}\} \tag{3.66}$$

where μ is the mean, and σ the standard deviation of the distribution.

Karlin and Altschul adapted this theory to the situation of ungapped local alignments [50]. Using the terminology of BLAST in Sect. 3.1.8, recall that a high-scoring segment pair (HSP), is an ungapped alignment of a segment where the score cannot be improved either by trimming or extending the segment. The distribution of HSP scores in *random sequences* is characterized by two parameters, λ and K. For two random sequences of lengths m and n, respectively, the probability of the local optimal alignment score S exceeding some value s, say, is given by the *P-value* of S

$$P = \mathbf{P}(S \ge s) \approx 1 - \exp\{-Knme^{-\lambda s}\} = 1 - e^{-E} \tag{3.67}$$

such that the parameters of the Gumbel distribution in (3.66) becomes $\mu = \log Kmn/\lambda$ and $\sigma = 1/\lambda$. The E is called the *E-value* for threshold s and corresponds to the expected number of HSPs having scores above s in the random model

$$E = E[\#\text{HSPs with score} \ge s] = Knme^{-\lambda s} \tag{3.68}$$

The parameters λ and K depend on the scoring scheme used, that is, the substitution matrix and the gap penalty model. Two restrictions are imposed on the scoring scheme for the theory to work; (1) the substitution matrix used to score residue pairs must include at least some positive values, and (2) the expected score per pair must be negative. If the latter condition is not fulfilled, the algorithm would score high for long alignments just based on their lengths, regardless of similarity [50].

To get rid of the dependence on the scoring system, or more specifically on the parameters λ and K, it is common to normalize the local alignment score into a "bit score"

$$S' = \frac{\lambda S - \log K}{\log 2} \tag{3.69}$$

The E-value corresponding to bit score S' is then approximated by

$$E \approx mn\,2^{-S'}. \tag{3.70}$$

BLAST reports the normalized score and the E-value in (3.69) and (3.68), respectively, where m is the length of the query sequence and n the length of the database (sum of all sequence lengths).

The statistical theory developed by Karlin and Altschul was meant for ungapped local alignments only, but there are strong indications that the same theory is applicable for gapped local alignments as well [4, 7, 70, 76, 84]. A main difference is that while parameters λ and K can be resolved analytically for ungapped alignments [50], they have to be estimated for gapped alignments.

3.2 Multiple Sequence Alignment

Multiple sequence alignment is the extension of pairwise alignments to multiple sequences. Multiple alignments provide more reliable information about sequence homology than pairwise comparisons, and have many uses in molecular biology, including:

- Identifying functionally important sites.
- Estimating the evolutionary relationship between sequences and constructing phylogenetic trees.
- Detecting weak but significant similarities.
- Predicting secondary structures of proteins.
- Predicting gene function.
- Designing primers for PCR experiments.

The problem of multiple sequence alignments is far from trivial due to its computational complexity. The field is growing fast and the algorithms improve continuously, but with the ever-increasing demand for faster and more accurate methods, we have probably only seen the beginning of this development. Here we give an overview of the most common approaches and directions. For a more thorough description, there are numerous good references giving detailed accounts of the underlying theory [11, 38, 81, 96] and surveying past and recent developments [14, 23, 34, 73, 93].

In Sect. 3.2.3, we discuss the extension of dynamic programming to multiple alignments. A dynamic programming approach would, similarly to its application to pairwise alignments, treat evolutionary events such as substitutions, insertions, and deletions simultaneously in all sequences considered. This problem is NP-complete, ,however, with the computational complexity growing exponentially in the number of sequences [94]. A possible solution, used for instance in the MSA package [61] (see Sect. 3.2.3), is instead to restrict the search space in some manner, and focus on a subset of multiple alignments that is likely to contain the optimal solution.

Unfortunately, the number of sequences that can be considered this way is still very limited.

Another popular approach is to use *progressive* methods, described in Sect. 3.2.4, where sequences are not aligned simultaneously, but instead included progressively into the multiple alignment in order of their evolutionary relationships. Sequences or groups of sequences are then aligned to the growing multiple alignment in a pairwise manner, by using sequence *profiles* to represent the subalignments. Profiles are discussed in Sect. 3.2.9.

One major problem with progressive methods is the high dependence on a correct initial alignment, since once errors are introduced they will be propagated through the entire process. To salvage this, many progressive methods include an *iterative refinement* strategy, where the multiple alignment is recomputed in various manners during its progressive build-up. Iterative methods are presented in Sect. 3.2.5.

An emerging field is that of *structure alignments*. As the tertiary structure of proteins tends to be more conserved than the primary sequence, the inclusion of structure information in the algorithm improves the quality of the multiple alignment immensely. Moreover, proteins with highly similar functions, but with too weak sequence similarity to be detected by sequence-based methods, may often be clearly distinguishable by structural alignment methods. This field shows great promise, and will continue to grow as more and more protein structures become available. However, the methods differ greatly from the sequence analysis tools described here, and is therefore outside the scope of this book. For references, see for instance [100].

3.2.1 Scoring Schemes

Similarly to pairwise alignments, a multiple alignment is obtained by placing the sequences on top of each other, and inserting gaps in various places and numbers in the sequences to increase the overall similarity. A multiple alignment can thus be seen as a two-dimensional lattice, where the sequences constitute the rows and each column represents a sequence position of common ancestry.

Ideally, we would score a multiple alignment based on the phylogenetic tree it induces. Unfortunately, we rarely have enough data to parametrize such a complex model, and a number of simplifications are necessary. As for pairwise alignments, it is common to assume independence between sequence positions and score each column separately. However, now that we have multiple characters in each column, it is not obvious how to do that. For instance, a direct generalization from pairwise alignments would be to score an alignment of three sequences

$$E\ A\ A\ S$$
$$V\ A\ -\ S$$
$$G\ C\ A\ -$$

as the sum of the column scores

$$S = s(E, V, G) + s(A, A, C) + s(A, -, A) + s(S, S, -), \qquad (3.71)$$

where each column score is built on the probability of observing the three residues together

$$s(a, b, c) = \log \frac{q_{abc}}{p_a p_b p_c}. \qquad (3.72)$$

However, the amount data required to achieve reliable estimates of the three-way probabilities q_{abc} makes this scoring scheme intractable already for three sequences.

Another problem is that the evolutionary history of the sequence family under analysis is not known, but needs to be inferred from the sequence alignment itself. As a result, our ability to determine a "correct" alignment will vary with the relatedness of the sequences, which makes it difficult to unambiguously produce a single correct alignment. It is therefore common to focus on subsets of columns corresponding to core structural elements that can be aligned with more confidence.

If we assume independence between columns, the scoring function of a multiple alignment M can be written

$$S(M) = \sum_C S(C) \qquad (3.73)$$

where C is a column in the multiple alignment. There are several possible ways to define the column score $S(C)$. One of the first, and probably the most popular, methods is the *Sum-of-Pairs* (SP) method [14], where the score of a multiple alignment column is simply the sum of all pairwise scores in that column. Several variations of this method exists, including the Weighted Sum-of-Pairs [5], which takes into account the phylogenetic relationships between the sequences. Other approaches, such as minimum entropy and maximum likelihood scores [34], use background distributions of the column residues.

Sum-of-Pairs (SP)

The Sum-of-Pairs (SP) method [14] is one of the most popular methods for scoring multiple alignments. It scores each column as the sum of all pairwise scores, such that a column $C = (a, b, c)^T$ gets score

$$S(C) = \log\left(\frac{q_{ab}}{p_a p_b}\right) + \log\left(\frac{q_{ac}}{p_a p_c}\right) + \log\left(\frac{q_{bc}}{p_b p_c}\right) = \sum_{i<j} s(a_i, a_j) \qquad (3.74)$$

where a_i and a_j are residues i and j, respectively, in the column. The corresponding score for the entire multiple alignment M is then as in (3.73).

There are several problems with the SP score, however, and while it makes sense in the pairwise case, it has no theoretical foundation in the multiple case. For instance, highly correlated sequences will tend to outvote other, more distantly related sequences that may carry more information [5]. In other words, single mutation events gets overweighted, and, more troubling, the relative score due to a single mutation *decreases* as the number of sequences increases. An illustrative example of this is given in [19].

Weighted Sum-of-Pairs (WSP)

One way to correct for the bias in SP scores is to introduce sequence weights between all sequence pairs [5], resulting in a kind of *weighted* SP score (WSP) such that the score of alignment column C can be written

$$S(C) = \sum_{i<j} w_{ij} s(a_i, a_j) \tag{3.75}$$

where w_{ij} is the weight between sequences i and j. By weighting the sequences, the contribution of highly similar sequences is decreased and the sequence bias gets less pronounced.

Minimum Entropy

Shannon entropy, or just *entropy*, in effect measures the level of variability, or randomness, of a variable or a process, and the idea of the minimum entropy score is to minimize the entropy of each column in the alignment. The entropy measure can be written as

$$H(X) = -\sum_{x} p(x) \log p(x) \tag{3.76}$$

where $p(x)$ is the density function of some random variable X, and the sum runs over all possible values of X. If we were to assume independence, not only between the columns in a multiple alignment, but between sequences as well, the probability of a column C could be written

$$\mathbf{P}(C) = \prod_{a} q_a^{f_a} \tag{3.77}$$

where q_a is the probability of residue a, and f_a is the proportional count of a in the column. The negative logarithm of this becomes exactly the minimum entropy score of the column

$$S(C) = -\sum_{a} f_a \log q_a. \tag{3.78}$$

The more conserved a column is, the smaller the score, and a completely conserved column yields score 0. Thus, the optimal alignment using minimum entropy scoring, is the one that minimizes the sum of column scores.

The minimum entropy score makes more biological sense than the SP score, since it directly emphasizes the conservation of a column. The assumption of independence between sequences is reasonable as long as the sequences aligned are not too biased. Various tree-based weighting schemes have been proposed as solutions to this problem.

3.2.1.1 Gap Costs

How to extend the gap models of pairwise alignments to multiple alignments is not immediately clear. Just as for pairwise alignments, the simplest model is to assume that each indel residue is independent of everything else. In that case, the gap cost can be included as en extra residue in the substitution matrix, and the inclusion in the alignment score is straightforward. For instance, the SP score is easy to extend to include gaps:

$$s(a_i, -) = -\delta. \tag{3.79}$$

Although, such a gap model typically generates multiple alignments with many isolated single-residue gaps, and, thus, serves as a bad imitation of the evolutionary process.

Affine gap costs, with separate gap opening and gap extension penalties (see Sect. 3.1.4) generate more sensible alignments. However, as the indel length grows the affine model becomes more and more similar to the linear. Moreover, the extension of the affine gap model to SP-scored multiple alignments, sometimes called a *natural* gap cost, is not straightforward and is computationally expensive. Altschul introduced the *quasi-natural* gap model, which, by modifying the affine gap model slightly, avoids the explosion in computational complexity that the natural gap cost model suffers from [2]. The quasi-natural gap cost is simply the natural gap cost with an additional cost for gaps that are opened or closed within another gap. This does not affect the resulting multiple alignment, but it reduces the information that needs to be stored by the dynamic programming algorithm significantly.

3.2.2 Phylogenetic Trees

The construction of phylogenetic trees is closely related to that of multiple alignments, as several phylogenetics methods require an initial multiple alignment to construct the tree. In particular, progressive alignment techniques use a phylogenetic tree to determine the order of the progressive adding of sequences.

The two basic method areas for phylogenetic analysis are *distance* methods and *character-based* methods. In distance methods the phylogenetic tree is constructed from estimated evolutionary distances between sequences (or organisms). The key assumption is that the pairwise distances are additive in the corresponding tree. Although this rarely holds true when using estimated distance data, coming for instance from sequence comparisons, the goal of distance methods is to find the phylogenetic tree that most closely fits the additivity assumption. Examples of distance-based methods are the UPGMA [85] and the Neighbor-Joining method [80]. Examples of character-based methods are parsimony methods and maximum likelihood methods. Parsimony methods search for the most "parsimonious" tree, i.e., the tree that requires the fewest number of evolutionary events to move between the sequences. Maximum likelihood methods search for the tree that gives the highest likelihood of the observed data under a given model of sequence evolution.

Phylogenetic data can be displayed in many different ways. For instance, the resulting trees can be rooted or unrooted. In a rooted tree, the distance to a common ancestor is included, while in an unrooted tree no assumptions about a common ancestor is made. A common notion in phylogenetic analysis is the Operational Taxonomic Units (OTUs). These constitute the actual objects, sequences, or molecules that we want to relate in a phylogenetic tree, and that will appear as external nodes (leafs). The internal nodes in the tree represent hypothetical evolutionary events, or ancestral units.

There are many different algorithms for constructing phylogenetic trees. Here we briefly describe the Neighbor-Joining method which is used both by MSA [61] and CLUSTALW [91].

3.2.2.1 The Neighbor-Joining Method

The neighbor-joining (NJ) method [80] is a distance-based clustering method used to construct phylogenetic trees. Unlike its simpler predecessor, UPGMA, [85], evolutionary rates are allowed to vary between lineages. Instead, the distances are assumed to be additive, meaning that the pairwise distance between two organisms can be achieved by adding the distances of each branch in the tree. However, even when the distances used are nonadditive distances, NJ produces reasonable phylogenetic trees. A basic concept of NJ is the concept of "neighbors." A pair of OTUs are neighbors if they are connected by a node in an unrooted, bifurcating tree.

Assume that we have a distance matrix of N OTUs where d_{ij} denotes the distance between OTUs i and j according to some measure. The NJ algorithm is initialized by placing the OTUs in a *star tree* with center node X, and proceeds as follows:

1. Given the distance matrix, compute a new matrix M of values

$$M_{ij} = (N - 2)d_{ij} - D_i - D_j \qquad (3.80)$$

where D_i represents the total dissimilarity from OTU S_i and the rest, and is given by

$$D_i = \sum_{j=1}^{N} d_{ij}. \qquad (3.81)$$

Fig. 3.13 The
Fitch-Margoliash method
can be used to determine the
branch lengths a, b, and c in
a star tree of three OTUs

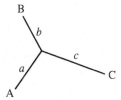

2. Choose the OTU pair with the smallest value in M and make them "neighbors" by creating a new node Y between this pair and the node X.
3. Temporarily collapse all other OTUs into one group and calculate the branch lengths between the neighbor pair and node Y using the Fitch-Margoliash method [29] for a three-branches star tree (see below).
4. Calculate the branch lengths of all other OTUs outside this pair to the new node Y, using the Fitch-Margoliash method iteratively on the pair and each outside OTU.
5. Consider the joined neighbors as a single, composite OTU and repeat steps 1–5 until only three OTUs remain.
6. Use the Fitch-Margoliash method on the final three (composite) OTUs.

Fitch-Margoliash

The Fitch-Margoliash method [29] can be used to calculate the branch lengths in a star tree of three OTUs, by using the additive assumption of the distance matrix. That is, in a system of three OTUs, A, B, and C say (see Fig. 3.13), the individual branch lengths are achieved by solving the following equation system

$$\begin{cases} d_{AB} = a + b \\ d_{AC} = a + c \\ d_{BC} = b + c \end{cases} \iff \begin{cases} a = (d_{AB} + d_{AC} - d_{BC})/2 \\ b = d_{AB} - a \\ c = d_{AC} - a \end{cases} \tag{3.82}$$

One problem with the neighbor-joining algorithm is that in the attempt to represent the data in an additive tree, branches may be assigned negative lengths. Since branch lengths provide an estimate of the number of substitutions that have occurred, negative measures become a problem. A common solution is to turn the negative length to zero, and transfer the difference to an adjacent branch so that the total distance between adjacent pairs remains unchanged. That way the topology of the tree is preserved.

3.2.3 Dynamic Programming

In order to guarantee that an optimal solution is found, ideally we would like to extend the dynamic programming approach of pairwise alignments to multiple alignments.

A naive implementation of aligning K sequences, however, requires a K-dimensional dynamic programming matrix. In fact, the problem of finding a global optimum for multiple alignments grows exponentially with the number of sequences and has been found to be NP-complete [94]. The algorithm in the MSA package, described next, still attempts to utilize the strengths of dynamic programming but by limiting the search space to the area most likely to harbor the optimal solution.

3.2.3.1 The MSA Package

The program MSA [61] (Multiple Sequence Alignment) extends the dynamic programming approach for pairwise alignments to multiple alignments by first reducing the search space by means of pairwise alignments and the Sum-of-Pairs (SP) method [14] described above. The idea is that a multiple alignment imposes constraints on the search space of each of the pairwise alignments and vice versa. Therefore, when projecting a multiple alignment onto a pairwise alignment space, it is possible to bound the number of points through which the projection can pass, which in turn limits the number of possible points in the original multiple space. The intersection of all these pairwise subsets of the search space is then thought to contain the optimal alignment. The MSA algorithm proceeds as follows:

1. Calculate the scores of all pairwise alignments.
2. Estimate a phylogenetic tree based on the pairwise scores.
3. Calculate sequence pair weights according to the estimated evolutionary relationships.
4. Produce a heuristic multiple alignment.
5. Calculate lower bounds of each sequence pair.
6. Compute the reduced search space.
7. Construct the final, optimal multiple alignment.

The pairwise alignment scores can be calculated using full pairwise dynamic programming, but is typically performed using a faster heuristic method such as FASTA [77]. Based on these scores, a phylogenetic tree is constructed using the neighbor-joining method [80] described above. To circumvent the problem of a biased sequence set giving too big importance to groups of near-similar sequences, MSA weights the sequence pairs using either of the methods described in [5].

MSA uses a clever algorithm for reducing the volume of the multidimensional dynamic programming matrix [14]. First, a heuristic multiple alignment is constructed, using progressive method similar to that of Feng and Doolittle [26]. Based on this, MSA calculates bounds on each of the pairwise alignments in order to reduce the search space. To illustrate, assume that we have a heuristic multiple alignment of N sequences with SP score

$$S = \sum_{k<l} S(a^{kl}),$$

(3.83)

where a^{kl} is the imposed pairwise alignment of sequences k and l and the sum runs over all such pairs $1 \leq k < l \leq N$. If, among all possible pairwise alignments of sequences k and l, $\hat{a}^{kl}$ is the optimal one, it holds that

$$S(\hat{a}^{kl}) \geq S(a^{kl}). \tag{3.84}$$

MSA then determines a lower bound β^{kl} for each sequence pair, and only considers pairwise alignments of scores higher than this bound

$$S(a^{kl}) \geq \beta^{kl}. \tag{3.85}$$

More specifically, a cell in the dynamic programming matrix is included in the reduced search space if the optimal pairwise alignment going through that cell scores higher than the bound β^{kl} (see Fig. 3.14). The default is to use the score of the pairwise alignment imposed by the heuristic multiple alignment as lower bound for each pair of sequences. The optimal multiple alignment is then searched for using dynamic programming in the intersection of all reduced pairwise matrices.

Since the choice of gap penalties influences the resulting optimal alignment heavily, it seems reasonable to choose a model that uses the same rationale as substitution scores. That is, a natural extension of the SP method would be to penalize gaps in a multiple alignment as the sum of all pairwise gap penalties. However, this has turned out to lead to problems, and instead MSA uses an affine gap model presented in [2], where gaps that span several columns are penalized less than the sum of the individual gap costs. As a result, columns are not scored independently, but depend on whether the previous column includes a gap or not.

In the original setting using the algorithm in [14], MSA could be used to align up to six sequences. Although the time and memory requirements have been substantially improved [37], the MSA method still becomes impractical when aligning more than

Fig. 3.14 A reduced search space of the MSA program. A cell of the dynamic programming matrix is included in the search space if the optimal pairwise alignment going through that cell has a score higher than the specified lower bound

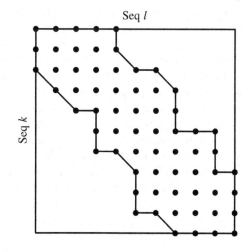

20 sequences or so. Because of the computational complexity, dynamic programming algorithms for multiple alignments are typically only used when benchmarking new heuristic methods, or when an extremely high-quality alignment is needed for a very small number of sequences. Moreover, the quality of the resulting alignment is questioned, as Gupta and colleagues have shown that the MSA package rarely produces a provable optimal alignment [37].

3.2.4 Progressive Alignments

While being a direct extension of the pairwise approach, the dynamic programming method for multiple alignments can be used only for very few, and relatively short sequences. Progressive methods are the most widely used approach when constructing multiple alignments. Although being heuristic, meaning that the optimal solution is not guaranteed to be found, they produce reasonable alignments of large sequence sets, and at a much lower computational cost. Progressive methods build-up the multiple alignment by adding sequences progressively according to their evolutionary relationships. The algorithms typically consist of two main steps: the construction of an evolutionary tree, called a *guide tree*, and the progressive building of the multiple alignment. The general approach for most progressive alignment methods goes as follows:

1. Produce pairwise alignments between all sequence pairs.
2. Calculate a distance measure for each pair.
3. Construct a guide tree based on these distances.
4. Combine the sequences into a multiple alignment in the order given by the guide tree, starting with the closest pair.

The progressive methods vary in how they treat these steps. The pairwise alignments can be constructed using dynamic programming, in which case the optimal similarity scores are used, but since this becomes computationally expensive for large sequence sets, it is common to use a fast heuristic also in this step. The guide tree is constructed by clustering the pairwise alignment scores using methods such as the UPGMA [85] or neighbor-joining method [80] described above.

In the final step, the guide tree is used to determine the order in which to incorporate new sequences. Starting with the closest pair of sequences, new sequences or groups of sequences are added progressively until all sequences have been included. If two or more sequences are joined by a common ancestor prior to their converging to the growing multiple alignment, we first align this subgroup, and then align that alignment to the multiple alignment. It is not obvious how to score such a procedure, however. If we use SP scores, a common approach when aligning two alignments is to use the average pairwise score between the two sequence groups.

Fig. 3.15 Alignment of alignments

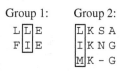

For instance, if we were to align the two groups in Fig. 3.15 the score of aligning the second column in group 1 to the first in group 2, say, would be

$$S = \frac{1}{6}\Big(s(L, L) + s(L, I) + s(L, M) + s(I, L) + s(I, I) + s(I, M)\Big),$$

where $s(L, L)$ is the score of aligning residue L in the first group to residue L in the second and so on. Regarding the gaps, a common policy is "once a gap, always a gap." That is, when adding a sequence to the growing multiple alignment, gaps are maintained throughout the progressive build-up.

The most widely used set of progressive methods is the Clustal family, with the weighted variant CLUSTALW [91], described in the next section, at the frontier. The Clustal suite offered great improvement in both sensitivity and speed when it was first introduced, and has inspired the construction of several methods that today perform better than their predecessor. One such method is T-Coffee [75], which is among the most accurate methods to date [23]. With T-Coffee came the novelty of using *consistency-based* scoring in progressive alignments, which has been adopted in several methods since then. The basic idea behind consistency-based scoring is to keep the pairwise alignments in a multiple alignment consistent. That is, given three sequences Y, Z, and W, and three pairwise alignments (Y, Z), (Z, W), and (Y, W), two of the alignments (Y, Z) and (Z, W) implicitly induce an alignment between sequences Y and W that may differ from the computed alignment (Y, W). Consistency-based scoring algorithms seek a multiple alignment that keeps the pairwise alignments consistent.

Example 3.4 CLUSTALW
The progressive alignment method CLUSTALW [91] is one of the most commonly used progressive methods. It extends the original algorithm CLUSTAL [44] to, among other things, include sequence weights according to similarity to avoid the problem of sequence bias. The procedure of CLUSTALW goes as follows:

1. Pairwise align all sequences.
2. Calculate a distance matrix of all sequence pairs based on their similarities.
3. Construct a guide tree from the distance matrix.
4. Progressively align the sequences in the order given by the guide tree.

The pairwise alignments are constructed using dynamic programming by default, but CLUSTALW offers as an option—the fast heuristic alignment method that was used in the original CLUSTAL program [98]. The guide tree is constructed from the pairwise alignments scores using the neighbor-joining method [80] described above.

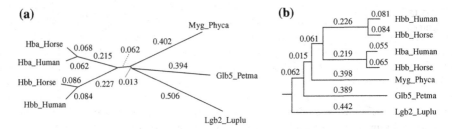

Fig. 3.16 The CLUSTALW guide tree is used to calculate sequence weights and determine the order of the progressive alignment. **a** illustrates a star tree and **b** a rooted tree. Reproduced from [91] by permission of Oxford University Press

Table 3.6 Distance matrix giving the number of mismatches per residue

	A	B	C	D	E	F	G
A - Hbb_Human	–	0.17	0.59	0.59	0.77	0.81	0.87
B - Hbb_Horse		–	0.60	0.59	0.77	0.82	0.86
C - Hba_Humam			–	0.13	0.75	0.73	0.86
D - Hba_Horse				–	0.75	0.74	0.88
E - Myg_Phyca					–	0.80	0.93
F - Glb5_Petma						–	0.90
G - Lgb2_Luplu							–

The guide tree determines both the progressive order of the multiple alignment and the weights assigned to the sequences. The sequences are weighted according to their distance to the root in the guide tree, and sequences of a common branch share the weights for that branch (see Fig. 3.16).

To illustrate the CLUSTALW procedure, we borrow the example in [91]. In this example, seven sequences are aligned: Hba_Human (human α-globin), Hba_Horse (horse α-globin), Hbb_Human (human β-globin), Hbb_Horse (horse β-globin), Myg_Phyca (sperm whale myoglobin), Glb5_Petma (lamprey cyanohaemoglobin), and Lgb2_Luplu (lupin leghaermoglobin). The sequences (and their names) come from the UniProt database [92]. The score between each sequence pair i, j is calculated as

$$d(i, j) = \frac{\# \text{ matches}}{\# \text{ ungapped positions}}. \tag{3.86}$$

This score is sometimes referred to as a *fractional identity* score. The corresponding distance measure is simply $1 - d(i, j)$, giving the average number of differences per residue. The resulting distance matrix is given in Table 3.6. The neighbor-joining method is then used to construct the unrooted tree illustrated in Fig. 3.16a. The corresponding rooted tree in Fig. 3.16b is produced by placing the root "in the middle" of the unrooted tree, such that the average branch lengths are equal on each side of the root [90]. The weights given in Fig. 3.16b depend on the distance from the root, and sequences with a common branch split the weight. Since Lgb2_Luplu does not share branches with anyone else, its weight becomes 0.442, while, for instance, the weight for Hbb_Human is

$$w(\text{Hbb\_Human}) = \frac{0.062}{6} + \frac{0.015}{5} + \frac{0.061}{4} + \frac{0.226}{2} + 0.081 = 0.223.$$

$\square$

One major problem with progressive alignments is that of local optima. Due to the greedy nature of the algorithm, there is no guarantee that the reported optimal solution is a global optimum. Errors introduced in intermediate steps become fixed and are propagated through the process, such that the result depends heavily on the correctness of the initial alignment and the quality of the guide tree. Thus, while computing optimal pairwise alignments between all sequences easily become too heavy computationally, using heuristic methods for determining the distances may compromise the guide tree. Another problem is the choice of scoring scheme. Since we might be dealing with sequences at very different evolutionary distances, using the same substitution matrix for all distances may be inappropriate. Some methods therefore use several substitution matrices based on how evolutionary related the individual sequences are.

3.2.5 Iterative Methods

The quality of progressive alignments depend heavily on the initial alignment. If the sequences to be aligned are evolutionary distant their alignments become less reliable, and errors made in the initial alignment are propagated through the multiple alignment. Iterative methods attempt to correct this by recalculating the multiple alignment through the progressive build-up. The refinement is typically produced by splitting the multiple alignment into subgroups that are iteratively realigned and joined back into a multiple alignment. The way of splitting into subgroups varies; it can be guided by a phylogenetic tree, certain sequences may be separated from the rest, or the subgroups may even be chosen at random. There is still no guarantee, however, that the resulting multiple alignment is the optimal solution.

An early example of iterative multiple alignments was presented in the software package PRRN/PRRP [33]. PRRN/PRRP uses a randomized iterative refinement strategy, known as *hill climbing*, to improve the initial multiple alignment [10]. Another iterative approach is introduced in DIALIGN (DIagonal ALIGNment) [67, 68], where local, gap-free segments of high similarity are combined into a multiple alignment. The gap-free segments correspond to diagonals in the pairwise alignment matrices, and are scored according to their degree of similarity as well as their overlap with diagonals in other sequence pairs. DIALIGN generally produces high-quality multiple alignments, but at the cost of computational complexity. To speed things up, the local alignment tool CHAOS [13] has been included as a preprocessing step. CHAOS is used to identify an initial set of highly similar regions in the input sequences that are then used as *anchor points* by DIALIGN. The anchor points constitute pairs of equally lengthed sequence stretches which have to be matched in the multiple alignments. These anchor points thus guide the multiple alignment and reduce the search space of possible solutions.

More recent approaches include MUSCLE [22] and MAFFT [51], which both use variants of *tree-dependent restricted partitioning* [45] as their iterative refinement method (see Example 3.5). A novelty in MAFFT is the use of a fast Fourier transform (FFT) to rapidly identify homologous segments between sequences or groups of sequences. The algorithm detects peaks in the correlation between the chemical properties of the amino acids compared, which then provide the basis for the guide tree.

Example 3.5 MUSCLE

MUSCLE [22] is a fairly recent progressive alignment method that employs iterative refinement. It uses strategies similar to that of PRRN [33] and MAFFT [51] and offers great improvements in both computational speed and alignment accuracy. The MUSCLE procedure consists of three main steps, each producing a multiple alignment that is built progressively using profile-to-profile alignments. The steps are detailed below, but to summarize, in the first step, a rough multiple alignment is generated using a fairly crude guide tree. A distance matrix is calculated based on the fraction of conserved k-mers (short exact matches of fixed length k) between pairs of sequences. In the second step, the initial alignment is improved by generating a more accurate guide tree based on the initial alignment. The pairwise similarity scores are recalculated using the same "fractional identity" score as in CLUSTALW (3.86). To reduce computational complexity, the multiple alignment is calculated only for those branches of the tree that differ from the original guide tree. The second step can be iterated if necessary, and in the final step the multiple alignment is further refined using a variant of the tree-dependent restricted partitioning method [45]. The MUSCLE procedure can be summarized as follows:

I. Draft progressive alignment:

    1. Calculate the k-mer similarity score for all pairs of sequences.
    2. Construct a distance matrix corresponding to the similarity scores.
    3. Construct a guide tree based on the distance matrix.
    4. Progressively align the sequences according to the guide tree.

II. Improved progressive alignment:

    1. Calculate the fractional identity score d of pairwise alignments imposed by the multiple alignment in the previous step.
    2. Calculate the Kimura distance [55] described in Sect. 3.1.2 for all sequence pairs

$$d_K = -\log(1 - d - d^2/5). \tag{3.87}$$

    3. Construct a new guide tree using the new distance matrix.

4. If the new tree differs from the previous a new progressive alignment is built. To reduce computational complexity, subalignments are produced only for the changed subtrees.
5. Either iterate this step or go to the next. The process is considered to have converged if the number of changed nodes does not decrease between iterations.

III. Iterative refinement (tree-dependent restricted partitioning):

1. Go through the edges in the order of the guide tree, starting with the edge most distant from the root.
2. Remove the edge, resulting in two disjoint subtrees.
3. Generate profiles for each subtree using the multiple alignment from the previous step. Discard columns containing only gaps.
4. Realign the profiles of the two disjoint subtrees.
5. Compare the SP scores of the old and the new alignment. If the score is increased, keep the new alignment, and discard it otherwise.
6. Iterate until all edges can be visited without changing the original alignment, or until reaching a user-defined maximum number of iterations.

The default method in MUSCLE for constructing the guide tree is UPGMA [85], but Neighbor-Joining [80] is implemented as a user option (see Sect. 3.2.2). A common scoring function for aligning profiles is a profile version of the SP score (PSP) used in both CLUSTALW and MAFFT. The PSP score for aligning columns y and z in two respective profiles is defined as

$$PSP(y, z) = \sum_i \sum_j f_i(y) f_j(z) S_{ij} \qquad (3.88)$$

where $f_i(y)$ is the observed frequency of residue i in column y and S_{ij} is the substitution matrix score for aligning residues i and j. MUSCLE uses this score along with a kind of position-specific affine gap penalty when assessing a new alignment in the refinement step. The substitution matrices used are the PAM JTT 200 [48] and PAM VTML 240 [71] matrices. In addition, MUSCLE implements a *log-expectation* (LE) score defined as

$$LE(y, z) = (1 - f_G(y))(1 - f_G(z)) \sum_i \sum_j f_i(y) f_j(z) \frac{p_{ij}}{p_i p_j}, \qquad (3.89)$$

where $f_G(y)$ is the observed frequency of gaps in column y. The probabilities p_i, p_j, and p_{ij}, corresponding to the background frequencies and joint alignment probability, respectively, of residues i and j, are taken from the PAM VTML 240 matrix. Since the substitution matrix score in (3.88) is in the form

$$S_{ij} = \log \frac{p_{ij}}{p_i p_j}, \qquad (3.90)$$

the two scoring functions are very similar, with the main difference being the two $(1 - f_G)$ factors in (3.89). These factors can be seen as "occupancy" factors of a column, and are included to counteract unnecessary inclusions of gaps.

The affine gap model used to align a gap of length k in sequence Y to residues $z_{m_1}, \ldots, z_{m_k}$ in sequence Z is given by

$$\gamma(k) = b(z_{m_1}) + t(z_{m_k}) + \varepsilon k, \tag{3.91}$$

where $b(\cdot)$ and $t(\cdot)$ are position-specific gap-opening and gap closing penalties, and ε is an extension penalty that does not vary with position. MUSCLE offers several different strategies for weighting the sequences according to their evolutionary relationships (including no weighting), but is using the CLUSTALW approach as default (see Example 3.4).   □

3.2.6 Hidden Markov Models

A hidden Markov model (HMM) is also a kind of iterative method, since the model is updated with each additional sequence that is included in the alignment. However, unlike other iterative methods, the parameters of an HMM are trained in parallel with the construction of the multiple alignment. Multiple alignments based on HMMs can be modeled by an acyclic graph of a linearly arranged set of vertices corresponding to the columns in the alignment, and where each multiple alignment corresponds to a path through the model. Recall from Sect. 2.1 that an HMM is composed of two interrelated process: a hidden process and an observed process. In multiple alignments, the observed process constitutes the aligned sequence columns, and the hidden process constitutes the underlying sequence of states where the state space consists of matches, insertions, and deletions, just as in PHMMs in Sect. 3.1.7. Typically, HMMs are trained using the Baum–Welch algorithm [79], but other methods such as gradient descent [8] or simulated annealing [15, 57] can also be used. These and other training methods are described in Chap. 6.

SAM—Sequence Alignment and Modeling

The program SAM [47] uses an HMM for multiple alignments. The state space, illustrated in Fig. 3.17, mimics pairwise sequence alignments with affine gaps [47]. The alignment columns are laid out linearly, and each position has three states: match, insertion, or deletion. In terms of adding a new sequence, a match state corresponds to matching a residue of the new sequence to the multiple alignment column. In the insertion state, the residue of the new sequence is "skipped" by the HMM, resulting in a completely new column in the multiple alignment where the inserted residue is matched to gaps in all previous sequences. In the deletion state, the multiple alignment column is skipped by the new sequence instead, resulting in a gap in the new sequence. In this fashion, each residue of the new sequence either belongs to a match or to an insertion state.

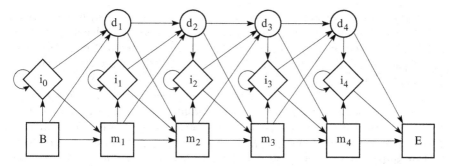

Fig. 3.17 The state space of an HMM for multiple alignment. Reproduced from [47] by permission of Oxford University Press

Just as in standard HMMs, the SAM model is composed of a hidden process and an observed process. The hidden process, denotes it $X_1^L = X_1, \ldots, X_L$, is a Markov chain that jumps between states in a state space $S = \{s_1, .., s_N\}$. The observed process now corresponds to the new sequence Y_1^L that we want to align to the model. Just as in pair HMMs (see Sect. 3.1.7), an alignment can be specified by a path through the state space. The probability of aligning sequence Y_1^L to the model can be written

$$\mathbf{P}(Y_1^L, X_1^L) = \prod_{l=1}^{L} a_{X_{l-1}, X_l} b_{X_l}(Y_l) \qquad (3.92)$$

where a_{ij} denotes the transition probability between states i and j, and $b_{X_l}(Y_l)$ denotes the emission probability of observing residue Y_l in state X_l. Note that both the observed and the hidden sequences are of the same length L here. This notation is possible if we only include match and insert states in the path X_1^L. The delete states are included implicitly in the calculations such that if two consecutive states are not directly connected in the state space, the transition probability between these states include the transition via one or more delete states. For instance, if a match state in column 1 (m_1) is followed directly by an insert state in column 4 (i_4) in the path, the transition probability is given by

$$a_{m_1, i_4} = a_{m_1, d_2} a_{d_2, d_3} a_{d_3, d_4} a_{d_4, i_4}. \qquad (3.93)$$

Note that this notation makes it easy to implement affine gap penalties. A transition from a match or insertion state to a deletion state correspond to a gap opening penalty, while transitions between deletion states correspond to a series of gap extension penalties.

The SAM parameters are trained using a modified version of the Baum–Welch algorithm in Sect. 6.5, called *maximum a posteriori estimation* (MAP). Typically, the parameters are estimated by using proportional counts of the corresponding events in training set. If the training set is small or biased, however, there is always a risk of *overfitting*, meaning that the model fits only the training set, and any new sequence will appear very unlikely under this model. SAM solves this problem by using a *regularizer* in the reestimation step of the Baum–Welch algorithm. A regularizer can be compared to the use of pseudocounts in parameter training (see Sect. 6.2) in which a constant or random amount is added to the estimate to avoid overfitting.

One advantage of HMMs over other multiple alignment approaches is that HMMs can be trained on unaligned sequences, while profiles, for instance, are constructed from an existing initial multiple alignment. Moreover, the probabilistic foundation of HMMs enables a meaningful interpretation of its parameters and scores. However, HMMs suffer from several problems as well. First, the training set needs to be rather large. Second, it is not possible to capture correlations between nonadjacent positions. This is a problem since, in reality, amino acids that are physically far apart in a polypeptide may become very close as the protein folds. Chemical and electrical interactions between such amino acids are not captured in a linear model. Also, residues are implicitly assumed to be independent in the model. This is something that is not always true biologically, since some amino acids are more likely to appear in near vicinity of each other than others. Third, like progressive alignments, the result is affected by the order in which sequences are included. HMMs cannot explicitly incorporate sequence weights, however.

3.2.7 Genetic Algorithms

Genetic algorithms, introduced by J.H. Holland in the 1960s [46], constitute a special kind of search techniques in machine learning, and can be used in various optimization problems. Although not originally developed for sequence analysis, the technique is inspired by the evolutionary process and Darwin's "survival of the fittest" by simulating a large set of solutions that "evolve" through inheritance, selection, mutations, and recombinations. The popularity of genetic algorithms in optimization problems is due to their ability to find the solution of rather complicated systems without having to explicitly model complex interactions. Here we give a brief overview of the technique applied to multiple alignments. For more details on the general theory, see for instance [66].

A genetic algorithm starts out by generating a random initial population, where each member is a potential solution to the optimization problem. A number of individuals are then selected for "reproduction." Individuals are selected based on their "fitness." The better the fitness the higher the chance of being selected. The fitness is measured by some kind of *objective function*, such as the Sum-of-Pairs (SP) measure for multiple alignments. The selected individuals are then modified in some manner, mimicking the mutation process, to generate a new generation. The most common modification operators are *crossover* and *mutation*. In a crossover operation, two parents are combined to generate the offspring, while a mutation operation only involves the modification of a single parent. Generation sizes are typically kept constant and the process is iterated either until a specified number of populations have been generated, or until a convergence criterion has been met. To ensure that the highest scoring solutions are not lost, it is common to let a fraction of the parent population move on unmodified to the next generation, while the mutation process is applied to the rest to generate the remaining offsprings of that new generation.

An advantage of genetic algorithms over other optimization methods is that they can move very abruptly; the offspring generated from one or two parents may be radically different from its parents. Thus, the method is less likely to fall into a local optimum. A disadvantage is that fit individuals tend to reproduce more quickly than less fit, resulting in a large fraction of highly similar individuals in the population. Such a reduced diversity will slow the algorithm down, unless the optimal solution is among this over-represented group. Therefore it is important to choose an objective function that is biased toward the more fit individuals, while still preserving the diversity of the population.

Example 3.6 SAGA
SAGA (Sequence Alignment by Genetic Algorithm) [74] is a multiple alignment software that uses a genetic algorithm to search for an optimal solution. The method attempts to mimic the evolutionary process by evolving a population of multiple alignments, while gradually improving the fitness of the population. The fitness is measured by an objective function, which in SAGA typically is a weighted Sum-of-Pairs score using the PAM250 matrix and an affine gap penalty. Given a set of sequences, the SAGA procedure can be summarized as follows:

1. *Generate an initial population of multiple alignments.* Each alignment is constructed by choosing a set of random offsets, one for each sequence, move the sequences to the right in the multiple alignment according to these offsets, and fill the beginning and end with gaps to achieve rows of equal lengths. Generate n such multiple alignments using a unique set of offsets for each alignment (typically $n = 100$).
2. *Evaluate the fitness of each member of the population.* Each member (multiple alignment) of the current population, G_n say, is assigned a fitness score using the objective function.
3. *Select individuals for reproduction.* The half of the current population G_n with the best fitness is sent directly to the next generation G_{n+1}, without modifications. The other half of G_{n+1} are generated by selecting parents from G_n and modifying them.
4. *Create the remaining half of the next generation.* The members in G_n are used for reproduction as follows:

   a. *Select parents from the current population.* The expected number of offsprings θ for each individual is estimated for each member in G_n (typically $\theta \in \{0, 1, 2\}$), and the selection probability of an individual is simply

   $$\mathbf{P}(\text{Individual } i \text{ is selected}) = \frac{\theta_i}{\sum_j \theta_j}.$$

b. *Select reproduction operator.* SAGA consists of a set of reproduction operators, each being either of crossover or mutation type. In the first round, all operators have equal probability of being selected, but as the process goes along these probabilities are dynamically improved based on their recent efficiency (typically based on 10 generations back). That is, if an operator is used to generate an offspring that gets better fitness than its parents, the operator (and preceding operators) gets a credit that will improve its selection probability.

c. *Generate an offspring.* A crossover involves the combination of two parent alignments, and results in two potential children. However, only the best-scoring child alignment is considered. A mutation involves the modification of a single parent. If the child is kept, the expected number of offsprings θ is reduced by one in each of the parents.

d. *Check for duplicates.* Duplicates are not allowed, thus a newly generated child is only kept if it is unique in G_{n+1} and is otherwise discarded.

e. *Iterate.* Continue until the new generation has been filled with n individuals.

5. *Evaluate the fitness of the new population.* If the end condition is met, return the highest scoring member as the solution. Otherwise, return to step 3 and construct a new generation of offsprings.

To ensure that not only the best fit members of a population are used for reproduction, each member get assigned an expected number of offspring value θ. During reproduction each parent is chosen with a probability proportional to this value, and each time a member is used for reproduction, its expected offspring value is decreased by 1. As a result, its probability of being selected as parent for the next child is reduced, corresponding to a variant of selection without replacement. This process is continued until all the parents have been chosen.

The operators used for reproduction of new generations belong to one of two groups: crossovers or mutations. Mutation operators modify the alignment of one parent in a number of different ways; by inserting gaps, by shuffling blocks of sequences or blocks of gaps around, or by locally rearranging the alignment within a block. A crossover involves splitting the two parental multiple alignments and rejoining the subalignments with the other parent. The split can be made in two ways, one-point or uniform. In the one-point variant, the multiple alignment of the first parent is cut vertically at a completely random position of the alignment, and the second parent alignment is cut in a staggered manner such that when recombining the subalignments between the parents, the individual sequences are preserved in both new alignments (see Fig. 3.18 for an illustration). One problem with this method is that it can be very disruptive, especially around the cut position. Therefore, the uniform variant is implemented in SAGA. The first step in this variant is to identify consistent positions between the two parent alignments, meaning columns that are completely conserved between the two. Then blocks between these consistent positions can be swapped freely between the two parents. Thus, a child alignment is constructed by fixing these consistent positions and then inserting blocks between them from either parent. The blocks can be chosen in a *stochastic* or in a *semi-hill climbing* manner.

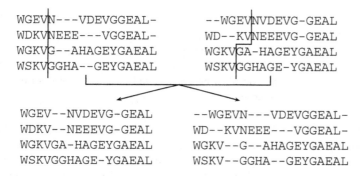

Fig. 3.18 One-point crossover. The first parent is cut vertically at some random position. The corresponding cut in the second parent becomes staggered in order to preserve the sequences. Reproduced from [74] by permission of Oxford University Press

In the stochastic variant, each block is chosen randomly from one of the parents, while in the semi-hill climbing variant the best-scoring sequence of blocks is used. □

3.2.8 Simulated Annealing

Physical *annealing* is used to shape metals or to make glass or crystals less brittle and more workable. The process consists of three stages: heating the material to a desired temperature, holding at that temperature, and then a slow cooling down to room temperature. In terms of thermodynamics, annealing is a process of obtaining low-energy states of a metal or a crystal. The heating makes the atoms "unquench" from the current state, and as the material is cooled down the atoms recrystallize in a more ordered fashion, until reaching the state with the minimum internal energy, the "frozen" state. However, if the initial temperature is too low the metal may become quenched in a metastable state, and if the cooling is done too fast, the internal stress may induce warping or cracking.

Simulated annealing, first introduced in [15, 57], is a *Monte Carlo sampling technique* used to obtain an approximate solution to large optimization problems. It has, for instance, been successfully applied to the *Traveling Salesman Problem* [15], which is a classical, NP-complete optimization problem. Simulated annealing attempts to mimic the annealing process in thermodynamic systems by means of heating, holding, and cooling the system. The correspondence between the two processes is rather straightforward. The states in the thermodynamic system, representing the position of the atoms, translates to different solutions to the optimization problem. The internal energy function is analogous to the objective function, and the final frozen state translates to the global minimum. There is no obvious analogy for the temperature, however, other than that it corresponds to a control parameter that represents the "willingness" of the system to jump to a state that is worse (has higher energy) than the current.

A simulated annealing algorithm starts out with a high temperature τ (can be infinity) that is gradually decreased until reaching $\tau = 0$ or to a preset termination value. The annealing schedule must be constructed in a way that ensures that the process reaches this termination value. At each temperature step, the algorithm runs a number of iterations where the current state is replaced by a random neighboring state. The iteration is a descendant to a Monte Carlo simulation method called the *Metropolis algorithm* [64], in which the choice of neighbor depends on the current temperature and on the difference in energies between the states. In the algorithm, a new state is chosen by making a slight modification to the current state. If the energy of the new state is lower than for the current one, the new state is kept. If the energy is higher, a *Boltzmann acceptance probability* is used to determine whether the state is accepted or rejected. Basically, the acceptance criterion states that if the energy difference is small enough, the state change is likely to be accepted, while if the difference is too large the state is likely to be rejected. In this formulation, the algorithm always chooses to go "downhill" when possible, but this is not a necessary condition for the method to work. One side effect of the Metropolis scheme is that both very good and very bad moves are excluded, but since the bad moves tend to outnumber the good ones, it is a reasonably efficient strategy.

The jumps between solutions are almost random for high temperatures, while lowering the temperature results in a gradually smaller search space of suboptimal solutions. The acceptance criterion for states of higher energy allows for some moves in the "wrong" direction, which, in contrast to more greedy algorithms, makes the method less prone to fall into local optima. As the temperature goes down, however, the process becomes more and more reluctant to move toward higher energies. Therefore, analogously to the physical process, too fast cooling will risk the process of getting stuck in a local minimum. The success of the process thus depends on the annealing schedule, which means the choice of initial temperature, the number of iterations at each temperature step, and how fast the temperature is decreased between steps.

Simulated annealing can both be used to train parameters, such as in the HMM-based multiple alignment software HMMER [20] and as a refinement method of already existing alignments such as in MSASA [52]. The application of training HMMs with simulated annealing is described in Sect. 6.10.

Example 3.7 MSASA
Simulated annealing is similar to genetic algorithms in that it produces high-scoring multiple alignments by iterative rearrangements. The main difference is that instead of producing a random population of multiple alignments, simulated annealing moves between the possible solutions using a probabilistic transition rule, in order to find the optimal solution. This approach has been implemented in the program MSASA (Multiple Sequence Alignment using Simulated Annealing) [52].

MSASA begins with a heuristic multiple alignment, which is then iteratively refined using a set of transition and acceptance rules. Let $\mathbf{M} = \{M_1, M_2, \ldots, M_n\}$

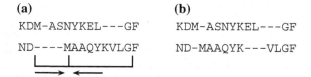

Fig. 3.19 The *Swap*(2, 4, 6, *right*) operation swaps in the second sequence the gaps to the right and including position 4 with the six residues to the right of these gaps

be the state space of possible multiple alignments, and $C(M_i)$ a real-valued cost function assigned to each state. The goal is to find the multiple alignment $M^* \in \mathbf{M}$ that minimizes the cost. The cost function constitutes the usual alignment score, including substitution costs (or scores) and gap costs. The scoring scheme in MSASA uses the Sum-of-Pairs score and natural gap costs described in Sect. 3.2.1.

The simulated annealing algorithm operates by generating a new alignment M_{new} from the current one M_{cur} using a transition rule that swaps the positions of null characters (gaps) and residues.

Transition Rule

- *Swap*(i, j, k, dir): where i is the sequence, j the position (column) in the multiple alignment, k the number of residues to be swapped, and dir is the swapping direction, which is either *left* or *right*.

If the swapping direction is *right*, the operator takes all consecutive nulls to the *right* of, and including, position j and swaps them with the k residues to the *right* of these nulls. For instance, the operation *Swap*$(2, 4, 6, right)$ alters the alignment (a) into alignment (b) in Fig. 3.19. The new multiple alignment M_{new} is produced by applying the *Swap*-operator to each sequence in M_{cur} that has a null in column j. Note that the *Swap*-operations may result in null columns in the new alignment, i.e., columns with gaps in all sequences. These cannot be removed during the simulated annealing process, however, since reducing the length of the alignment reduces the search space of possible solutions and thus risks excluding the optimal alignment. Null columns are removed only after the annealing schedule has finished.

Acceptance Rule

The new alignment M_{new} is either accepted or rejected using the Metropolis acceptance scheme [64] as follows. For $\Delta E = C(M_{\text{new}}) - C(M_{\text{cur}})$ if

$$\Delta E : \begin{cases} \leq 0 & \text{accept the new state } M_{\text{new}}, \\ > 0 & \text{accept the new state with probability } \mathbf{P}(\Delta E) = e^{-\Delta E/\tau}. \end{cases}$$

The second criterion enables moves to states of higher cost and thereby prevents the algorithm from getting stuck in a local minimum. The temperature τ controls the rearrangement rate and the likelihood of each new state.

The MSASA algorithm proceeds as follows:

1. *Generate an initial multiple alignment*: the initial alignment can be arbitrary, but in order to speed things up MSASA is initiated by an alignment produced by the MSA program [61] described in Sect. 3.2.3.
2. *Choose an initial temperature* τ: the temperature is initially set high and then gradually decreased towards zero.
3. *Create a new alignment from the current*: the transition rule is applied as follows:

    a. Choose a column j at random in the current alignment M_{cur}.
    b. Choose a random direction $dir \in \{left, right\}$, and a random integer $k \in [1, 10]$ and apply the $Swap(i, j, k, dir)$ operator to each sequence in M_{new} that has a gap in column j.

4. *Accept or reject the new alignment*: if the new alignment M_{new} has a lower cost than the current, accept it. Otherwise, accept it with probability $\mathbf{P}(\Delta E) = e^{-\Delta E/\tau}$.
5. *Decrease the temperature*: $\tau = \gamma\tau$ for some preset constant $0 < \gamma < 1$.
6. *End condition*: if the temperature is below the final temperature $\tau_f = 1/\log k$, remove all null columns and return the final alignment. Otherwise, return to step 3 and construct a new alignment. $\qquad\qquad\qquad\qquad\qquad\qquad\qquad\qquad\qquad$ □

3.2.9 Alignment Profiles

In the early days of protein sequence analysis it was observed that some proteins contained long segments that were very similar to other proteins, while the rest of the sequence had no detectable similarity to any known protein at all. Today, we know that most functional sequences come in families having diverged from a common ancestor either through duplications within the genome, or through speciation into different organisms. We know that proteins are composed of *domains*, which are sequence segments that are responsible for the structure and function of the protein, and that are often separated by sequence stretches that have little or no impact on the activities of the protein. There are today several databases [27, 43, 82] that keep track of known domains, which proteins that are involved in which processes, and that store multiple sequence alignments of relevant segments in the protein sequences. These databases allow for analyses of new sequences in terms of which families they belong to, and what domains they contain, and thereby inferring their potential function.

We can use a sequence family to search a database for new members, or search for family membership of a new sequence in a database of families. A naive way to do this would be to perform pairwise searches for each sequence in the family, but then we might miss distantly related sequences matching features appearing in the multiple alignment but not in the pairwise alignment. In order to identify new family members, or family membership of a new sequence, we want to utilize features common for the family as a whole, rather than distinct similarities in a pairwise comparison. *Profile analysis* is about identifying and using such features for multiple alignments.

3.2.9.1 Standard Profiles

The notion of *profiles* was introduced by Gribskov et al. [35] as means for detecting more distantly related sequences than pairwise comparisons would allow in a database search. The idea was straightforward and very efficient. A profile can be constructed from a set of multiply aligned sequences, or from a single sequence that has been equipped with additional structural information. The result is a *position-specific scoring matrix* (PSSM) that scores the probability of observing a given amino acid, or a gap, in a specific position in the multiple alignment. A PSSM typically consists of 20+ columns and L rows, where L is the length of the underlying multiple alignment, and the columns represent the 20 amino acids, and possibly additional columns for gaps and unknown residues. The gap score can be split into two columns, one for a gap opening and one for a gap extension penalty. Thus, note that while the rows in a multiple alignment correspond to the sequences in the alignment, the representation is rotated in a profile (see Fig. 3.20).

An entry M_{ka} in the profile represents the score of observing amino acid a in position k in the underlying multiple alignment. The method used by Gribskov et al. [35] is commonly referred to the *average score method*. In this method each profile entry is on the form

$$M_{ka} = \sum_{b=1}^{20} \frac{c_{kb}}{N} s(a, b) \tag{3.94}$$

where $s(a, b)$ is the substitution matrix score (PAM or BLOSUM) for amino acids a and b, c_{kb} is the count of amino acid b in column k in the multiple alignment, and N is the number of aligned sequences. The average method suffers from the same problem as the Sum-of-Pairs scoring method for multiple alignments; if the sequence set is biased, some sequences will be given too big importance in relation to more distantly related sequences. Just as in multiple alignments, this problem can be resolved by appropriate weighting of the sequences [62, 90]. An overview of different sequence weights is found in [19].

Another problem, which in a sense is on the other side of the same coin, is that with the average score method the score of a completely conserved column will be equal to that of a single sequence. This is counter-intuitive since having a completely conserved column in a multiple alignment should give that residue more weight the more sequences that were included. The use of sequence weights will unfortunately not resolve this problem completely. What we would like is some kind of *probability distribution* over the residues in each position. This problem gets a natural solution, however, in the application of hidden Markov models to profile analysis, described in the next section.

(a)

```
S S Q S L L D S G D G N T Y L
G D S L R - - - - - G Y D A
A S G F T F S - - - - A N D M
A T G Y T F S S - - - - Y E L
```

(b)

Pos					Cons	A	C	D	...	Y	Gap
1	A	A	G	S	A	10	3	4	...	-4	9
2	T	S	D	S	S	4	3	5	...	-4	9
3	G	G	S	Q	G	5	1	6	...	-6	9
4	Y	F	L	S	F	-1	2	-4	...	7	9
5	T	T	R	L	T	1	-2	0	...	-2	9
6	F	F	.	L	F	-2	-3	-6	...	8	4
7	S	S	.	D	S	3	2	5	...	-3	4
8	S	.	.	S	S	2	3	1	...	-2	4
9	.	.	.	G	G	2	0	2	...	-2	4
10	.	.	.	D	D	1	-1	4	...	-1	4
11	.	.	.	G	G	2	0	2	...	-2	4
12	.	A	G	N	A	6	0	4	...	-3	4
13	Y	N	Y	T	Y	0	5	0	...	6	9
14	E	D	D	Y	D	2	-2	9	...	0	9
15	L	M	A	L	L	3	-5	-3	...	0	9

Fig. 3.20 A part of the profile example given in [35]. **a** The multiple alignment underlying the profile. **b** The leftmost column, with numbers, gives the profile positions, the next four columns represent the rotated multiple alignment, followed by a column representing the consensus of the alignment. The corresponding profile is given in the framed box. Each column gives the score for a specific amino acid at each position in the profile, and the rightmost column gives the gap penalty for each position

3.2.9.2 Profile HMMs

Profile HMMs, first introduced in [40, 58], constitute a sophisticated version of the type of position-specific scoring matrices described in the previous section, and possess several advantages over standard profiles. Just as in all other HMMs, the underlying model is probabilistic, resulting in a consistent and mathematically interpretable treatment of evolutionary events. The model parameters are estimates of true frequencies rather than observed, which means that a profile of only 10–20 sequences can be of as good quality as a profile of 50 sequences. Moreover, profile HMMs require less skill and manual tuning in general, in comparison to the construction of standard profiles.

The underlying state space in a profile HMM consists of a series of nodes, roughly corresponding to the columns in a multiple alignment. We say "roughly," because insertions and deletions get a slightly different treatment than in standard profiles. If we ignore the occurrence of gaps for a moment, the model would consist of a linear sequence of *match* states with transition probabilities equal to 1 between them. Just as in pair HMMs, "match" only means that a residue is matched to the current column, but it may very well be a "mismatch."

If we consider an ungapped multiple alignment of length T, the probability of adding a new sequence Y_1^T can be written

$$\mathbf{P}(Y_1^T) = \prod_{t=1}^{T} b_{M_t}(Y_t), \qquad (3.95)$$

where b_{M_t} is the emission probability of match state M_t. However, we want to relate the probability of observing a given residue a in the current position t, to the background probability q_a of that residue. Therefore, it is common to use the log-odds ratio instead

$$S(Y_1^T) = \sum_{t=1}^{T} \log \frac{b_{M_t}(Y_t)}{q_{Y_t}}. \qquad (3.96)$$

The summed log-odds scores are similar to those used in substitution matrices, except that we now score the pairing of a residue and a profile *position*, rather than the pairing of two residues.

The occurrence of gaps are modeled by including an *insert* and a *delete* state along with each match state. Figure 3.21 illustrates the full model of a profile HMM. For simplicity, we can view the model as a sequence generating machine, or, analogously, as a means to score a new observed sequence according to an existing multiple alignment. Each match state and each insert state has their own sets of transition and emission probabilities. The match states score the match of a residue in the observed sequence to the model, while the insert states only "process" residues of the observed sequence that do not match any position in the profile. The delete states correspond to inserting gaps in the observed sequence and they only have state-specific transition

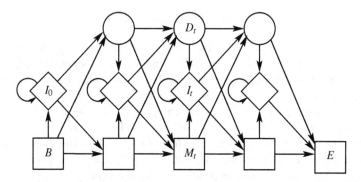

Fig. 3.21 The state space and transitions of a profile HMM. *Boxes* represent match states, *diamonds* insert states, and *circles* are delete states. B and E represent the silent begin and end states, respectively. Reprinted from [58] with permission from Elsevier

probabilities connected to them. Note how only the insert states are equipped with self-transitions. The actual profile starts with the first match state M_1 or delete state D_1, but to allow for the possibility that the beginning of the observed sequence does not match the profile, we add a silent begin state B with a possibility jump to an initial insert state I_0.

Having state-specific model parameters has the advantage that gaps can be penalized differently depending on where they occur in the profile. Also, we can assign different probabilities to self-transitions and transitions out of an insert state, such that they correspond to the gap opening and gap extension penalties in an affine gap model.

Scoring a New Sequence

The main purpose of profiles is to search for homologous sequences to a given protein family or protein domain. This can be done in profile HMMs by using a slightly modified version of the Viterbi algorithm. Moreover, since the intrinsic gap model is affine, we use the same trick as in Sect. 3.1.5.1 to improve computational complexity. That is, instead of using just one dynamic programming matrix we implement three, one for each kind of state. The modified Viterbi algorithm for a profile of length L and a new sequence Y_1^T to be matched to the model is given by

$$\delta_{M_j}(t) = \log \frac{b_{M_j}(Y_t)}{q_{Y_t}} + \max \begin{cases} \delta_{M_{j-1}}(t-1) + \log a_{M_{j-1}M_j}, \\ \delta_{I_{j-1}}(t-1) + \log a_{I_{j-1}M_j}, \\ \delta_{D_{j-1}}(t-1) + \log a_{D_{j-1}M_j}. \end{cases} \tag{3.97}$$

$$\delta_{I_j}(t) = \log \frac{b_{I_j}(Y_t)}{q_{Y_t}} + \max \begin{cases} \delta_{M_j}(t-1) + \log a_{M_j I_j}, \\ \delta_{I_j}(t-1) + \log a_{I_j I_j}, \\ \delta_{D_j}(t-1) + \log a_{D_j I_j}. \end{cases} \tag{3.98}$$

$$\delta_{D_j}(t) = \max \begin{cases} \delta_{M_{j-1}}(t) + \log a_{M_{j-1}D_j}, \\ \delta_{I_{j-1}}(t) + \log a_{I_{j-1}D_j}, \\ \delta_{D_{j-1}}(t) + \log a_{D_{j-1}D_j}. \end{cases} \tag{3.99}$$

Note that when we allow for gaps, the length T of the new sequence may differ from the profile length L. We initiate by naming the silent begin state M_0 and set $\delta_{M_0}(0) = 0$. Similarly, we name the end state M_{L+1}, and calculate $V_{M_{L+1}}(T)$ using the equations above, but without the emission term.

Besides scoring new sequences or searching a database for homologies to the current protein family or domain, profile HMMs can be used as a tool for multiple alignments. By training the model on *unaligned* sequences, using for instance the EM-algorithm described in Chap. 6, we will in effect produce an optimal multiple alignment of these sequences. This additional feature and others are included in the profile HMM package HMMER [21].

References

1. Alexandersson, M., Bray, N., Pachter, L.: Pair hidden Markov models. In: Jorde, L.B., Little, P., Dunn, M., Subramanian, S. (eds.) Encyklopedia of Genetics, Genomics, Proteomics and Bioinformatics, Ch. 4.2 (17) (2005)
2. Altschul, S.F.: Gap costs for multiple alignments. J. Theor. Biol. **138**, 297–309 (1989)
3. Altschul, S.F.: Amino acid substitution matrices from an information theoretic perspective. J. Mol. Biol. **219**, 555–565 (1991)
4. Altschul, S.F., Gish, W.: Local alignment statistics. Methods Enzymol. **266**, 460–480 (1996)
5. Altschul, S.F., Carroll, R.J., Lipman, D.J.: Weights for data related by a tree. J. Mol. Biol. **207**, 647–653 (1989)
6. Altschul, S.F., Gish, W., Miller, W., Myers, E.M., Lipman, D.J.: Basic local alignment search tool. J. Mol. Biol. **215**, 403–410 (1990)
7. Altschul, S.F., Madden, T.L., Schäffer, A.A., Zhang, J., Zhang, Z., Miller, W., Lipman, D.J.: Gapped BLAST and PSI-BLAST: a new generation of protein database search programs. Nucleic Acids Res. **25**, 3389–3402 (1997)
8. Baldi, P., Chauvin, Y., Hunkapiller, T., McClure, M.A.: Hidden Markov models of biological primary sequence information. Proc. Natl. Acad. Sci. USA **91**, 1059–1063 (1994)
9. Benson, D.A., Karsch-Mizrachi, I., Lipman, D.J., Ostell, J., Wheeler, D.L.: GenBank. Nucleic Acids Res. **36**, D25–D30 (2008)
10. Berger, M.P., Munson, P.J.: A novel randomized iterative strategy for aligning multiple protein sequences. Comput. Appl. Biosci. **7**, 479–484 (1991)
11. Bishop, M.J., Rawlings, C.J. (eds.): DNA and Protein Sequence Analysis. Oxford University Press, Oxford (1997)
12. Boeckmann, B., Bairoch, A., Apweiler, R., Blatter, M.-C., Estreicher, A., Gasteiger, E., Martin, M.J., Mochoud, K., O'Donovan, C., Phan, I., Pilbout, S., Schneider, M.: The SWISS-PROT protein knowledgebase and its supplement in 2003. Nucleic Acids Res. **31**, 365–370 (2003)
13. Brudno, M., Chapman, M., Göttgens, B., Batzoglou, S., Morgenstern, B.: Fast and sensitive multiple alignment of large genomic sequences. BMC Bioinform. **4**, 66 (2003)
14. Carrillo, H., Lipman, D.: The multiple sequence alignment problem in biology. SIAM J. Appl. Math. **48**, 1073–1082 (1988)
15. Černý, V.: Thermodynamical approach to the traveling salesman problem: an efficient simulation algorithm. J. Optim. Theor. Appl. **45**, 41–51 (1985)
16. Dayhoff, M.O.: Atlas of Protein Sequence and Structure. National Biomedical Research Foundation, Washington (1969)
17. Dayhoff, M.O., Schwartz, R.M.: Matrices for detecting distant relationships.In: Dayhoff, M.O. (ed.) Atlas of Protein Sequence and Structure, vol. 5, pp. 353–358 (1978)
18. Dayhoff, M.O., Schwartz, R.M., Orcutt, B.C.: A model of evolutionary change in proteins. In: Dayhoff, M.O. (ed.) Atlas of Protein Sequence and Structure, vol. 5, pp. 345–352 (1978)
19. Durbin, R., Eddy, S., Krogh, A., Mitchinson, G.: Biological Sequence Analysis: Probabilistic Models of Proteins and Nucleic Acids. Cambridge University Press, Cambridge (1998)
20. Eddy, S.R.: Multiple alignment using hidden Markov models. Proc. Int. Conf. Intell. Syst. Mol. Biol. **3**, 114–120 (1995)
21. Eddy, S.R.: Profile hidden Markov models. Bioinformatics **14**, 755–763 (1998)
22. Edgar, R.C.: MUSCLE: a multiple sequence alignment method with reduced time and space complexity. BMC Bioinform. **5**, 113 (2004)
23. Edgar, R.C., Batzoglou, S.: Multiple sequence alignment. Curr. Opin. Struct. Biol. **16**, 368–373 (2006)
24. Felsenstein, J.: Evolutionary trees from DNA sequences: a maximum likelihood approach. J. Mol. Evol. **17**, 368–376 (1981)
25. Feng, D.F., Johnson, M.S., Dolittle, R.F.: Aligning Amino Acid sequences: comparison of commonly used methods. J. Mol. Evol. **21**, 112–125 (1985)
26. Feng, D.F., Doolittle, R.F.: Progressive sequence alignment as a prerequisite to correct phylogenetic trees. J. Mol. Evol. **25**, 351–360 (1987)

27. Finn, R.D., Tate, J., Mistry, J., Coggill, P.C., Sammut, S.J., Hotz, H.-R., Ceric, G., Forslund, K., Eddy, S.R., Sonnhammer, E.L.L., Bateman, A.: The Pfam protein families database. Nucleic Acids Res. **36**, D281–D288 (2008)
28. Fitch, W.M.: Random sequences. J. Mol. Biol. **163**, 171–176 (1983)
29. Fitch, W.M., Margoliash, E.: Construction of phylogenetic trees. Science **155**, 279–284 (1967)
30. Gibbs, A.J., McIntyre, G.A.: The diagram, a method for comparing sequences. Its use with amino acid and nucleotide sequences. Eur. J. Biochem. **16**, 1–11 (1970)
31. Gonnet, G.H., Cohen, M.A., Benner, S.A.: Exhaustive matching of the entire protein sequence database. Science **256**, 1443–1445 (1992)
32. Gotoh, O.: An improved algorithm for matching biological sequences. J. Mol. Biol. **162**, 705–708 (1982)
33. Gotoh, O.: Significant improvement in accuracy of multiple protein sequence alignments by iterative refinement as assessed by reference to structural alignments. J. Mol. Biol. **264**, 823–838 (1996)
34. Gotoh, O.: Multiple sequence alignments: algorithms and applications. Adv. Biophys. **36**, 159–206 (1999)
35. Gribskov, M., McLachlan, A.D., Eisenberg, D.: Profile analysis: detection of distantly related proteins. Proc. Natl. Acad. Sci. USA **84**, 4355–4358 (1987)
36. Gumbel, E.J.: Statistics of Extremes. Columbia University Press, New York (1958)
37. Gupta, S.K., Kececioglu, J.D., Schäffer, A.A.: Improving the practical space and time efficiency of the shortest-paths approach to sum-of-pairs multiple sequence alignment. J. Comput. Biol. **2**, 459–472 (1995)
38. Gusfield, D.: Algorithms on Strings, Trees and Sequences: Computer Science and Computational Biology. Cambridge University Press, Cambridge (1997)
39. Hasegawa, M., Kishino, H., Yano, T.: Dating of human-ape splitting by a molecular clock of mitochondrial DNA. J. Mol. Evol. **22**, 160–174 (1985)
40. Haussler, D., Krogh, A., Mian, I.S., Sjölander, K.: Protein modeling using hidden Markov models: analysis of globins. In: HICSS-26, vol. 1, pp. 792–802 (1993)
41. Henikoff, J.G., Greene, E.A., Pietrokovski, S., Henikoff, S.: Increased coverage of protein families with the blocks database servers. Nucleic Acids Res. **28**, 228–230 (2000)
42. Henikoff, S., Henikoff, J.G.: Amino acid substitution matrices from protein blocks. Proc. Natl. Acad. Sci. USA **89**, 10915–10919 (1992)
43. Henikoff, S., Henikoff, J.G., Pietrokovski, S.: Blocks+: a non-redundant database of protein alignment blocks derived from multiple compilations. Bioinformatics **15**, 471–479 (1999)
44. Higgins, D.G., Sharp, P.M.: CLUSTAL: a package for performing multiple sequence alignment on a microcomputer. Gene **73**, 237–244 (1988)
45. Hirosawa, M., Totoki, Y., Hoshida, M., Ishikawa, M.: Comprehensive study on iterative algorithms of multiple sequence alignment. Comput. Appl. Biosci. **11**, 13–18 (1995)
46. Holland, J.H.: Adaptation in Natural and Artificial Systems. University of Michigan Press, Ann Arbor (1975)
47. Hughey, R., Krogh, A.: Hidden Markov models for sequence analysis: extension and analysis of the basic method. Comput. Appl. Biosci. **12**, 95–108 (1996)
48. Jones, D.T., Taylor, W.R., Thornton, J.M.: The rapid generation of mutation data matrices from protein sequences. Comput. Appl. Biosci. **8**, 275–282 (1992)
49. Jukes, T.H., Cantor, C.R.: Evolution of protein molecules. In: Munro, H.N. (ed.) Mammalian Protein Metabolism, pp. 21–123. Academic Press, New York (1969)
50. Karlin, S., Altschul, S.F.: Methods for assessing the statistical significance of molecular sequence features by using general scoring schemes. Proc. Natl. Acad. Sci. USA **87**, 2264–2268 (1990)
51. Katoh, K., Misawa, K., Kuma, K., Miyata, T.: MAFFT: a novel method for rapid multiple sequence alignment based on fast Fourier transform. Nucleic Acids Res. **30**, 3059–3066 (2002)
52. Kim, J., Pramanik, S., Chung, M.J.: Multiple sequence alignment using simulated annealing. Comput. Appl. Biosci. **10**, 419–426 (1994)

53. Kimura, M.: A simple method for estimating evolutionary rates of base substitutions through comparative studies of nucleotide sequences. J. Mol. Evol. **16**, 111–120 (1980)

54. Kimura, M.: Estimation of evolutionary distances between homologous nucleotide sequences. Proc. Natl. Acad. Sci. USA **78**, 454–458 (1981)

55. Kimura, M.: The Neutral Theory of Molecular Evolution. Cambridge University Press, Cambridge (1983)

56. Kimura, M., Ohta, T.: On the stochastic model for estimation of mutational distances between homologous proteins. J. Mol. Evol. **2**, 87–90 (1972)

57. Kirkpatrick, S., Gelatt, C.D., Vecchi, M.P.: Optimization by simulated annealing. Science **220**, 671–680 (1983)

58. Krogh, A., Brown, M., Mian, I.S., Sjölander, K., Haussler, D.: Hidden Markov models in computational biology: applications to protein modeling. J. Mol. Biol. **235**, 1501–1531 (1994)

59. Kruskal, J.B.: An overview of sequence comparison: time warps, string edits, and macromolecules. SIAM Rev. **25**, 201–237 (1983)

60. Lipman, D.J., Pearson, W.R.: Rapid and sensitive protein similarity searches. Science **227**, 1435–1441 (1985)

61. Lipman, D.J., Altschul, S.F., Kececioglu, J.D.: A tool for multiple sequence alignment. Proc. Natl. Acad. Sci. USA **86**, 4412–4415 (1989)

62. Lüthy, R., Xenarios, I., Bucher, P.: Improving the sensitivity of the sequence profile method. Protein Sci. **3**, 139–146 (1994)

63. Maizel, J.V., Lenk, R.P.: Enhanced graphic matrix analysis of nucleic acid and protein sequences. Proc. Natl. Acad. Sci. USA **78**, 7665–7669 (1981)

64. Metropolis, N., Rosenbluth, A.W., Rosenbluth, M.N., Teller, A.H., Teller, E.: Equations of state calculations by fast computing machines. J. Chem. Phys. **21**, 1087–1092 (1953)

65. Meyer, I.M., Durbin, R.: Comparative ab initio prediction of gene structures using pair HMMs. Bioinformatics **18**, 1309–1318 (2002)

66. Mitchell, M.: An Introduction to Genetic Algorithms. The MIT Press, Cambridge (1998)

67. Morgenstern, B., Frech, K., Dress, A., Werner, T.: DIALIGN: finding local similarities by multiple sequence alignment. Bioinformatics **14**, 290–294 (1998)

68. Morgenstern, B.: DIALIGN 2: improvement of the segment-to-segment approach to multiple sequence alignment. Bioinformatics **15**, 211–218 (1999)

69. Morrison, D.R.: PATRICIA—practical algorithm to retrieve information coded in alphanumeric. J. ACM **15**, 514–534 (1968)

70. Mott, R.: Maximum-likelihood estimation of the statistical distribution of Smith-Waterman local sequence similarity scores. Bull. Math. Biol. **54**, 59–75 (1992)

71. Müller, T., Spang, R., Vingron, T.: Estimating Amino Acid substitution models: a comparison of Dayhoff's estimator, the resolvent approach and a maximum likelihood method. Mol. Biol. Evol. **19**, 8–13 (2002)

72. Needleman, S.B., Wunsch, C.D.: A general method applicable to the search for similarities in the Amino Acid sequence of two proteins. J. Mol. Biol. **48**, 443–453 (1970)

73. Notredame, C.: Recent evolutions of multiple sequence alignment algorithms. PLoS Comput. Biol. **3**, e123 (2007)

74. Notredame, C., Higgins, D.G.: SAGA: sequence alignment by genetic algorithm. Nucleic Acids Res. **24**, 1515–1524 (1996)

75. Notredame, C., Higgins, D.G., Heringa, J.: T-Coffee: a novel method for fast and accurate multiple sequence alignment. J. Mol. Biol. **302**, 205–217 (2000)

76. Pearson, W.R.: Empirical statistical estimates for sequence similarity searches. J. Mol. Biol. **276**, 71–84 (1998)

77. Pearson, W.R., Lipman, D.J.: Improved tools for biological sequence comparison. Proc. Natl. Acad. Sci. USA **86**, 2444–2448 (1988)

78. Pustell, J., Kafatos, C.: A high speed, high capacity homology matrix: zooming through SV40 and polyoma. Nucleic Acids Res. **10**, 4765–4782 (1982)

79. Rabiner, L.R.: A tutorial on hidden Markov models and selected applications in speech recognition. Proc. IEEE **77**, 257–286 (1989)

80. Saitou, N., Nei, M.: Neighbor-joining method: a new method for reconstructing phylogenetic trees. Mol. Biol. Evol. **4**, 406–425 (1987)

81. Sankoff, D., Kruskal, J.B.: Time Warps, String Edits, and Macromolecules: The Theory and Practice of Sequence Comparison. Addison-Wesley, New York (1983)

82. Sigrist, C.J.A., Cerutti, L., Hulo, N., Gattiker, A., Falquet, L., Pagni, M., Bairoch, A., Bucher, P.: PROSITE: a documented database using patterns and profiles as motif descriptors. Brief. Bioinform. **3**, 265–274 (2002)

83. Smith, T.F., Waterman, M.S.: Comparison of biosequences. Adv. Appl. Math. **2**, 482–489 (1981)

84. Smith, T.F., Waterman, M.S., Burks, C.: The statistical distribution of nucleic acid similarities. Nucleic Acids Res. **13**, 645–656 (1985)

85. Sneath, P.H.A., Sokal, R.R.: Numerical Taxonomy. Freeman, San Francisco (1973)

86. Staden, R.: An interactive graphics program for comparing and aligning nucleic acid and amino acid sequences. Nucleic Acids Res. **10**, 2951–2961 (1982)

87. Steinmetz, M., Frelinger, J.G., Fisher, D., Hunkapiller, T., Pereira, D., Weissman, S.M., Uehara, H., Nathenson, S., Hood, L.: Three cDNA clones encoding mouse transplantation antigens: homology to immunoglobulin genes. Cell **24**, 125–134 (1981)

88. Tamura, K., Nei, M.: Estimation of the number of nucleotide substitutions in the control region of mitochondrial DNA in humans and chimpanzees. Mol. Biol. Evol. **10**, 512–526 (1993)

89. Tavare, S.: Some probabilistic and statistical problems in the analysis of DNA sequences. In: Lectures on Mathematics in the Life Sciences, vol. 17, pp. 57–86 (1986)

90. Thompson, J.D., Higgins, D.G., Gibson, T.J.: Improved sensitivity of profile searches through the use of sequence weights and gap excision. Comput. Appl. Biosci. **10**, 19–29 (1994)

91. Thompson, J.D., Higgins, D.G., Gibson, T.J.: CLUSTAL W: improving the sensitivity of progressive multiple sequence alignment through sequence weighting position-specific gap penalties and weight matrix choice. Nucleic Acids Res. **22**, 4673–4680 (1994)

92. The UniProt Consortium: The universal protein resource (UniProt) 2009. Nucleic Acids Res. **37**, D169–D174 (2009)

93. Wallace, I.M., Blackshields, G., Higgins, D.G.: Multiple sequence alignments. Curr. Opin. Struct. Biol. **15**, 261–266 (2005)

94. Wang, J., Jiang, T.: On the complexity of multiple sequence alignment. J. Comput. Biol. **1**, 337–448 (1994)

95. Waterman, M.S., Smith, T.F., Beyer, W.A.: Some biological sequence metrics. Adv. Math. **20**, 367–387 (1976)

96. Waterman, M.S.: Introduction to Computational Biology: Maps, Sequences and Genomes. Chapman & Hall/CRC, London (1995)

97. Wilbur, W.J., Lipman, D.J.: Rapid similarity searches of Nucleic Acid and protein data banks. Proc. Natl. Acad. Sci. USA **80**, 726–730 (1983)

98. Wilbur, W.J., Lipman, D.J.: The context dependent comparison of biological sequences. SIAM J. Appl. Math. **44**, 557–567 (1984)

99. Yang, Z.: Estimating the pattern of nucleotide substitution. J. Mol. Evol. **39**, 105–111 (1994)

100. Zaki, M.J., Bystroff, C.: Protein structure prediction. In: Zaki, M.J., Bystroff, C. (eds.) Methods in Molecular Biology, vol. 413. Humana Press, New Jersey (2008)

Chapter 4
Comparative Gene Finding

In the previous chapter we presented various alignment techniques in order to shed light on questions around evolutionary relationships. In the context of comparative gene finding, sequence alignments can help in pinpointing regions of importance, by highlighting evolutionary preserved segments. Single species gene finding algorithms suffer from several problems, the main one being the tendency to overpredict. The new generation of gene finders attempt to salvage this problem by automatizing the integration of gene prediction and sequence alignment. Since the introduction of comparative gene finders a decade or so ago, it has been proved beyond any doubt that comparative-based gene finding works and has a considerable number of advantages over its single species predecessors. These advantages include much higher accuracy in the predictions, the ability to annotate a variety of features that have previously eluded computational approaches, and, more fundamentally, a new view of gene finding and alignment that explains how the two problems are intimately related. In this chapter we exemplify the main areas of comparative gene finding, ranging from similarity-based techniques, to pair HMMs and generalized pair HMMs, to gene mapping. Last but not least we present the first attempts to extend pairwise approaches to multiple sequence gene finding. We expect to see much more developments in this area in the near future, if only we can come to terms with the vast increase in computational complexity that it imposes.

4.1 Similarity-Based Gene Finding

The first attempts to extend the single species algorithms go under the label of similarity-based, or homology-based, gene finding. Methods under this label do not fully integrate the homology information such as in pairwise alignments, but use it to strengthen the signal of conserved elements in the genome. In addition to strengthening the coding signal, incorporating database matches can provide useful information about the gene structure, sort out false positives, and improve boundary prediction.

© Springer-Verlag London 2015
M. Axelson-Fisk, *Comparative Gene Finding*, Computational Biology 20,
DOI 10.1007/978-1-4471-6693-1_4

A straightforward approach of integrating external homology data is to include database match scores from protein, cDNA, or EST databases, and use such scores as additional indications of coding sequences, besides the intrinsic patterns of the input sequence itself. Methods that apply this approach can crudely be divided into two classes: those that make use of mRNA and cDNA sequence information, and those aligning homologous proteins to the target sequence. Implementations of the former class include GMAP [36], BLAT [17] and ECgene [18]. These methods apply local alignment techniques to match the cDNAs of related species to the target genome. A common problem for such an approach is that they need to account for a high error rate in ESTs, and are not really useful with the homology information is too distant. For example, the evolutionary distance of human and mouse is too great.

The method of integrating protein homology has been applied in a long line of implementations, including GeneParser [35], Genie [21, 22], GRAIL [37, 38], SGP-2 [30], HMMgene [20], and GenomeScan [39]. GenomeScan, which is an extension of Genscan [9] in Sect. 2.2.4, is described in detail in the next section. Typically, in this line of methods, a set of candidate exons are extracted from the target sequence using various de novo methods (see Chap. 2), and then matched to a database for further information.

A third type of similarity-based gene finding methods is implemented in the informant-based software Twinscan [19]. Twinscan is also an extension of Genscan, but in contrast to GenomeScan, the external homology is composed of a database of sequences from a specific *informant genome*. The informant sequences are locally aligned to the target sequence, and the conserved segments are scored and combined with the Genscan framework to produce a final prediction. Twinscan is described in a little more detail in Sect. 4.1.2.

Other methods, such as Procrustes [12] and GeneWise [6], have chosen to match the genomic sequence to an informant protein directly, rather than extracting homology information from a protein database. GeneWise uses a spliced alignment approach, where a predicted protein from the target sequence is aligned to a known protein to improve gene structure accuracy. Procrustes uses a combinatorial approach, where potential exon blocks in the target sequence are assembled into various possible protein predictions, which then are matched against known proteins.

4.1.1 GenomeScan: GHMM-Based Gene Finding Using Homology

Genscan [9], described in Sect. 2.2.4, is a very popular and powerful tool for single species gene finding. It exhibits a high sensitivity in its gene predictions, meaning that it detects a high proportion of the genes in a sequence. However, along with most single species gene finders, it suffers from a high rate of false positives, leading to a fairly low specificity (see Sect. 7.3 for details on accuracy measures). In order to improve this, GenomeScan [39] extends the GHMM in Genscan by matching the target sequence to a protein database, using the database search tools BLASTP [2] and BLASTX [13], which are variants of the BLAST algorithm described in Sect. 3.1.8. The GenomeScan procedure can be summarized as follows:

1. Mask repeats in the genomic sequence using RepeatMasker [34].
2. Run Genscan on the masked sequence, and match the predicted peptides against a protein database using BLASTP. Use matches with E-values above an appropriate threshold (default is $E > 10^{-5}$).
3. Compare all open reading frames (ORFs) in the genomic sequence to the database hits retrieved in the previous steps, using BLASTX in a more sensitive search with increased gap penalties and a relaxed E-value cutoff.
4. Run GenomeScan on the masked sequence using homology information of the hits in the previous step as input.

Recall from Sect. 2.1 that an HMM is composed of two interrelated random processes, a hidden process which is Markov and jumps between the states in a state space, and an observed process that generates outputs depending on the underlying state sequence. We denote the state space as $S = \{s_1, \ldots, s_N\}$ as before, the hidden sequence generated by the hidden process as X_1^L, and the observed sequence as Y_1^T, where the notation Y_a^b corresponds to a sequence stretching between indices a and b, respectively. GenomeScan uses the same state space as Genscan with the main states being the intergenic, intron, and exon states (see Fig. 2.11 for a simplified version). Recall from Sect. 2.2.4 that Genscan uses a GHMM to determine the optimal path through the state space. It is mainly the exon states that are generalized in this model, meaning that unlike a standard HMM where the output in each step is a single residue, an exon state first chooses a state dependent duration from a generalized distribution, and then outputs a sequence of residues of that length. To facilitate this, we attach a sequence of state durations d_1^L to the state sequence, one for each state, and recall the state sequence and the corresponding state durations is called a *parse* of the observed sequence Y_1^T.

To solve the gene finding problem we would like to find the state sequence and the corresponding duration sequence that maximize the probability of the parse given the observed sequence $\mathbf{P}(X_1^L, d_1^L | Y_1^T)$. We recall that this probability is maximized in the same point as the joint probability $\mathbf{P}(Y_1^T, X_1^L, d_1^L)$. The extension in GenomeScan is to maximize the conditional probability of the hidden and observed data $\mathbf{P}(Y_1^T, X_1^L, d_1^L | \Gamma)$, given some additional similarity information Γ. The similarity information is retrieved from a protein database in two steps. First, the peptides predicted by Genscan are matched against a protein database using a regular protein-to-protein search (BLASTP), and the matching proteins are selected from the database. Next all open reading frames (ORFs) in the genomic sequence are matched against this reduced set of proteins using BLASTX, which translates the ORFs in all reading frames before the matching. The highest scoring region of a BLASTX hit is termed the *centroid* of the protein. Proteins with several high scoring regions (multimodal), or with a BLASTX hit extending over 100 codons, are separated into several single-modal hits with a centroid each.

Let Γ denote the resulting set of separated BLASTX hits, each with a defined centroid. GenomeScan only considers parses that have at least one exon overlapping a centroid in Γ. Let $\phi_i = (X_1^L, d_1^L)$ denote a given parse, and let $\Phi_\Gamma = \{\phi_i\}$ denote the set of parses with at least one exon overlapping a centroid. Since a BLAST hit not necessarily infers homology, we let P_H denote the probability that the BLASTX hit is truly homologous, while $P_A = 1 - P_H$ denotes the probability that the hit is artefactual. The joint probability of a given parse and the genomic sequence, given the homology information, is then defined as

$$
\mathbf{P}(\phi_i, Y_1^T | \Gamma) = \begin{cases} \left(\dfrac{P_H}{\mathbf{P}(\Phi_\Gamma)} + P_A \right) \mathbf{P}(\phi_i, Y_1^T) & \text{if } \phi_i \in \Phi_\Gamma \\ P_A \, \mathbf{P}(\phi_i, Y_1^T) & \text{if } \phi_i \notin \Phi_\Gamma. \end{cases} \tag{4.1}
$$

$\mathbf{P}(\phi_i, Y_1^T)$ is the Genscan probability of the path and the observed sequence, while $\mathbf{P}(\Phi_\Gamma)$ is the probability that Φ_Γ contains the true parse. As a result, the probability in (4.1) favors parses that are consistent with the similarity information, but does not completely rule out inconsistent parses. $\mathbf{P}(\Phi_\Gamma)$ can be derived using a procedure similar to the forward-backward algorithm described in Sect. 6.5. The probability P_A that a hit is artefactual is related to the BLASTX hit score P_B using a heuristic $P_A = (P_B)^{1/r}$, where r is a small integer (default value $r = 10$). Using the formula in (4.1) as basis, gene prediction is performed running the forward, backward and Viterbi algorithms as in Sect. 2.2. For details, see [39].

4.1.2 Twinscan: GHMM-Based Gene Finding Using Informant Sequences

Twinscan [19] is another extension of Genscan, in which the target sequence is compared to an informant sequence, which is added to boost or reduce the probability of a potential gene component. The informant sequence is not used directly in the gene finding process. Instead, the homology found between the target sequence and the informant sequence, is represented by a *conservation sequence*, obtained using a gapped version of BLAST [2] (see Sect. 3.1.8). The conservation sequence matches one of three possible symbols to each symbol in the target sequence:

  . unaligned
  | match
  : mismatch.

The unaligned symbol corresponds to a gap in the informant sequence, the match symbol to identical residues, and the mismatch symbol to differing residues in the informant sequence. Gaps in the target sequence are uninteresting in the current context and are therefore ignored. As a result the target sequence and the conservation sequence will be of the same lengths.

The procedure in Twinscan is similar to that of Genscan, except that instead of outputting a single sequence, each state generates a pair of outputs corresponding to the target sequence and the conservation sequence. For instance, a typical output would be of the form:

```
                      10          20          30
                      |           |           |
        Target:  AGGAAGTCCTCATCAGGCTCTTTAAGGGTCAC
        Conserv: ...||:|||:|:|||||:||:|||::||....
        Inform:  -AACTCCACGTCAGGGTCATTACCGG--
```

As before we let X_1^L and d_1^L denote the underlying state path and the corresponding state durations, respectively. Furthermore, we let Y_1^T denote the target sequence, and let Z_1^T correspond to the conservation sequence. Given the underlying state path, Twinscan assumes conditional independence between Y and Z, such that

$$\mathbf{P}(Y_1^T, Z_1^T | X_1^L, d_1^L) = \mathbf{P}(Y_1^T | X_1^L, d_1^L)\mathbf{P}(Z_1^T | X_1^L, d_1^L). \tag{4.2}$$

The first probability on the right-hand side of (4.2) is simply the probability of the observed sequence under the Genscan model

$$\mathbf{P}(Y_1^T | X_1^L, d_1^L) = \prod_{l=1}^{L} b_{X_l}(Y_{d_{l-1}+1}^{d_l} | Y_1^{d_{l-1}}) \tag{4.3}$$

where $b_i(Y_a^b)$ is the emission probability of sequence Y_a^b in state i in the GHMM. The second probability in (4.2) is similar to the first, but operates on the three-symbol conservation sequence. The model used in Twinscan for the conservation sequence is a fifth-order Markov model. That is, if we split the conversation sequence according to the state sequence

$$\mathbf{P}(Z_1^T | X_1^L, d_1^L) = \prod_{l=1}^{L} \mathbf{P}(Z_{d_{l-1}+1}^{d_l} | X_l), \tag{4.4}$$

each of the probabilities in the right hand product will be of the form

$$\mathbf{P}(Z_1^{d_1} | X_1) = \prod_{t=1}^{d_1} \mathbf{P}(Z_t | Z_{t-5}^{t-1}, X_1). \tag{4.5}$$

The gene prediction algorithm then follows the same route as in Genscan, but with the addition of the conservation sequence in the probability model. That is, the optimal path determined by the Viterbi algorithm is the state sequence and state durations that maximize the joint probability $\mathbf{P}(Y_1^T, Z_1^T, X_1^L, d_1^L)$. The procedure of Twinscan can be summarized as follows:

1. Mask repeats in the target sequence using RepeatMasker [34].
2. Create a conservation sequence by aligning the target and the informant sequences using WU-BLASTN (W. Gish, unpublished).
3. Run Genscan on the target and the conservation sequences, using the extended model in (4.2).

Integrating the information sequence improves the Genscan accuracy significantly, in particular when it comes to reduce the amount of false positives. Later in this chapter we will see a further extension of the Twinscan model, to N-SCAN [14] where multiple informant sequences are used to infer homology. Twinscan was one of the gene finders used by the Mouse Genome Sequencing Consortium [26], along with SLAM [1] and SGP-2 [30], in the initial comparison of the human and mouse genomes.

4.2 Heuristic Cross-Species Gene Finding

The first implementations of cross-species gene finding are probably in the softwares ROSETTA [5] and CEM [4]. Both programs take two homologous genomic sequences as input and produce gene predictions simultaneously in both. ROSETTA begins by producing a rough alignment map, focusing on highly similar k-mers, and then refining the alignment successively in the regions in between these k-mers. The gene prediction is then produced by combining various coding measures within the alignment of the two sequences. CEM is similar, but uses TBLASTX [13] to identify candidate exons. High-scoring BLAST-hits are extended (or shrunk) to include putative splice sites, and then the candidate exons are chained together to produce a complete gene structure.

4.2.1 ROSETTA: A Heuristic Cross-Species Gene Finder

If the gene structure of a specific gene is known in one organism, it is fairly straightforward to identify the homologous counterpart in a related genome, given that the gene exists and that the two organisms are not too distantly related. A far more challenging task is to perform de novo gene prediction in the two organisms directly. ROSETTA [5], named after the *Rosetta stone*, is among the first softwares that have implemented a full cross-species gene prediction, in which two homologous sequences are annotated simultaneously. ROSETTA operates by first aligning regions of high similarity, and then searching for coding exons in these regions. The alignments are performed by a program named GLASS (Global Alignment

SyStem), which uses a hierarchical alignment approach to produce a heuristic global alignment.

The motivation for constructing GLASS was that neither standard dynamic programming algorithms such as the Needleman-Wunsch algorithm [27], nor faster heuristics such as BLAST [2], were satisfactory for the task at hand. Dynamic programming algorithms are not well adapted to identifying short stretches of highly conserved regions interspersed by long regions of low or no similarity. On the other hand, although heuristic search tools such as BLAST perform very well at identifying local similarities, they usually focus on perfect matches, and may therefore miss relevant regions.

GLASS starts out with a rough alignment consisting of long stretches of high identity, which is then iteratively improved by aligning shorter and shorter regions in between these stretches. In the final step, the remaining regions are aligned using standard approaches. GLASS proceeds as follows:

1. Choose an initial value k and find all matching k-mers in the two sequences.
2. Treat each unique k-mer as a specific object, and convert both sequences into strings of such matching objects.
3. Align the two object strings as follows: for each object align the flanking regions before and after (e.g. 12 bp before and after) using dynamic programming, and assign the object the combined scores of these two regions. Mismatches and gaps receive score 0.
4. Identify the pairs of the k-mers that score above a given threshold T.
5. From this list of k-mers, remove inconsistent pairs. For instance, if two k-mers overlap in one sequence but not the other, they are inconsistent.
6. Fix the alignment between the two sequences for the remaining k-mers.
7. Recursively repeat steps 1–6 for the regions between the k-mers, by successively decreasing k. For instance, $k = 20, 15, 12, 9, 8, 7, 6, 5$.
8. After all recursions are completed, extend each aligned segment in both direction using dynamic programming.
9. Finish by aligning all remaining unaligned segments using dynamic programming.

ROSETTA parses the sequences into exons, introns, and intergenes, where three types of exons are identified: initial, internal, and terminal exons. The gene recognition is performed using a dynamic programming approach, where each parse is scored according to the sum of scores of the individual exons in the parse. The submodels used for scoring an exon are a splice site model, codon usage, amino acid similarity, and exon length. The splice sites are scored using a combination of the maximal dependence decomposition (MDD) algorithm [7, 8] described in Sect. 5.4.3, and a dictionary-based approach [29]. The codon usage score of an exon is computed by adding the log-odds ratio for each codon, using species-specific codon frequencies. The amino acid similarity of a human-mouse exon is computed using the PAM20 matrix (see Sect. 3.1.3). The exon length score combines the agreement with known exon length distributions and a penalty for differing lengths in the exon pair.

ROSETTA was applied to a set of 117 orthologous human and mouse sequences in [5], and proved beyond any doubt that comparative gene finding improves the prediction accuracy significantly.

4.3 Pair Hidden Markov Models (PHMMs)

Pair hidden Markov models (PHMMs) extend the theory of HMMs to model pairs of observed sequences. The underlying hidden process still jumps between different states in a state space, but instead of emitting a single residue as in standard HMMs (Sect. 2.1), or a sequence of residues of generalized length as in GHMMs (Sect. 2.2), the output is an aligned pair of residues. While mainly used for pairwise alignments as in Sect. 3.1.7, the PHMM theory has also been successfully applied to gene finding in the comparative software DoubleScan [24] presented in the next section. Here the standard PHMM state space, typically consisting of a match state, an insertion state, and a deletion state, is extended to include the labeling into different gene features. The resulting model is a complex composition of several PHMMs, representing different types of exons, introns and intergenic regions in the gene model, and the output of the model is two labeled sequences along with their pairwise alignment.

4.3.1 DoubleScan: A PHMM-Based Comparative Gene Finder

DoubleScan [24] is a comparative gene finder that predicts complete gene structures in two homologous DNA sequences. In addition to predicting complete, multiple, or partial genes, DoubleScan can detect homologous genes that have diverged through exon-splitting or exon-fusion. The output of DoubleScan is the simultaneous annotation of the two input sequences, along with their alignment.

The DoubleScan model is based on a PHMM, and uses two methods to retrieve the gene predictions; the Viterbi algorithm and a heuristic method called the *stepping stone algorithm*. To save on memory complexity, a linear-space implementation of the Viterbi is used, called the *Hirschberg algorithm* [16].

If the Viterbi algorithm has memory complexity $O(NTU)$, where N is the number of states and T and U are the lengths of the observed sequences, respectively, the Hirschberg algorithm linearizes the complexity to $O(N \min\{T, U\})$ and at most doubles the time requirement of the Viterbi. The stepping stone algorithm reduces the time requirement of the Viterbi to near linear, but the optimal solution is no longer guaranteed to lie within the search space.

The State Space

The DoubleScan model consists of 54 states, a simplified version is illustrated in Fig. 4.1. Each state outputs an aligned pair of symbols, along with one of three class labels: exon, intron, or intergene. The *match* states output sequence residues in both

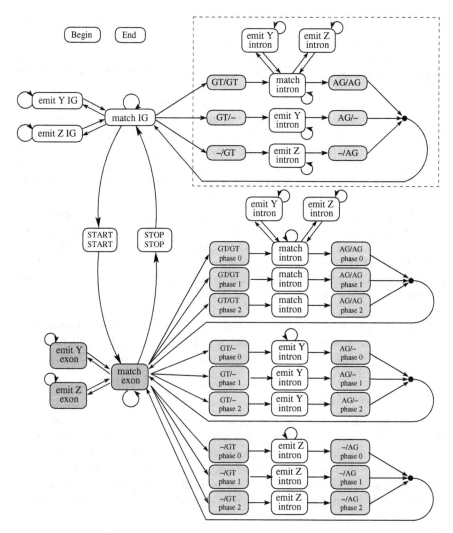

Fig. 4.1 The DoubleScan state space. In reality, extra states are included for each splice site and each intron, to account for the three different frames the exons can be in. The *dashed box* at the *top* represent UTR introns. Reproduced from [24] with permission from Oxford University Press

sequences, while the *emit* states represent insertion and deletion events and output a residue in one sequence and a gap in the other. All paths through the state space start and end in the silent states *begin* and *end*, respectively. These states are connected to all other states except themselves and each other, allowing for the sequence output to begin and end in any state.

Initial and single exons begin with the START START state, which outputs the start codon ATG in both sequences, and then jumps to the *match exon* state. The *match exon* state and the two *emit exon* states all output complete codons in each step, each coding for an amino acid. That is, exons can produce internal, in-frame ATG-codons, but not any stop codons. The stop codon, in the end of a single or a terminal exon is produced by the STOP STOP state, which outputs a stop codon in each sequence.

Introns belong to one of three phases, representing the position in the codon where the coding sequence is spliced (see Sect. 5.1.1). The *phase* states surrounding the introns are responsible for producing the (aligned) sequences of spliced codons, along with the corresponding splice site sequences, where splice sites are modeled to adhere to the GT-AG consensus of donor and acceptor sites (see Sect. 5.1.2). Intergenic states appear in between complete gene structures, and are modeled with the least constraints among all states, forming a regular PHMM.

UTR exons are not modeled explicitly, but are invoked in and labeled by the intergenic states. However, the model allows for UTR splicing by including a separate set of UTR introns in the state space. The reasoning for this is that the inclusion of UTR exons and introns improves the detection of the start and stop codons of the translated portion of the genes [24]. Introns within UTRs have no phase, but are surrounded by splice sites following the GT-AG consensus just as the introns in the coding region.

DoubleScan predicts genes on the forward strand only, the reverse-complemented strand requires a separate run. This could be handled, however, by adding a mirror image to the state space, doubling the number of states, much like the Genscan state space in Sect. 2.2.4.

The Stepping Stone Algorithm

When dealing with long pairs of sequences the memory and space requirements quickly become overwhelming. Here we present the stepping stone algorithm used by DoubleScan to reduce the search space in which the Viterbi algorithm searches for an optimal path. The idea is to narrow the search around regions of high similarity by assuming that the optimal path passes near such regions. Regions of high similarity are detected by running BLASTN [2] on the pair of input sequences. The stepping stone procedure goes as follows:

1. Start with the highest scoring BLASTN match and use its middle point in the alignment matrix (x, y) as reference.
2. Find the next highest scoring match which is compatible with existing reference points and add its middle point to the set. Continue until no more high scoring matches can be added.
3. Extend the submatrices formed between two subsequent reference points (constituting the bottom-left and top-right corners of the sub-matrix) by a 15 bp region around each middle point.

A match is *compatible* with existing reference points if all (x, y) pairs can be ordered simultaneously. That is, in the ordered set of reference points, each point should appear above and to the right of the previous one in the alignment matrix (see Fig. 4.2). A variant of the Viterbi algorithm is then run on the reduced search space, where

Fig. 4.2 The search space
reduced by the stepping
stone algorithm

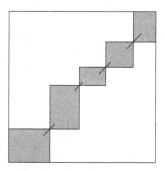

only the necessary values are kept to continue the calculation. Each new submatrix
calculation is initiated by the last values of the previous submatrix. Once the top-
right corner is reached the optimal state path is retrieved by recalculating the Viterbi
algorithm backwards through the search space. By only keeping the necessary Viterbi
values, the time requirements double since the Viterbi algorithm has to run twice,
but the memory requirements are decreased to a minimum.

4.4 Generalized Pair Hidden Markov Models (GPHMMs)

The GHMM has proved to be very successful for gene finding, and the PHMM a
suitable model for sequence alignment, and for quite some time the two problems
were tackled separately. This is not surprising, since a priori there was no fundamental
reason to believe that the two problems were related. However, it has become more
and more clear that solving one of the problems can aid in the solution of the other.
For instance, knowing the alignment of two related sequences can greatly improve
the gene finding, and vice versa. The first results suggesting the effectiveness of
comparative methods for gene finding included ad hoc studies of dot plots and more
sophisticated PIP plots [32] that show that exons and regulatory domains tend to be
in conserved regions [3, 15].

4.4.1 Preliminaries

A *generalized pair* HMM (GPHMM) [28] is a seamless merging of a GHMM and a
PHMM model, where the problems of gene finding and alignment are unified to be
solved both at once. Each state is (possibly) generalized as in the GHMM, but now
outputs an aligned pair of observations, just as in the PHMM.

We keep all the previous notation used in the GHMMs in Sect. 2.2 and in the
PHMMs in Sect. 3.1.7. As usual the HMM is composed of a hidden Markov process
and an observed (typically not Markov) process. The hidden process jumps between

the states in a state space $S = \{s_1, \ldots, s_N\}$, and generates a a state path $X_1^L = X_1, \ldots, X_L$.

The hidden process is initiated by the initial distribution $\pi = \{\pi_1, \ldots, \pi_N\}$ and progresses according to the transition probabilities

$$a_{ij} = \mathbf{P}(X_{l+1} = j | X_l = i), \quad i, j \in S. \tag{4.6}$$

Just as in PHMMs the observed process generates an aligned pair of sequences, Y_1^T and Z_1^U, each taking values in some alphabet $V = \{v_1, \ldots, v_M\}$, and just as in PHMMs we need to associate state durations and partial sums with the observed sequences to keep track of their respective indices. We let d_1^L and e_1^L denote the durations, with corresponding partial sums $p_l = \sum_{k=1}^{l} d_k$ and $q_l = \sum_{k=1}^{l} q_k$, for sequence Y and Z, respectively. The main difference from PHMMs is that now the durations can be generalized as in GHMMs, and the duration pairs (d_l, e_l) are chosen from some joint generalized length distribution $f_{X_l}(d_l, e_l)$. Thus, when in state X_l the observed process first chooses a pair of durations (d_l, e_l) and then generates aligned output sequences according to the joint emission distribution $b_{X_l}(Y_{p_{l-1}+1}^{p_l}, Z_{q_{l-1}+1}^{q_l} | Y_1^{p_{l-1}}, Z_1^{q_{l-1}})$. In non-generalized states, to account for insertions and deletions in the aligned output, we allow one (but not both) of the durations (d_l, e_l) to be zero. Recall that in order to keep the notation consistent we let Y_a^b (or Z_a^b) correspond to a gap for $a > b$. For simplicity we assume that we $p_L = T$ and $q_L = U$, meaning that all of the observed output generated by the final state X_L is included in the observed sequences. Inserting all this into Eq. (2.64) gives us the joint probability of hidden and observed data for the GPHMM

$$\mathbf{P}(Y_1^T, Z_1^U, X_1^L, d_1^L, e_1^L) =$$
$$= \prod_{l=1}^{L} a_{X_{l-1}, X_l} f_{X_l}(d_l, e_l) b_{X_l}(Y_{p_{l-1}+1}^{p_l}, Z_{q_{l-1}+1}^{q_l} | Y_1^{p_{l-1}}, Z_1^{q_{l-1}}), \tag{4.7}$$

where X_0 represent the silent begin state with

$$a_{X_0, X_1} = \pi_{X_1}. \tag{4.8}$$

4.4.1.1 The Forward, Backward, and Viterbi Algorithms

As before, the forward algorithm is used to calculate the probability (or the likelihood) of the observed data $\mathbf{P}(Y_1^T, Z_1^U)$ under the given model. As in GHMMs we let D denote the maximum possible duration in any state and define the forward variables as

$$\alpha_i(t, u) = \mathbf{P}\left(Y_1^t, Z_1^u, \{\text{some hidden state } i \text{ ends at } (t, u)\}\right) \tag{4.9}$$

$$= \mathbf{P}\left(Y_1^t, Z_1^u, \{\bigcup_{l=1}^{L}(X_l = i, p_l = t, q_l = u\}\right)$$

$$= \sum_{j=1}^{N}\sum_{d=1}^{D}\sum_{e=1}^{D} \alpha_j(t - d, u - e)\, a_{ji}\, f_i(d, e)\, b_i(Y_{t-d+1}^t, Z_{u-e+1}^u | Y_1^{t-d}, Z_1^{u-e}).$$

We initialize the process in a silent begin state X_0 with

$$\begin{aligned} \alpha_i(0, 0) &= \pi_i, \\ \alpha_i(t, 0) &= 0 \text{ if } t > 0, \\ \alpha_i(0, u) &= 0 \text{ if } u > 0, \end{aligned} \tag{4.10}$$

and terminate in a silent end state X_{L+1} with

$$\alpha_i(T + 1, U + 1) = \sum_{j \in S} \alpha_j(T, U)\, a_{ji}. \tag{4.11}$$

As before, we can now compute the likelihood of the observed data using

$$\mathbf{P}(Y_1^T, Z_1^U) = \sum_{i=1}^{N} \alpha_j(T + 1, U + 1). \tag{4.12}$$

The backward algorithm is the probability of all observed data after time (t, u), given the observed data up to that time and given that the last hidden state ended at (t, u). That is, the backward variables are given by

$$\beta_i(t, u) = \mathbf{P}\left(Y_{t+1}^T, Z_{u+1}^U | Y_1^t, Z_1^u, \bigcup_{l=1}^{L}(X_l = i, p_l = t, q_l = u)\right)$$

$$= \sum_{j=1}^{N}\sum_{d=1}^{D}\sum_{e=1}^{D} \beta_j(t + d, u + e)\, a_{ji}\, f_i(d, e)\, b_i(Y_{t+1}^{t+d}, Z_{u+1}^{u+e} | Y_1^t, Z_1^u),$$

$$\tag{4.13}$$

and are initialized by

$$\beta_i(T + 1, U + 1) = 1, \quad i \in S. \tag{4.14}$$

As before the backward algorithm terminates upon calculation of $\beta_i(0, 0)$. The Viterbi algorithm, which is used to determine the optimal path through the state space under the given model, uses maxima instead of sums but is otherwise analogous to the forward algorithm. The Viterbi variables are thus given by

$$\delta_i(t, u) = \max_{j,d,e} \delta_j(t - d, u - e) \, a_{ji} \, f_i(d, e) \, b_i(Y_{t-d+1}^t, Z_{u-e+1}^u | Y_1^t, Z_1^u). \quad (4.15)$$

The initialization and termination is the same as for the forward algorithm with

$$\delta_i(0, 0) = \pi_i,$$
$$\delta_i(t, 0) = \delta_i(0, u) = 0 \quad \text{for } t, u > 0, \quad\quad\quad (4.16)$$
$$\delta_i(T + 1, U + 1) = \max_j \delta_j(T, U) \, a_{ji}.$$

During the Viterbi computation we record the optimal previous state along with the corresponding durations for each observed sequence. In PHMMs the durations are given by the specific state directly, but here the durations may be generalized, and the backtrack thus needs to know three values

$$\big(\psi_i(t, u), \phi_i(t, u), \xi_i(t, u)\big) \quad\quad\quad\quad\quad\quad\quad\quad (4.17)$$
$$= \operatorname*{argmax}_{j,d,e} \Big\{ \delta_j(t - d, u - e) \, a_{ji} \, b_i(Y_{t-d+1}^t, Z_{u-e+1}^u | Y_1^{t-d}, Z_1^{u-e}) \Big\}.$$

The probability of the optimal path is given by the largest element in the silent state $\max_i \delta_i(T + 1, U + 1)$, and the backtrack starts in the corresponding state

$$i^* = \operatorname*{argmax}_i \delta_i(T + 1, U + 1). \quad\quad\quad\quad (4.18)$$

The backtrack proceeds by following the pointers to the previous state, starting with $\psi_{i^*}(T + 1, U + 1)$ which has durations $\phi_{i^*}(T + 1, U + 1)$ and $\xi_{i^*}(T + 1, U + 1)$, respectively.

4.4.2 SLAM: A GPHMM-Based Comparative Gene Finder

The GPHMM theory was first introduced in [28], and later implemented in a the cross-species gene finder SLAM [1]. Instead of treating the problems of gene finding and pairwise sequence alignment as separate, the implementation of GPHMMs places the two problems on an equal footing. It can be used to annotate genomic sequences by using the added signal strength in conserved regions, or it can be used as a global alignment tool which takes advantage of the statistical features of biologically functional regions to improve the alignment accuracy. SLAM has been used in several large genome projects including the initial analysis of the mouse genome [26] (along with Twinscan [19] and SGP-2 [30]) and the comparative analysis of the rat genome [11, 31].

The State Space

It is an interesting observation that we can use the same state space for the GPHMM as for the GHMM, the difference resides in the output of the states. Thus, the SLAM state space can be illustrated just as the Genscan state space in Fig. 2.11. The figure only shows the model for the forward strand, and the full state space includes a mirror image to account for genes on the reverse strand as well, resulting in a state space of 27 states; 13 for each strand and the joint intergenic strands. There are four types of exons on each strand: single, initial, internal, and terminal exons. Single exons are bounded by a start and a stop codon, initial exons by a start codon and a donor site, internal exons by a donor and an acceptor site, and terminal exons by an acceptor site and a stop codon. The introns are separated according to phase, corresponding to where in the sequence it occurs between two exons. Intron $_i$, $i = 0, 1, 2$ means that the previous exon was spliced after i bases into its last codon. Similarly, exons are indexed by the phases of the surrounding introns, where exon $E_{i,j}$ is preceded by an intron of phase i and succeeded by an intron of phase j, and $i = I$ stands for initial, and $j = T$ for terminal exons, respectively. As in Genscan it is mainly the exon states that need to be generalized, and we partition the state space as before into E-states (exons) and I-states (introns and intergene).

The E- and I-states in SLAM are themselves rather complex models, illustrated in Fig. 4.3. The E-states in Fig. 4.3a include the entire exon sequences, with possibly spliced codons, as well as the corresponding boundary models for start or stop codons and splice sites. An exon output is generated by the model as follows:

1. Choose a pair of durations from a joint generalized length distribution generating lengths of the correct phase. That is, the length includes a random number of codons, plus the extra bases needed for the spliced codons (if any) at the beginning and the end of the exon.
2. Generate the pair of initial boundaries (start codon or acceptor site) corresponding to the exon type, using a specified boundary model.
3. Generate extra initial bases (if any) according to exon phase.
4. Use a PHMM that outputs complete codons to generate two aligned exon sequences according to the chosen lengths.
5. Generate extra terminal bases (if any) according to exon phase.
6. Generate the pair of terminal boundaries (donor site or stop codon) corresponding to the exon type, using a specified boundary model.

The I-states in Genscan are modeled by standard HMM states, where a single residue is generated in each step, resulting in geometrically distributed duration lengths. Thus, a direct translation to the GPHMM would be to model the I-states in SLAM by regular PHMMs. An intrinsic property of PHMMs, however, is that they generate sequence pairs of on average the same length. While this holds true for conserved, functional regions, introns and intergenes tend to differ in length quite notably even between fairly closely related species. As an example, intron lengths in human and mouse differ on average by a factor of 1.5 [26]. To account for this the SLAM I-states generate sequences of high conservation interspersed by long stretches of

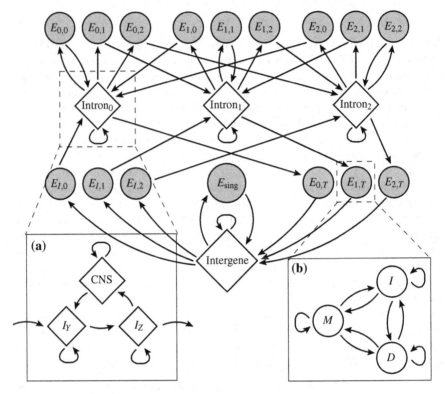

Fig. 4.3 The SLAM state space. **a** Each I-state consists of three parts, a PHMM for conserved noncoding regions, and two independent stretches of unconserved I-sequence. **b** Exons are modeled as regular PHMMs in codon-space

independent sequences with geometrically distributed lengths. The SLAM I-states thus consist of two parts (Fig. 4.3b): a pair of independent I-states for each observed sequence, modeling long unrelated noncoding regions, and a conserved noncoding sequence (CNS) state for modeling non-coding conserved regions. The CNSs are modeled by regular PHMMs, but the interspersing I-states are modeled independently, which allows for very different lengths of the resulting introns or intergenes in the two observed sequences. Details on the submodels and an overview of the implementation of SLAM is given in Chap. 7.

Reducing Computational Complexity

A major issue with GPHMMs is the computational complexity, both in memory usage and running time. The required storage for the full forward algorithm in (4.9) is $O(TUN)$, where T and U are the observed sequence lengths, respectively, and N is the number of states in the model. The running time depends very much upon how long it takes to evaluate the emission distribution $b_i(Y_{t-d+1}^t, Z_{u-e+1}^u | Y_1^{t-d}, Z_1^{u-e})$. A natural way to model the output is as a PHMM, which would require $O(de)$

calculations, where d and e are the durations of the output sequences, respectively. Then, if D is the maximal duration of a state, the number of computations needed for each forward variable would become $O(ND^4)$, and the total number of calculations for the forward algorithm becomes $O(TUN^2D^4)$. However, since SLAM uses the same overall state space as Genscan, we can use the same tricks to reduce computational complexity as described in Sect. 2.2.4.2. That is, we can partition the state space in E- and I-states and store the forward, backward and Viterbi variables for the I-states only.

To illustrate how this is done let N_I and N_E denote the number of I- and E-states respectively, such that $N = N_I + N_E$. Using the forward variable as an example, we can break up the expression in (4.9) into sums over the two state classes

$$\alpha_i(t, u) =$$

$$= \sum_{j \in I} \sum_{d,e} \alpha_j(t - d, u - e) \, a_{ij} \, f_i(d, e) \, b_i(Y^t_{t-d+1}, Z^u_{u-e+1} | Y^{t-d}_1, Z^{u-e}_1) \tag{4.19a}$$

$$+ \sum_{j \in E} \sum_{d,e} \alpha_j(t - d, u - e) \, a_{ij} \, f_i(d, e) \, b_i(Y^t_{t-d+1}, Z^u_{u-e+1} | Y^{t-d}_1, Z^{u-e}_1). \tag{4.19b}$$

The sum in the first class (4.19a) requires α-variable values from previous I-states only. However, recall that the only possible transition from an I-state to an I-state is via a self-transition, which means that the first sum over I-states only has one positive term. Moreover, the duration pairs (d, e) in I-states run over three values only, denote it $B = \{(1, 1), (1, 0), (0, 1)\}$ (corresponding to a match, an insertion or a deletion in Y, respectively). Thus (4.19a) becomes

$$(4.19a) = a_{ii} \sum_{(d,e) \in B} \alpha_i(t - d, u - e) \, f_i(d, e) \, b_i(Y^t_{t-d+1}, Z^u_{u-e+1} | Y^{t-d}_1, Z^{u-e}_u). \tag{4.20}$$

Recall that Y^b_a denotes a gap if $a > b$. To make the second class (4.19b) to sum over previous I-states only we need to extend the backward dependence one more hidden state. We recall that for an ordered pair of I-states (k, j) with an intervening E-state, the E-state is uniquely defined $E_{k,j}$, and once in an E-state there is only one possible transition out of it. We use these properties and let the sum in the second class (4.19b) extend over two states back to get the (computationally) simplified form of the forward algorithm

$$\alpha_i(t, u) = \sum_{(d,e)\in B} f_i(d, e) \, b_i(Y^t_{t-d+1}, Z^u_{u-e+1} | Y^{t-d}_1, Z^{u-e}_1) \Bigg[\alpha_i(t - d, u - e) \, a_{ii} +$$

$$+ \sum_{k\in I} \sum_{d'=1}^{D} \sum_{e'=1}^{D} \alpha_k(t - d' - d, u - e' - e) \, a_{k, E_{k,i}} \, f_{E_{k,i}}(d', e')$$

$$\cdot b_{E_{k,i}} (Y^{t-d}_{t-d-d'+1}, Z^{u-1}_{u-e-e'+1} | Y^{t-d-d'}_1, Z_{u-e-e'}) \Bigg]. \qquad (4.21)$$

The first part is thus the self-transition of the current I-state $i \in I$, which sums over the three possible durations of (d, e) in B. The second part extends two states back $k \to E_{k,i} \to i$, to the previous I-state k, via the intervening E-state, which has generalized durations denoted (d', e'). The initiation and termination is the same as in Sect. 4.4.1.1, and the backward and Viterbi algorithms are simplified in the same manner.

If we store the variables for the I-states only, the memory usage of the forward algorithm reduces to $O(TUN_I)$. SLAM consists of $N = 39$ states in the full model in Fig. 4.3, not counting the states in the boxes separately, but only $N_I = 7$ I-states which results in an 82 % reduction in memory usage, and nearly a 97 % reduction in the required number of operations.

The main computational issue, however, is the lengths of the input sequences. To handle this we need to limit the search space in some manner, for instance, as in the stepping stone algorithm in DoubleScan in Sect. 4.3.1. SLAM uses something called an *approximate alignment*, which is based on a global alignment of the input sequences, which is then relaxed in various ways to ensure that it (most likely) contains the optimal solution. The approximate alignment in SLAM is discussed further in Sect. 7.2.3.

4.5 Gene Mapping

With the ever-increasing amount of sequence data available, it becomes more and more common that with every newly sequenced genome, there is an evolutionary-related genome already sequenced. If that homologous genome is fairly well anno-tated, a *gene mapping* approach such as used by Projector [25] and GeneMapper [10] might be more appropriate to annotate the new sequence.

Gene mapping traditionally means constructing a *genetic map* by assigning DNA sequences (possibly fragmented) to specific chromosome positions. This is typically done by either *genetic mapping* or *physical mapping*. Genetic mapping uses *linkage analysis* to determine the *relative* position between genes on the chromosome, while physical mapping uses molecular biology techniques to analyze the DNA molecules directly in order determine the *absolute* position of the genes on the chromosome.

The gene mapping methods we describe in this section are performed completely 'in silico' (biological experiments carried out on computer only) by means of mapping the genes of a well-annotated reference genome onto a newly sequenced target genome. This idea is similar to that of using external homology such as cDNA/EST, mRNA or protein sequences, described in Sect. 4.1. The main difference is that similarity-based methods only use external evidence to boost the signal of coding regions, while the exon–intron structure of the underlying gene is not known. The gene mapping methods described in this section utilizes not only the sequence similarity in homologous genes, but incorporates the extra information about splice sites as well in order to predict complete gene structures.

4.5.1 Projector: A Gene Mapping Tool

Projector [25] is a gene mapping tool that is based on the PHMM model used in DoubleScan, described in Sect. 4.3.1. However, instead of comparing two unannotated genomes and predicting genes in both, Projector uses an annotated sequence as *informant* sequence and "projects" its genes onto the unannotated target sequence. In HMM terminology, instead of determining the globally optimal state path through the model, Projector chooses the most probable state path among those that coincide with the gene annotation of the informant sequence. This can be done by simply modifying the Viterbi algorithm in (3.64) slightly.

Let Y_1^T denote the informant sequence with the known genes, and let Z_1^U correspond to the target sequence to be annotated with the genes in the informant sequence. Furthermore, let $X_1^{(Y)}, \ldots, X_T^{(Y)}$ denote the sequence of (known) state labels for Y, where the state space is the same as for DoubleScan (see Fig. 4.1). Note that this label sequence is not exactly the same as the the state sequence X_1^L that the PHMM predicts, as it does not take insertions and deletions into account. (Actually, the sequence $\{X_t^{(Y)}\}_{t=1}^T$ is the same as X_1^L when all deletion (D) states have been removed.) The modified Viterbi can be written as

$$\delta_i(t, u) = \max_{j,d,e} \left\{ \delta_j(t - d, u - e) \, a_{ji} \, b_i(Y_{t-d+1}^t, Z_{u-e+1}^u | Y_1^{t-d}, Z_1^{u-e}) \, \mathbb{I}\{X_t^{(Y)} = i\} \right\}$$

$$(4.22)$$

where $\mathbb{I}$ is the indicator function and $(d, e) \in \{(1, 1), (1, 0), (0, 1)\}$ indicating a match, an insertion, and a deletion in the alignment in sequence Y, respectively. Thus, the only difference in the Viterbi formulation is the factor $\mathbb{I}\{X_t^{(Y)} = i\}$, which is added to ensure that we stay within the path dictated by the gene annotation of Y. All paths violating this restriction get probability zero. The result of the model is then the corresponding gene prediction in Z along with an alignment of the two sequences (see Fig. 4.4).

Projector and DoubleScan use position-dependent transition probabilities and utilizes an external program, called StrataSplice [23], to capture candidate splice sites

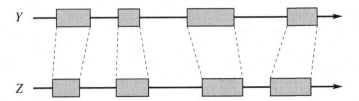

Fig. 4.4 The gene prediction in Projector of the target sequence is produced within the bounds of the known annotation in the informant sequence

in the sequence. This is done in a preprocessing step, where StrataSplice processes each sequence separately and assigns a score to each potential donor, acceptor and translation start site. The chosen splice sites are then used both to reduce the search space of the dynamic programming algorithm, and to modify the transition probabilities into translation starts or splice sites in the HMM. If the signal score of the potential exon boundary is high, the transition probability is unmodified, while if the signal score is low, the transition probability into that site is reduced.

4.5.2 GeneMapper—Reference-Based Annotation

A gene mapping software that falls into the same category as Projector is the tool GeneMapper [10]. Similar to Projector, GeneMapper takes an already annotated genomic sequence and maps the coding exons to a homologous target sequence. The pairwise version uses dynamic programming to map the exons, while a multiple species version of the program works with profiles, similar to protein profiles (see Sect. 3.2.9), which enables the mapping over fairly large evolutionary distances. Both GeneMapper and Projector can only work properly on conserved gene structures, however. Deletions or insertions of whole exons in either sequence will complicate the prediction task and compromise the results.

While Projector uses a heuristic threshold of the splice sites scores to reduce the search space of the algorithm, GeneMapper uses a subroutine called ExonAligner to map the annotated exons onto the target sequence. ExonAligner implements a Smith-Waterman-like approach (see Sect. 3.1.6) to locally align each exon to the target sequence. The alignment is constrained by the location of the potential splice sites, which, like in Projector, are scored using StrataSplice [23]. The dynamic programming algorithm in GeneMapper allows for both single nucleotide sequencing errors as well as frameshifts during mapping, but since these events are expected to be rare, the transitions into such events are heavily penalized. The procedure of constructing a reference annotation in GeneMapper can be summarized as follows:

1. Only the most conserved exons are mapped onto the target sequence:

    (i) The approximate location of the exon in the target genome is found using tBLAST and extending the best hit (only significant hits are used).
    (ii) The exact ortholog in the target sequence is predicted using ExonAligner.
    (iii) The exon alignment is tested for significance using a likelihood ratio test, and only the most highly conserved exons are let through.

    These mappings are then used to provide an outline for the entire gene structure.
2. Less conserved exons are now mapped using tBLAST and ExonAligner as above, but by being restricted by the already mapped exons.
3. Events of exon splitting and exon fusion are searched for:

    (i) Introns are required to have a minimum length, thus exons separated by a too short intron are fused together.
    (ii) Exon alignments with gaps greater than the minimum intron length and with splice sites at the gap ends are split into two.

There is a multiple species version of GeneMapper, in which the annotated gene is mapped onto multiple target sequences using profiles. The gene gets mapped onto the target sequences according to evolutionary distance, starting with the genome closest to the annotated sequence, and updating the profile after each mapping. By using profiles in this manner, genes can be mapped between fairly distant sequences, as long as the gene structure is conserved (still allowing for exon splitting and fusion).

4.6 Multiple Sequence Gene Finding

With the ever-increasing amount of sequence data available, the need for more efficient and more accurate methods to analyze them is larger than ever. Increasing the number of species in a sequence comparison naturally improves the gene finding accuracy further. Moreover, with the aid of several annotated genomes at various evolutionary distances, more and more remotely related sequences can be characterized. However, the availability of computer softwares that manage to utilize the immense flow of data is still lacking. There are several possible explanations to this. One is that until very recently, not that many sequences were available. The sequencing development has exploded during the last decade. Another explanation is the that the computational complexity grows fast with every new sequence added. For instance, the SLAM method presented in Sect. 4.4.2 becomes infeasible already for three sequences. A third explanation is that the problem of multiple alignments is still a difficult one. The more distantly related sequences that are compared, the more problems arise with genome rearrangements such as insertions and deletions, sequence transitions and inversions. With that said, methods are emerging that attempt to utilize multiple sequences in the gene finding to improve the prediction accuracy. One such method, presented next, is N-SCAN [14], which uses a multiple alignment as reference when annotating a related, but uncharacterized sequence.

4.6.1 N-SCAN: A Multiple Informant-Based Gene Finder

The program *N-SCAN* [14] is a further development of Twinscan [19] described in
Sect. 4.1.2, which in turn is an extension of the GHMM-based gene finder Genscan [9]
described in Sect. 2.2.4. While Genscan is a single species gene finder, Twinscan
uses a homologous *informant* sequence to boost or decrease the probabilities of the
various gene components. N-SCAN is a further extension of this, where, instead of
using a single informant sequence, the target sequence is compared to a multiple
alignment of evolutionary related informant sequences. N-SCAN only annotates one
target sequence at a time, however.

We let Z_1^T denote the target sequence as before, and now we let $\mathbf{Y}_t = \{Y_t^{(1)}, \ldots,$
$Y_t^{(N)}\}$, $t = 1, \ldots, T$ denote the column at position t in the multiple alignment of
the N informant sequences. We assume that we use a kth-order model for the target
sequence and an mth-order model for the informant sequences. Then the probability
of outputting a column in the alignment between the informant sequences and the
target sequence, given the previous k bases in Z and the previous m bases in the
informants, is given by

$$\mathbf{P}(\mathbf{Y}_t, Z_t | \mathbf{Y}_{t-1}, \ldots, \mathbf{Y}_{t-m}, Z_{t-1}, \ldots, Z_{t-k}) =$$
$$= \mathbf{P}(\mathbf{Y}_t | \mathbf{Y}_{t-1}, \ldots, \mathbf{Y}_{t-m}, Z_{t-1}, \ldots, Z_{t-k}) \, \mathbf{P}(Z_t | Z_{t-1}, \ldots, Z_{t-k}). \quad (4.23)$$

This equality comes from the assumption that $\mathbf{Y}_t$ is independent of Z_t, and that
Z_t given the previous k bases is independent of the m positions of the informant
sequences. The second part $\mathbf{P}(Z_t | Z_{t-1}, \ldots, Z_{t-k})$ of (4.23) is calculated as for a nor-
mal GHMM, while the first part $\mathbf{P}(\mathbf{Y}_t | \mathbf{Y}_{t-1}, \ldots, \mathbf{Y}_{t-m} Z_{t-1}, \ldots, Z_{t-k})$ is computed
by N-SCAN using Bayesian networks similar to that in [33]. A brief introduction to
Bayesian networks is given in Sect. 5.4.7 in the framework of modeling splice sites.

Consider a phylogenetic tree, where the leaf nodes represent sequences from
existing species, and the internal nodes represent common ancestors to the branch-
ing subtrees. Such a tree can be seen as a Bayesian network of the probability distribution of the columns in the multiple alignment. The nodes then represent the
specific residues in the different rows of a column, and the edges correspond to
the dependencies between the nodes. Moreover, given the state of an ancestral node,
the probability distributions of the child nodes are conditionally independent of the
rest of the tree. Thus, the joint probability of a column can be factorized as

$$\mathbf{P}(\mathbf{Y}_t) = \mathbf{P}(Y_t^{(1)}, \ldots, Y_t^{(N)}) = \prod_{i=1}^{n} \mathbf{P}(Y_t^{(i)} | \mathbf{Y}_{\text{pa}(i)}) \quad (4.24)$$

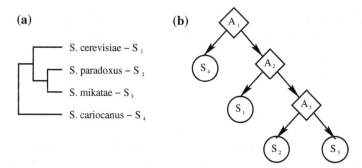

Fig. 4.5 **a** A phylogenetic tree of sequences from *S. cerevisiae* (S_1), *S. paradoxus* (S_2), *S. mikatae* (S_3), and *S.cariocanus* (S_4). **b** The phylogenetic tree in (**a**) seen as a Bayesian network with ancestral nodes A_1, A_2, and A_3

where $\mathbf{Y}_{\mathrm{pa}(i)}$ denotes the parent nodes of $Y_t^{(i)}$. For instance, assume that we have homologous sequences from the four Saccharomyces species, *S. cerevisiae* (S_1), *S. paradoxus* (S_2), *S. mikatae* (S_3), and *S. cariocanus* (S_4) as in Fig. 4.5, with internal ancestral nodes A_1, A_2, and A_3. The Bayesian network corresponding to this tree could then be factorized as

$$\mathbf{P}(S_1, S_2, S_3, S_4, A_1, A_2, A_3) =$$
$$= \mathbf{P}(A_1)\,\mathbf{P}(S_4|A_1)\,\mathbf{P}(A_2|A_1)\,\mathbf{P}(S_1|A_2)\,\mathbf{P}(A_3|A_2)\,\mathbf{P}(S_2|A_3)\,\mathbf{P}(S_3|A_3). \quad (4.25)$$

The real multiple alignment only consists of the sequences in the leaf nodes, and during the training process the ancestral sequences are therefore treated as missing data. However, instead of using the phylogenetic tree directly, N-SCAN transforms it to make the target sequence the root node. This is done by simply reversing the edges between the root and the target node. Assume for instance that *S. cerevisiae* (S_1) is the target in Fig. 4.5. By reversing the edges between the root node and the S_1 leaf node, we get a new tree, as given in Fig. 4.6. The new tree represents the same probability distribution, but the factorization now becomes

$$\mathbf{P}(S_1, S_2, S_3, S_4, A_1, A_2, A_3) =$$
$$= \mathbf{P}(S_1)\,\mathbf{P}(A_2|S_1)\,\mathbf{P}(A_1|A_2)\,\mathbf{P}(A_3|A_2)\,\mathbf{P}(S_4|A_1)\,\mathbf{P}(S_2|A_3)\,\mathbf{P}(S_3|A_3). \quad (4.26)$$

The purpose of this transformation is to be able to factor out the target sequence S_1, without conditioning it on any other sequence. Finally, any ancestral node with only one child is removed and the edge goes directly from the parent to the child of the removed node. In Fig. 4.6 this means that node A_2 is removed and an edge is drawn from A_1 to S_4 directly. Since A_1 represents an unknown ancestral sequence this will not affect the results of the model.

The Bayesian network described above represents a model where the columns of the multiple alignment are assumed to be independent. The model can be extended to higher orders, though, following the recipe in [33]. Instead of treating each node

Fig. 4.6 The transformed
tree achieved by reversing
the edges between the root
node and the S_1-node in the
tree in Fig. 4.5

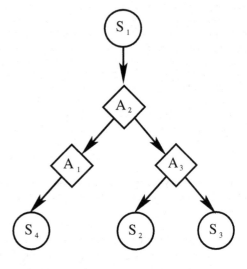

in the network as a univariate random variable, the nodes now represent a model of
order m, with each random variable depending on the previous m positions in the
multiple alignment.

References

1. Alexandersson, M., Cawley, S., Pachter, L.: SLAM: cross-species gene finding and alignment
   with a generalized pair hidden Markov model. Genome Res. **13**, 496–502 (2003)
2. Altschul, S.F., Gish, W., Miller, W., Myers, E.M., Lipman, D.J.: Basic local alignment search
   tool. J. Mol. Biol. **215**, 403–410 (1990)
3. Ansari-Lari, M.A., Oeltjen, J.C., Schwartz, S., Zhang, Z., Muzny, D.M., Lu, J., Gorrell, J.H.,
   Chinault, A.C., Belmont, J.W., Miller, W., Gibbs, R.A.: Comparative sequence analysis of a
   gene-rich cluster at human chromosome 12p13 and its syntenic region in mouse chromosome
   6. Genome Res. **8**, 29–40 (1998)
4. Bafna, V., Huson, D.H.: The conserved exon method for gene finding. Proc. Int. Conf. Intell.
   Syst. Mol. Biol. **8**, 3–12 (2000)
5. Batzoglou, S., Pachter, L., Mesirov, J., Berger, B., Lander, E.S.: Human and mouse gene
   structure: comparative analysis and application to exon prediction. Genome Res. **10**, 950–958
   (2000)
6. Birney, E., Clamp, M., Durbin, R.: Genewise and genomewise. Genome Res. **14**, 988–995
   (2004)
7. Burge, C.B.: Modeling dependencies in pre-mRNA splicing signals. In: Salzberg, S.L., Searls,
   D.B., Kasif, S. (eds.) Computational Methods in Molecular Biology, pp. 109–128. Elsevier
   Science B.V. (1998)
8. Burge, C.: Identification of genes in human genomic DNA. Ph.D. thesis, Stanford University,
   Stanford CA (1997)
9. Burge, C., Karlin, S.: Prediction of complete gene structures in human genomic DNA. J. Mol.
   Biol. **268**, 78–94 (1997)
10. Chatterji, S., Pachter, L.: Reference based annotation with GeneMapper. Genome Biol. **7**, R29
    (2006)

11. Dewey, C., Wu, J.Q., Cawley, S., Alexandersson, M., Gibbs, R., Pachter, L.: Accurate identification of novel human genes through simultaneous gene prediction in human, mouse, and rat. Genome Res. **14**, 661–664 (2004)
12. Gelfand, M.S., Mironov, A.A., Pevzner, P.A.: Gene recognition via spliced sequence alignment. Proc. Natl. Acad. Sci. USA **93**, 9061–9066 (1996)
13. Gish, W., States, D.J.: Identification of protein coding regions by database similarity search. Nat. Genet. **3**, 266–272 (1993)
14. Gross, S.S., Brent, M.R.: Using multiple alignments to improve gene prediction. J. Comput. Biol. **13**, 379–393 (2006)
15. Hardison, R.C., Oeltjen, J., Miller, W.: Long human-mouse sequence alignments reveal novel regulatory elements: a reason to sequence the mouse genome. Genome Res. **7**, 959–966 (1997)
16. Hirschberg, D.S.: A linear space algorithm for the computing maximal common subsequences. Comm. ACM **18**, 341–343 (1975)
17. Kent, W.J.: BLAT—the BLAST-like alignment tool. Genome Res. **12**, 656–664 (2002)
18. Kim, N., Shin, S., Lee, S.: ECgene: genome-based EST clustering and gene modeling for alternative splicing. Genome Res. **15**, 566–576 (2005)
19. Korf, I., Flicek, P., Duan, D., Brent, M.R.: Integrating genomic homology into gene structure prediction. Bioinformatics **17**, S140–S148 (2001)
20. Krogh, A.: Using database matches with HMMGene for automated gene detection in drosophila. Genome Res. **10**, 523–528 (2000)
21. Kulp, D., Haussler, D., Reese, M.G., Eeckman, F.H.: A generalized hidden Markov model for the recognition of human genes in DNA. Proc. Int. Conf. Intell. Syst. Mol. Biol. **4**, 134–142 (1996)
22. Kulp, D., Haussler, D., Reese, M.G., Eeckman, F.H.: Integrating database homology in a probabilistic gene structure model. Pac. Symp. Biocomput. **2**, 232–244 (1997)
23. Levine, A.: StrataSplice at http://www.sanger.ac.uk/Software/analysis/stratasplice/
24. Meyer, I.M., Durbin, R.: Comparative ab initio prediction of gene structures using pair HMMs. Bioinformatics **18**, 1309–1318 (2002)
25. Meyer, I.M., Durbin, R.: Gene structure conservation aids similarity based gene prediction. Nucleic Acids Res. **32**, 776–783 (2004)
26. Mouse Genome Sequencing Consortium: Initial sequencing and comparative analysis of the mouse genome. Nature **420**, 520–562 (2002)
27. Needleman, S.B., Wunsch, C.D.: A general method applicable to the search for similarities in the amino acid sequence of two proteins. J. Mol. Biol. **48**, 443–453 (1970)
28. Pachter, L., Alexandersson, M., Cawley, S.: Applications of generalized pair hidden Markov models to alignment and gene finding problems. J. Comput. Biol. **9**, 389–399 (2002)
29. Pachter, L., Batzoglou, S., Spitkovsky, V.I., Banks, E., Lander, E.S., Kleitman, D.J., Berger, B.: A dictionary based approach for gene annotation. J. Comput. Biol. **6**, 419–430 (1999)
30. Parra, G., Agarwal, P., Abril, J.F., Wiehe, T., Fickett, J.W., Guigó, R.: Comparative gene prediction in human and mouse. Genome Res. **13**, 108–117 (2003)
31. Rat Genome Sequencing Consortium: Genome sequence of the Brown Norway rat yields insights into mammalian evolution. Nature **428**, 493–521 (2004)
32. Schwartz, S., Zhang, Z., Frazer, K.A., Smit, A., Riemer, C., Bouck, J., Gibbs, R., Hardison, R., Miller, W.: PipMaker—a web server for aligning two genomic DNA sequences. Genome Res. **10**, 577–586 (2000)
33. Siepel, A., Haussler, D.: Phylogenetic estimation of context-dependent substitution rates by maximum likelihood. Mol. Biol. Evol. **21**, 468–488 (2004)
34. Smit, A.F.A., Hubley, R., Green, P.: RepeatMasker at http://www.repeatmasker.org
35. Snyder, E.E., Stormo, G.D.: Identification of protein coding regions in genomic DNA. J. Mol. Biol. **248**, 1–18 (1995)
36. Wu, T.D., Watanabe, C.K.: GMAP: a genomic mapping and alignment program for mRNA and EST sequences. Bioinformatics **21**, 1859–1875 (2005)
37. Xu, Y., Mural, R.J., Einstein, J.R., Shah, M.B., Uberbacher, E.C.: GRAIL: a multi-agent neural network system for gene identification. Proc. IEEE **84**, 1544–1552 (1996)

38. Xu, Y., Uberbacher, E.C.: In: Salzberg, S.L., Searls, D.B., Kasif, S. (eds.) Computational Methods in Molecular Biology, pp. 109–128. Elsevier Science B.V. (1998)
39. Yeh, R.F., Lim, L.P., Burge, C.B.: Computational inference of homologous gene structures in the human genome. Genome Res. **11**, 803–816 (2001)

Chapter 5
Gene Structure Submodels

A gene model algorithm integrates a wide range of scores, or signals, coming from the ingoing states of the model. These states are themselves complex submodels, which incorporate a number of *sensors* used to score the different characteristics of the submodel. Such sensors are traditionally divided into two groups: *content* sensors and *signal* sensors. Signal sensors model the transition between states, and attempt to detect the boundaries between exons and introns in the sequence, while content sensors score the content of a candidate region, such as the base composition or length distribution of a candidate exon or intron. In this chapter we describe some of the main submodels used in gene finding algorithms, and detail a number of different methods for the integrating the sensors the submodels incorporate. There are several different types of genes, exhibiting various kinds of characteristics, but in what follows we focus our attention on protein-coding genes.

5.1 The State Space

The main states of a gene finding model are the *exons*, *introns*, and *intergene* states. The intergene state corresponds to the long stretches of sequence in between genes, while the exons and introns constitute the main gene components of the protein-coding portion of the gene. The sequences surrounding the boundaries between exons and introns are called *splice sites*, with the *donor* site residing at the beginning of an intron, and the *acceptor* site at the end. Although being of utmost importance for the exact prediction of exons, the splice sites are typically not represented as separate states in the gene model, but are usually included in the exon submodel.

In addition to the three main types of states, a gene finding model can include various types of *regulatory* states. The *untranslated regions* (UTRs) harbor important *binding sites*, and are located right before the beginning and right after the end of the coding portion of the gene. We refer to them as the 5' and the 3' UTRs according to their location, respectively, with 5'UTR signifying the upstream region and 3'UTR

© Springer-Verlag London 2015
M. Axelson-Fisk, *Comparative Gene Finding*, Computational Biology 20,
DOI 10.1007/978-1-4471-6693-1_5

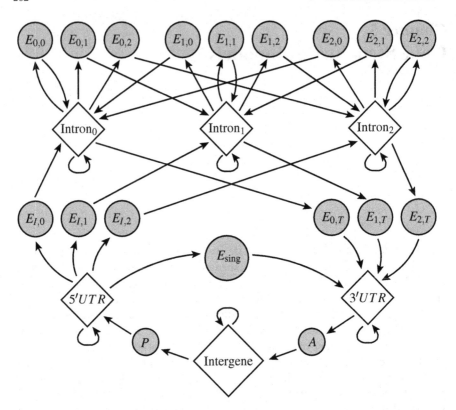

Fig. 5.1 The main states of a gene finding model: intergene, exons (E), introns, UTRs, promoter (P), and polyA-signal (A)

the downstream region of the gene. The UTRs are typically structured into exons and introns as well, but these exons are referred to as *noncoding*, as they are not translated into protein. The UTRs are typically not included in single species gene finding algorithms, mainly because their signal is rather weak and hard to detect, but also because not enough training data has been available in the past. In a cross-species setting, however, noncoding exons are often conserved to some degree, and tend to disturb the gene prediction if excluded from the model. Other common regulatory states include the *promoter* (P) and the *polyA-signal* (A). Although these elements correspond to very short sequence motifs, which makes them difficult to detect in a larger genomic sequence, their inclusion in a gene model may improve the prediction of the beginning and end of the gene. Figure 5.1 illustrates a state space incorporating the gene features that we discuss in this chapter.

5.1.1 The Exon States

There are four different types of protein-coding exons: single, initial, internal, and terminal exons. The different types exhibit fairly different properties, and benefit from separate parameter sets for their content sensors. An exon has both a *phase* and a *frame*. The frame refers to the reading frame with respect to the beginning of the sequence, such that an exon beginning at coordinate k in the sequence, is of frame (k mod 3). The phase has to do with the codon periodicity, and in a sense it is the introns that have the phase assignation, rather than the exons. An intron of phase 0 splits the exons right between two codons, while phase 1 introns splits a codon between the 1st and the 2nd base, and phase 2 introns between the 2nd and the 3rd codon position (see Fig. 5.2).

The subscripts of the exons in Fig. 5.1 indicate the exon type and the phase. A single exon is denoted E_{sing} and corresponds to an exon that begins in a start codon and ends in a stop codon. $E_{I,j}$ denotes an initial exon that is bounded by a start codon and an donor splice site, and ends with $j = 0, 1, 2$ extra bases after the last codon, to be completed in the beginning of the next exon. $E_{i,j}$ denotes an internal exon that is bounded by an acceptor and a donor site, and that has to finish the last codon of the previous exon with $[(3 - i) \bmod 3]$ extra bases, and ends with j extra bases after its last codon. Similarly, $E_{i,T}$ denotes a terminal exon, bounded by an acceptor site and a stop codon, and that begins by finishing off the last codon with $[(3 - i) \bmod 3]$ extra bases. The introns intervening the exons inherit the phase of the preceding exon and passes it on to the next. In this manner, by including a bunch of extra states that keep track of the phase, the process can be of first order, rather than carrying around memory about previous exons. Table 5.1 lists the boundaries connected with each exon type.

The characteristics of an exon, modeled in an exon submodel, include the codon (or dicodon) composition, the state length distribution, and the boundary models. The codon composition is a rather strong indicator for coding potential alone, as the sequence pattern within coding exons differ significantly from that of noncoding sequence. The prediction of the exact exon boundaries gets fuzzy, however, if codon composition were to be the only indicator. The inclusion of a generalized length distribution and good boundary models strengthen the exon signal and improves the prediction accuracy of the gene finding algorithm.

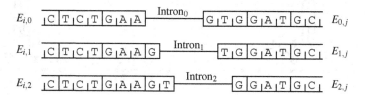

Fig. 5.2 Illustrating the notion of exon and intron phase. Intron$_j$ comes between exon $E_{i,j}$ and exon $E_{j,k}$, $j = 0, 1, 2$

Table 5.1 The boundaries connected to the four exon types

Exon	Notation	Left bdy (5′)	Right bdy (3′)
Single	E_{sing}	Start codon	Stop codon
Initial	$E_{I,j}$	Start codon	Donor site
Internal	$E_{i,j}$	Acceptor site	Donor site
Terminal	$E_{i,T}$	Acceptor site	Stop codon

5.1.2 Splice Sites

After the gene sequence has been transcribed into an RNA molecule, and before it is translated into protein, the introns are cut out of the RNA transcript. The cleavage process is called *splicing*, and the location of the exon–intron cuts are called *splice sites*. These sites are physical stretches of sequences exhibiting specific characteristics that enable the spliceosome to recognize and attach to it. The splice site at the beginning of an intron (5′ end) is called the *donor* site, and the one at the end (3′ end) is called the *acceptor* site (see Fig. 5.3).

Splice sites have a very specific sequence composition that can be illustrated for instance using *sequence logos* [56], such as in Fig. 5.4. Sequence logos are created from multiple alignments of sequence samples of the site in question, and show the level of conservation, or the level of *information*, to be found in each sequence position of the site. In a sequence logo, the residues observed in each column of the multiple alignment are stacked on top of each other in increasing order, with the

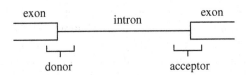

Fig. 5.3 The donor site is located at the exon–intron boundary at the beginning of an intron, and the acceptor at the intron–exon boundary at the end of the intron

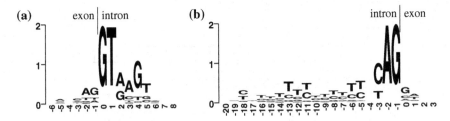

Fig. 5.4 Sequence logos of **a** human donor sites, and **b** human acceptor sites. The height of the letter in position t is given by $R_{seq}(t)$ in (5.2)

most frequent residue on top. The level of information in each position t is computed using the *Shannon entropy* $H(t)$, given by

$$H(t) = - \sum_{v \in V} f_t(v) \log_2 f_t(v) \tag{5.1}$$

where V is the set of possible residues in the alphabet in question, and $f_t(v)$ is the frequency of residue $v \in V$ at position t. The Shannon entropy measures the information content in "bits," using log base 2. For DNA sequences, with four possible residues in the alphabet ($V = \{A, C, G, T\}$), $H(t)$ takes values between 0 and 2, where $H(t) = 2$ is the maximum uncertainty indicating a uniform, or completely random, distribution over the residues, and $H(t) = 0$ indicates complete certainty with $f_t(v) = 1$ for one of the residues in V and 0 for all the others.

The height of a sequence logo indicates the amount of information available in each position, and is for DNA sequences calculated using

$$R_{seq}(t) = 2 - H(t). \tag{5.2}$$

Figure 5.4 illustrates sequence logos of human donor and acceptor sites, generated by the WebLogo software [24]. Directly we see the very characteristic invariant consensus dinucleotides 'GT' in donors and 'AT' in acceptors, appearing in the first two and the last two positions of the intron, respectively. In addition, we see that the positions surrounding these dinucleotides contain some level of information, but that the information content degrades rapidly as we move away from these positions. The aim of splice site models is to capture such sequence characteristics and position dependencies present in the splice site signals. As mentioned in Sect. 5.1.1, splice sites are typically not modeled as separate states, but are often included as part of the exon states, although their sequence extend into the adjacent intron as well.

5.1.3 Introns and Intergenic Regions

The *intergenic* and *intragenic* (*intron*) regions comprise an overwhelming portion of the genomes of higher organisms (e.g. over 95 % of the human genome). The intergenes are the long stretches of DNA located between genes, while the introns are the noncoding sequences separating the exons of a multi-exon gene. Introns and intergenic sequences have often been referred to as "junk-DNA" in the past [48], as there were no known function of these regions. This term is becoming outdated, however, as there is increasing indications that both intergenes and introns serve particular purposes, both regarding structure and regulation of the genome [7, 30].

While it may seem natural to denote the sequence separating two adjacent genes on the same strand as intergene, note that the intergene state in Fig. 5.1 is common to both strands, and can in our definition be interrupted by genes on *either strand* (see Fig. 5.5). In higher organisms, the intergenic stretches can be very long, and while

Fig. 5.5 We define an intergene as the sequence between two adjacent genes, regardless of the strand assignation of these genes

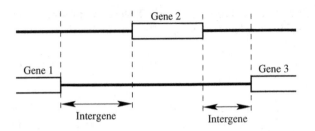

they contain various regulatory elements, they are for the most part long stretches of sequence of unknown function.

While rare or nonexistent in most prokaryotes, introns are common in eukaryotic genes. Besides harboring various regulatory elements, it is becoming more and more evident that the introns themselves have both structural and regulatory purposes, and are vital to the viability of the cells. In prokaryotes introns are mainly found in the tRNA (transfer) and rRNA (ribosome) coding genes. One possible explanation for this is that for smaller organisms such as bacteria, an efficient reproduction is key, and a streamlined genome is therefore more valuable than whatever added functions introns provide.

Introns are bounded by a donor and an acceptor splice site, and almost all introns begin and end with the consensus dinucleotides GT and AG, respectively (see the next section for more on this). In order to carry the phase information between exons, a common trick is to include numerous intron (and exon) states in the state space. In Fig. 5.1 the index of an Intron$_i$ state indicates that the previous exon generated $i = 0, 1, 2$ extra bases after its last completed codon. As a result, the succeeding exon needs to begin with $j = [(3 - i) \bmod 3]$ extra bases, respectively, before producing new codons.

The characteristics modeled in intron and intergene submodels are typically sequence composition and length distributions. While the position dependencies are much less than in exons, the intron and intergene sequences still show indication of structure, and the sequence model used for instance in SLAM [2] is a first-order, homogeneous Markov chain. The length distribution used is typically the geometric. Although the sheer length of these regions make any other model infeasible, the geometric distribution is in fact a reasonable length model for most eukaryotic noncoding regions.

5.1.4 Untranslated Regions (UTRs)

As described in Sect. 1.1 a protein-coding gene is expressed in several steps, the two main steps being the *transcription* and *translation*. In the transcription step, an RNA molecule is synthesized, using the DNA sequence between the transcription start and the transcription end of the gene as a template. The transcribed RNA molecule, also known as the *primary transcript*, consists of the protein-coding core of the gene, surrounded by *untranslated regions* (UTRs) both before and after. The 5'UTR is located upstream the core and extends from the transcription start to right before the start codon, and the 3'UTR extends downstream of the core from right after the stop codon to the transcription end. Before translation the introns are spliced out from the primary transcript, and the resulting mRNA molecule consists only of the UTRs and the coding exons. The UTRs are involved in the regulation and stabilization during translation, but are not part of the final protein product. The 5'UTR contains various regulatory binding sites, most notably the ribosomal binding site described in Sect. 5.3.2. The 3'UTR contains several regulatory sequences, such as the polyA-signal which is involved in preparing the RNA transcript for translation.

Just as the coding portion of the gene, the UTRs are structured into exons and introns as well, and undergo splicing between the transcription and the translation steps. The UTR exons are referred to as *noncoding* as they are not translated into protein during translation. UTRs are often excluded in single species gene finding models, as they are difficult to detect. Although the sequence composition is more structured than in intergenic sequences, it varies greatly, and, since the sequence is not translated, there is no clear codon structure or frame consistency to look for. On the other hand, when it comes to comparative gene finding it has been found that including the UTRs in some form is of vital importance to the accuracy of the model. It turns out that the UTR exons tend to be significantly conserved between species, and ignoring them will greatly compromise the gene prediction. As an example, the SLAM model [2], described in Sect. 4.4.2 includes a *conserved noncoding sequence* (CNS) state as part of the intergenic region, that accounts for any significantly conserved region other than coding. The CNS model in SLAM is a simple pair HMM (see Sect. 3.1.7), only characterized by the sequence composition and the pairwise similarity, while the state length is still allowed to be geometric. Although such a simple model, it improves the gene finding accuracy of SLAM vastly, as it enables the algorithm to distinguish between true exons and exon-similar conserved elements.

5.1.5 Promoters and PolyA-Signals

The promoter is located at the transcription start site (TSS) of a gene, and constitutes the docking site for the complex of enzymes and transcription factors needed to initiate transcription. Besides the TSS, the promoter contains a TATA-box, located about 30 bp upstream of the TSS, and a variety of regulatory binding sites for transcription modulators such as *enhancers* and *silencers* that boost or decrease the expression level of the gene, respectively. Identification of the promoter region is of huge interest, not only in order to direct gene prediction, but also to identify the regulatory regions of the gene. These regions hold the secret to how transcription occurs, what triggers it, and under what circumstances the gene is activated.

The problem of promoter prediction is a difficult one, however, complicated by numerous factors. For instance, there is no clear consensus sequence to go by. The binding sites typically follow position-specific patterns, but are too short to identify individually. Searching for whole groups of binding sites at once is a possible way around this problem, but since the regulatory machinery varies greatly between genes, this is not an easy task either. As regards to the TATA-box, some promoters have it and some do not. Some promoters occur in a CpG-island while some are not connected to such patterns at all. In addition, there is no one-to-one correspondence between the gene and the promoter. A gene can have several promoters, containing several TSSs each. The TSSs (in eukaryotes) can be located all over the gene, both upstream and downstream of the coding region, and in both exons and intron. Moreover, genes can share promoters, and promoter regions may overlap one another.

Identifying promoters as stand-alone objects is thus very challenging. However, by invoking the promoter as a submodel in a complete gene structure, and predicting it as part of that structure, has the potential of improving both the promoter prediction and the gene prediction. A wide range of methods have been applied to the modeling and classification of promoter sequences, including weight matrices, Markov chains, neural networks, k-tuple frequency models, interpolated Markov chains, and discriminant analysis models. These methods are described in the context of splice site detection in Sect. 5.1.2, but the methodology can easily be modified to model promoters in the same fashion.

5.2 State Length Distributions

Recall from Sect. 2.1.1 that the *duration* of a state in a standard Markov process, is the number of self-transitions into that state before leaving. In a first-order Markov chain, or in any other "memoryless" random process, the state duration follows a *geometric distribution*. This distribution arises naturally, and is the least expensive computationally, as no memory of the number of self-transitions made so far has to be kept. The geometric distribution is not always a good model for certain types of states, however, and forcing it onto a state duration when it is a bad fit will compromise prediction accuracy. Typically, the geometric distribution works well on gene features such as introns and intergenes, where there are rather few functional constraints on the lengths. This is rather fortunate since, due to the vast lengths of these features, a more general model would be infeasible. Empirical data shows, however, that exons follow a length distribution that is significantly different from the geometric. Although much shorter, the variation in exon lengths is still considerable, and some care has to be taken when designing more general distributions.

We begin by giving a brief encounter of the properties of the geometric distribution before we present a few of the most common approaches to exon length modeling.

5.2.1 Geometric and Negative Binomial Lengths

The *geometric distribution* is a discrete distribution constructed from a (possibly unbounded) number of *Bernoulli trials*. A Bernoulli trial is a random experiment where the outcome is one of two possible values, such as "success" or "failure," or 0 or 1. For instance, if we flip a coin, the outcome "heads" may be considered a success and "tails" a failure (or vice versa). A Bernoulli distributed random variable X has probability mass function

$$\mathbf{P}(X = x) = \begin{cases} p & \text{if } x = 1, \\ 1 - p & \text{if } x = 0, \end{cases} \tag{5.3}$$

where $0 \leq p \leq 1$. A *Bernoulli process* consists of a series of independent Bernoulli trials, and the distribution of the number Y of trials up to and including the first success is geometrically distributed with parameter p. That is, if the first success occurs after $Y = k$ trials, there must have been $k - 1$ preceding failures. The geometric probability mass function (for $p > 0$) becomes

$$\mathbf{P}(Y = k) = (1 - p)^{k-1} p, \quad k = 1, 2, 3, \ldots \tag{5.4}$$

The expected value and the variance of the geometric distribution are given by

$$E[Y] = \frac{1}{p}, \quad \text{Var}(Y) = \frac{1 - p}{p^2}. \tag{5.5}$$

The distribution is called geometric, because the probabilities form a *geometric sequence*: $p, pq^2, pq^3, \ldots$ with $q = 1 - p$, and the probability mass function decays exponentially (see Fig. 5.6).

Fig. 5.6 The probability mass function of the geometric distribution decays exponentially

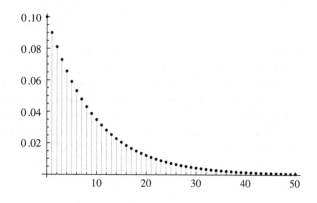

In terms of the duration of a state in memoryless random processes such as Markov chains, we are interested in the distribution of failures until the first success. Thus, the success is not included in the count, and this distribution is sometimes called a *shifted* geometric distribution. The probability mass function of the duration D of a memoryless state is thus

$$\mathbf{P}(D = k) = (1 - p)^k p, \quad k = 0, 1, 2, 3, \ldots \tag{5.6}$$

The variance is the same as for the standard geometric distribution, but the expected value becomes

$$E[D] = \frac{1 - p}{p}. \tag{5.7}$$

If m denotes the observed mean length of a specific gene feature, the parameter p can thus be estimated using

$$\hat{p} = \frac{1}{1 + m}. \tag{5.8}$$

A distribution related to the geometric is the *negative binomial* distribution. It is a generalization of the geometric distribution, and appears as the distribution of the number of Bernoulli trials needed to observe $r \geq 1$ successes. If Y denotes the number of trials up to and including the rth success, the probability mass function of the negative binomial distribution is given by

$$\mathbf{P}(Y = k) = \binom{k - 1}{r - 1} p^r (1 - p)^{k-r} \tag{5.9}$$

where

$$\binom{k - 1}{r - 1} = \frac{(k - 1)!}{(r - 1)!(k - r)!} \tag{5.10}$$

is the *binomial coefficient*. Note that with $r = 1$ the negative binomial distribution coincides with the geometric. While slightly more complex than the geometric distribution, the negative binomial is still fairly "cheap" computationally, since it can be implemented by linking r geometrically distributed states in sequence. By varying the number of states r, we can get varying shapes of the distribution, as illustrated in Fig. 5.7. A generalized way of implementing sequential geometric states to model state durations is used in *acyclic discrete phase-type distribution* described in Sect. 5.2.3.

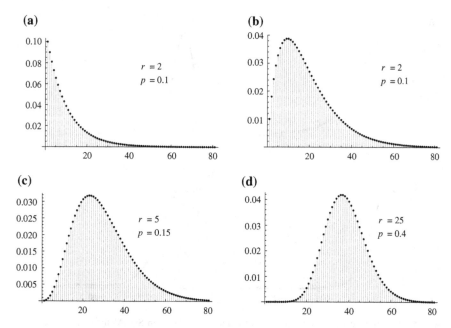

Fig. 5.7 Negative binomial density plots for the number of successes r and the probability of success p. With $r = 1$ we get the geometric distribution, while increasing r goes toward the normal distribution. **a** $r = 1$ and $p = 0.1$. **b** $r = 2$ and $p = 0.1$. **c** $r = 5$ and $p = 0.15$. **d** $r = 25$ and $p = 0.4$

5.2.2 Empirical Length Distributions

If $Y_1, \ldots, Y_n$ is an observed independent, identically distributed (i.i.d.) sample of a random variable Y with probability distribution $F_Y(y|\theta)$, where θ is a parameter characterizing the distribution, the *empirical distribution* of the sample is given by

$$F_n(y) = \frac{\#Y_i \leq y}{n} = \frac{1}{n} \sum_{i=1}^{n} \mathbb{I}(Y_i \leq y) \tag{5.11}$$

where $\mathbb{I}$ is the indicator function. The empirical distribution converges pointwise toward the true distribution and serves as a consistent, unbiased estimator of the distribution from which the sample was drawn. Thus, the length distribution of a gene feature can be estimated from a sample of such sequences. Figure 5.8 illustrates the empirical exon length distributions for human genes used in SLAM [2], with the exon length on the x-axis and the empirical probability on the y-axis. We see that these distributions clearly differ from the geometric, but also from each other with large differences in both mean and variance.

The advantage of using the empirical distribution over some characterized distribution is that no assumptions regarding the specific shape or behavior of the true

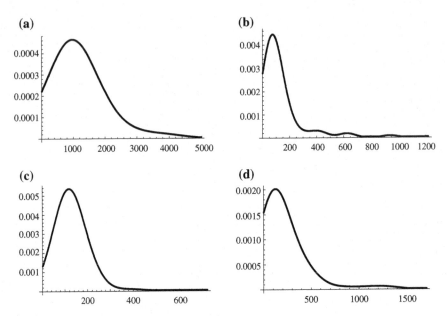

Fig. 5.8 The empirical length frequencies of human genes used in SLAM for **a** single exons, **b** initial exons, **c** internal exons, and **d** terminal exons

distribution has to be made. For a large enough sample, the empirical distribution will converge toward the true one, and will provide the best possible fit of the model to the training data. One disadvantage, however, is the added computational complexity. For instance, exon lengths vary from less than a hundred to several thousands of bases. Moreover, the empirical distribution is characterized simply by its sample, which means we need to store the probabilities of each possible duration. When searching for the optimal state path through a gene model we have to run through this long list of probabilities, something that is very expensive computationally and often forces the use of truncated distributions.

Another problem with using empirical distributions is the high dependence on the training data. For instance, just because a certain length is not represented in the given sample, it does not mean that the probability of that length is zero. Therefore, for small samples, one might want to smooth the empirical distribution in some manner. The approach taken by Burge in [13] is to simulate an evolution model of the exon lengths. Each exon length in the training sample is viewed as a representative of a whole population of exon lengths, diverging from the same ancestral exon. It can be argued (see [13]) that the length distribution of an observed exon of length k will be approximately normal with mean k and variance $2k$. This is used as motivation for the following smoothing procedure.

Assume we have an observed sample of exon lengths $\mathbf{n} = n_1, \ldots, n_m$, where n_k denotes the number of exons of length k, and m is some maximum length. Then instead of using the empirical densities $f_k = n_k/N$, where $N = \sum_{k=1}^{m} n_k$, the f_k is replaced by a "discretized" normal density with mean k and variance $2Ck/n_k m$, where C is a positive constant scaled to ensure a total mass of f_k. The division by n_k in the variance results in a distribution that is more smoothed in areas of sparse data, and that converges toward the unsmoothed empirical distribution as the training set increases.

5.2.3 Acyclic Discrete Phase-Type Distributions

A possible generalization of the geometric and the negative binomial distributions that still utilizes the efficiency of the geometric distribution, but that has more flexibility in its properties, is to link two or more geometrically distributed substates such as in Fig. 5.9. This can be done using *acyclic discrete phase-type distributions* (ADPH) [9], which for instance are used as state length distributions in the GHMM-based gene finder Agene [45].

Recall from Chap. 2 that a transient state is a state where the probability of returning is strictly less than 1. An *absorbing* state i is the opposite; the probability of exiting such a state is zero. Consider a discrete Markov chain, consisting of $m + 1$ states of which m states are transient and one is absorbing. The transition matrix of such a Markov chain can then be written as

$$A = \begin{pmatrix} A_0 & a \\ 0 & 1 \end{pmatrix} \tag{5.12}$$

where A_0 is an $(m \times m)$-matrix consisting of the transition probabilities between the transient states, a is an $(m \times 1)$-vector of transition probabilities from the transient states into the absorbing state, 0 a $(1 \times m)$-vector of zeros, and 1 is the probability of a self-transition in the absorbing state. Similarly, we let $\pi = (\Pi_0, \pi_{m+1})$ denote the initial probabilities of the states, where Π_0 is the vector of initial probabilities of the transient states and π_{m+1} is the initial probability of the absorbing state.

A *discrete phase-type distribution* (DPH) is the distribution of the time (the number of jumps) until the chain reaches the absorbing state. In this setting, the states are often referred to as *phases*. An *acyclic DPH* (ADPH) is a DPH where the underlying phases form an acyclic graph, such as in Fig. 5.11. As a result the transient phases can be ordered such that A_0 becomes an upper triangular matrix:

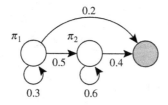

Fig. 5.9 An example of an ADPH with initial probabilities π_1 and π_2, and transition probabilities as given. Reprinted from [9] with permission from Elsevier

$$\mathbf{A_0} = \begin{pmatrix} a_{11} & a_{12} & a_{13} \\ 0 & a_{22} & a_{23} \\ 0 & 0 & a_{33} \end{pmatrix} \tag{5.13}$$

The possible transitions in each transient phase i is either a self-transition, or a transition to some phase $j > i$. As a result, the duration of each transient phase becomes geometrically distributed, with exit rate $1 - a_{ii}$, $i = 1, \ldots, m$.

If we let τ denote the time till absorption, we call τ a DPH random variable of order m and representation $(\Pi_0, \mathbf{A_0})$. This representation is in general nonunique, meaning that there might exist another representation defining the same distribution. Moreover, the matrix representation is very redundant, since there are $m^2 - 1$ free parameters to be estimated but only $2m - 1$ degrees of freedom in the DPH. It is therefore common to find a unique, minimal representation, called a *canonical* form of the DPH [9].

The time to absorption depends on the underlying path through the phase space. We define the *path* of the ADPH as the sequence of *unique* phases visited before absorption. That is, a phase is visited at most once in an ADPH, although the process can linger in that phase via a number of self-transitions. Such a path $v = (v_1, \ldots, v_n)$ then occurs with probability

$$\mathbf{P}(v) = \prod_{i=1}^{n} \frac{a_{v_i, v_{i+1}}}{1 - a_{v_i, v_i}} \tag{5.14}$$

The time to absorption τ can now be conveniently represented using the *generating function*

$$\mathscr{F}(z) = E[z^\tau] = \sum_{x} z^x \mathbf{P}(\tau = x). \tag{5.15}$$

More specifically, the generating function of τ via path v is given by

$$\mathscr{F}(z, v) = \prod_{i=1}^{n} \frac{(1 - a_{v_i, v_i})z}{1 - a_{v_i, v_i} z}. \tag{5.16}$$

If we let V_i denote all paths starting in state i, the generating function of τ, assuming that the initial state is i, is given by

$$\mathscr{F}_i(z) = \sum_{v \in V_i} \mathbf{P}(v)\mathscr{F}(z, v). \tag{5.17}$$

From this we get a very useful corollary [9].

Corollary 5.1 *The generating function of an ADPH is the mixture of the generating functions of its paths*

$$\mathscr{F}(z) = \sum_{i=1}^{m} \pi_i \sum_{v \in V_i} \mathbf{P}(v)\mathscr{F}(z, v). \tag{5.18}$$

The generating function of the paths of an ADPH can be decomposed further. First, we see in Corollary 5.1 that the generating function of τ does not depend on the order of the phases, and thus the phases can be reorganized such that the probabilities of a self-transition, i.e., the diagonal elements in $\mathbf{A}_0$, appear in decreasing order. Denote the ordered self-transition probabilities $q_1 \geq q_2 \geq \cdots \geq q_m$, and let $p_i = 1 - q_i$ denote the exit rate of phase i. The exit rates are thus ordered increasingly $p_1 \leq p_2 \leq \cdots \leq p_m$.

Now we can denote each path v as a binary vector over the ordered q_i's, with 1's for the phases visited by the path and 0's for the rest. This way each path has a unique representation, and a path of length k contains k 1's and $(m - k)$ 0's. We call a path of length k a *basic path* if it visits the k "fastest" phases, $q_{m-k+1}, \ldots, q_{m-1}, q_m$, and the corresponding binary vector, called the *basic vector*, thus contains $(m - k)$ initial 0's and k terminal 1's. With this representation we can express the following result.

Theorem 5.1 *The generating function of an ADPH is a mixture of the generating functions of its basic paths.*

As a direct consequence of Theorem 5.1 every ADPH can be uniquely represented as a mixture of basic paths. This leads to the simplest canonical form, which is when the only allowed jump in phase i, besides a self-transition, is a jump to the next phase $j = i + 1$, as in Fig. 5.10.

The transition probabilities of the transient phases can then be written as

$$\pi = (\pi_1, \ldots, \pi_m) \quad \mathbf{A}_0 = \begin{pmatrix} q_1 & p_1 & 0 & \cdots & 0 \\ 0 & q_2 & p_2 & \cdots & 0 \\ \vdots & \vdots & \vdots & & \vdots \\ 0 & 0 & 0 & \cdots & q_m \end{pmatrix} \tag{5.19}$$

Sometimes it is more convenient to have all the initial mass concentrated in one single phase, such that the initial phase is fixed. An alternative canonical form, leading to an equivalent distribution of the time τ to absorption, is then one where transitions

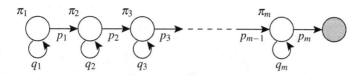

Fig. 5.10 In the simplest canonical form of an ADPH the only jumps allowed are a self-transition, or a transition to the next phase. Reprinted from [9] with permission from Elsevier

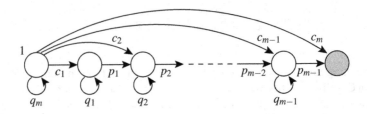

Fig. 5.11 A more convenient canonical form of an ADPH where the initial phase is fixed. Reprinted from [9] with permission from Elsevier

are possible from the initial phase to all other phases, including the absorbing phase, while the remaining transient phases still can only move to itself or to the next phase (see Fig. 5.11).

The ADPH representation is thus

$$\Pi_0 = (1, 0, 0, \ldots, 0) \quad \mathbf{A}_0 = \begin{pmatrix} q_m & c_1 & c_2 & \cdots & c_{m-1} \\ 0 & q_1 & p_1 & \cdots & 0 \\ \vdots & \vdots & \vdots & & \vdots \\ 0 & 0 & 0 & \cdots & q_{m-1} \end{pmatrix} \tag{5.20}$$

By introducing sufficiently many states, the ADPH can approximate any discrete (integer-valued) distribution. The trade-off is between computational complexity and the quality of the fit.

Example 5.1 Agene
Agene [45] is a GHMM-based, single species, eukaryotic gene finder that uses an ADPH as state length distribution for exons. Each ADPH consists of 15 phases that are fitted to an empirical length distribution, and the transition probabilities of the phases are estimated using the software PhFit [8].

The complexity of an ADPH depends on the number of fitted phases, but this number is constant in Agene, with the result that the computational complexity of the GHMM becomes linear in the sequence length. An example, borrowed from [45], of an ADPH and the corresponding length distribution plot is given in Fig. 5.12. The example corresponds to the special case where the length distribution is a mixture of a geometric and three negative binomial distributions.   □

Fig. 5.12 A length
distribution model in Agene

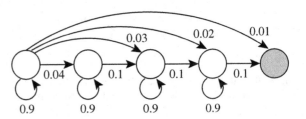

5.3 Sequence Content Sensors

Functional sequences are exposed to much higher evolutionary pressure than non-functional sequences, resulting in very different evolutionary rates. As a consequence there is a clearly distinguishable statistical bias between functional and nonfunctional sequences, and a wide variety of coding measures have been constructed with the aim to capture this bias. In a review of a large number of such measures [27], it was concluded that the most efficient coding measure, outranking a large number of more "sophisticated" measures, is to simply compare oligomer (k-tuple) counts in coding versus noncoding sequences. The distribution over the 64 different codons is significantly different in coding regions compared to noncoding regions, a feature that has proved to be a successful means to discriminate between coding and noncoding regions.

5.3.1 GC-Content Binning

Eukaryotic genomes tend to be a mosaic of *isochores*, which are long stretches of sequences (200–1000 kb) with fairly homogeneous sequence composition within, but with striking differences between different isochores [5, 6]. The isochores are typically divided into five groups characterized by the GC-content, signifying two 'light' groups (L), and three 'heavy' groups (H). The groupings vary slightly between references, but the Human Genome Project [39] used the following in their initial analysis of the human genome:

- L1: < 38 % GC-content.
- L2: 38 − 42 % GC-content.
- H1: 42 − 47 % GC-content.
- H2: 47 − 52 % GC-content.
- H3: > 52 % GC-content.

The structure of a genome, and in particular the density of its genes, varies significantly with the GC-content. For instance, although the H2+H3 groups together only make up about 12 % of the human genome, they harbor on the order of 54 % of all the genes [5]. Moreover, genes in different types of isochores exhibit clear differences in both structure and function, and ignoring these compositional variations would greatly affect the accuracy in gene prediction [60].

Gene model components such as base composition, length distributions, and transition probabilities are all affected by the GC-content surrounding the gene. For instance, generally the genes are more tightly packed in GC-rich regions, resulting particularly in shorter noncoding segments such as introns and intergenes. Exon lengths seem to be rather homogeneous over isochores, but instead the number of exons per gene are fewer, resulting in shorter gene products [39].

A novelty introduced in Genscan [15] was to "bin" the parameters according to the GC-content of the sequence to be analyzed. That is, based on the GC-content, four different parameter sets were produced: I (<43 % GC-content), II (43 − 51 %), III (51 − 57 %) and IV (>57 %). The first and second groups roughly corresponds to isochore groups L1+L2 and H1+H2 respectively, while group III and IV correspond to subsets of the most gene dense isochore group H3. One problem when splitting the parameter set is to decide where to draw the boundaries between different GC-groups in the input sequence, before applying the gene prediction algorithm. At the time when Genscan was constructed, the gene prediction task for the most part concerned sequences of up to *contig* length (100 kb). Thus, the parameter bin to be used was chosen based on a preprocessing step where the GC-content of the entire input sequence was determined. These days, when much larger quantities of contiguous sequences are considered, we need automatic means to determine isochore boundaries within an input sequence. The most hands-on approach is to move a sliding window of some length (e.g., 10–20 kb) across the sequence, register GC-content in each window, and then identify suitable sequence divisions. An automatized method for determining the sequence divisions is presented, for instance, in [49]. While it is not possible to test the accuracy of this algorithm, as only a few "true" isochore boundaries are known, the output seems more than satisfactory for the purpose of gene prediction.

5.3.2 Start Codon Recognition

Due to its very modest length, the start codon sequence appears spuriously in vast amounts throughout a genome, and even in when a strong coding potential indicates the corresponding exon, it might be difficult to locate the true translation start. A means to strengthen the signal of the translation start is to attempt to implement the consensus sequence of the ribosomal binding site occurring in direct vicinity of the start codon.

The start codon is typically ATG in eukaryotes, while in prokaryotes additional variations exist, such as GTG or TTG in *E. coli*, for instance. The ribosomal binding site (RBS) provides a signal for protein synthesis, and occurs about 6–7 bases upstream of the start codon in prokaryotic genes. Its consensus AGGAGG (or sometimes AGGAGGT), referred to as the *Shine-Dalgarno box* [59], can be used to strengthen the signal and facilitate the prediction of the translation start. The eukaryotic counterpart is called the *Kozak sequence* [37], with consensus sequence ACCATGG, or, more generally (GCC) RCCATGG where R stands for 'purine' (A or G) and the ATG is the actual start codon of the gene.

The authors of GeneMark.hmm [42] constructed a position-specific scoring matrix for the ribosome binding site in *E. coli* in order to sort out ambiguities of alternative start codons. They selected 325 *E. coli* genes that had annotated RBSs in Genbank, and collected the region between the 4 and 19 bp upstream of the start codon. The 16 bp subsequences were aligned, using simulated annealing (see Sect. 3.2.8), resulting in the consensus sequence AGGAG shown in Table 5.2. A scoring matrix, such as in

Table 5.2 Nucleotide frequencies for the RBS site in *E. coli*

Base	Position				
	1	2	3	4	5
A	**0.681**	0.105	0.015	**0.861**	0.164
C	0.077	0.037	0.012	0.025	0.046
G	0.077	**0.808**	**0.960**	0.043	**0.659**
T	0.161	0.050	0.012	0.071	0.115

Table 5.2, can be included in the gene finding model to boost the signal of potential start codons.

The inclusion of the Shine-Dalgarno or Kozak sequences strengthens the signal of translation start sites significantly. Unfortunately, no such sequences have (yet) been detected for the termination sites of translation. Besides the stop codon consensus, the sequences surrounding the stop codons seem to have very little in common.

5.3.3 Codon and Amino Acid Usage

Codon usage is by far the most important measure of coding potential in computational gene prediction, and it was early a main ingredient in gene prediction algorithms, in particular in the analysis of compact prokaryotic genomes such as the bacterium *E. coli* [52] and the yeast S. cerevisiae [3]. Codon usage measures hold the most discriminative power between coding and noncoding sequences [27], the reason being the difference in selective pressure acting upon coding and noncoding sequences. While mutations in functional regions are most often harmful and therefore selected against, the selection pressure in nonfunctional regions is more or less neutral, resulting in a significant differentiation in base composition. There are a number of different approaches to measure coding potential.

A direct way, implemented already in 1982 in [64], is to compare the codon usage of the sequence under analysis, to a precomputed codon distribution of the corresponding specie. Assume for instance that we have sequence

$$a_1 b_1 c_1 a_2 b_2 c_2 \cdots a_n b_n c_n a_{n+1} b_{n+1} c_{n+1}.$$

The probability of the sequence being in each frame is then given by

$$
\begin{aligned}
p_1 &= \pi_1 p(a_1 b_1 c_1) p(a_2 b_2 c_2) \cdots p(a_n b_n c_n), \\
p_2 &= \pi_2 p(b_1 c_1 a_2) p(b_2 c_2 a_3) \cdots p(b_n c_n a_{n+1}), \\
p_3 &= \pi_3 p(c_1 a_2 b_2) p(c_2 a_3 b_3) \cdots p(c_n a_{n+1} b_{n+1}),
\end{aligned}
\tag{5.21}
$$

where $p(a_i b_i c_i)$ is the probability of codon $a_i b_i c_i$ in the known codon distribution in frame i, and π_i is the probability that the coding frame is i. Typically we let $\pi_i = 1/3$

Table 5.3 Codon distribution of Leucine in *E. coli* O157:H7

Codon	Coding	Noncoding
CTA	0.038	0.076
CTC	0.101	0.116
CTG	0.493	0.238
CTT	0.110	0.164
TTA	0.134	0.214
TTG	0.125	0.191

for all frames. By sliding a window of size L across the sequence one triplet at a time, using the measure

$$P_i = p_i/(p_1 + p_2 + p_3) \ , \ i = 1, 2, 3, \tag{5.22}$$

it is possible to distinguish coding regions from noncoding regions.

Many codon usage measures focus on the distribution of *synonymous codons*. Codons that code for the same amino acid are said to be synonymous. Synonymous codons are codons that code for the same amino acid. For instance, in Fig. 1.3 we see that TTA, TTG, CTT, CTC, CTA and CTT all code for Leucine, and are thus synonymous. In random sequences there should be no particular preference for either codon in a group of synonymous codons, while in coding sequences this distribution is typically nonuniform. For instance, in O157:H7 the Leucine-coding codons is distributed as in Table 5.3. We see a clear preference for the codon CTG in coding sequences, while in noncoding sequences, the distribution is much more uniform. Examples of synonymous codon usage measures are the Codon Adaptation Index (CAI) [58], the Codon Bias Index (CBI) [4], the Effective Number of Codons (Nc) [66], and Frequency of Optimal Codons (Fop) [32].

What all these measures typically do is to measure the distance between the observed codon usage of a gene to the "preferred" codon usage, where only the most frequent among synonymous codons are used. That is, given the sequence of amino acids that the current gene codes for, the preferred codon usage is the sequence of most frequently used codons in the genome that would code for the same amino acid sequence. Since synonymous codon usage tend to be correlated with gene expression levels, synonymous codon measures can be used both to predict genes in DNA and to predict expression levels of genes [58].

5.3.4 K-Tuple Frequency Analysis

A natural extension of codon usage measures is to compare frequencies of longer "words" than triplets. In coding sequences, for instance, empirical data show dependencies not only within codons, but between adjacent codons as well. A direct approach to capture such dependencies is taken by *k-tuple frequency analysis* methods. K-tuple frequency analysis has many applications, such as discrimination between coding and noncoding regions, determination of coding frame, prediction of splice sites, and detection of binding sites and promoters. Basically, the frequencies of all "words" of length k in the sequence are calculated and compared. The idea is to utilize the statistical bias in k-tuple distributions to separate different features in the sequence. There are several different k-tuple analysis methods available, but they differ mainly in how the k-tuple information is used to make the separation.

A very straightforward approach is taken by Claverie and colleagues [21]. The method presented is used to discriminate between coding and noncoding sequences in the genome. Two frequency tables are constructed for the purpose, representing the k-tuple distribution for each feature, and the discriminant measure for a k-tuple x is given by

$$d(x) = p(x)/(p(x) + q(x)) \qquad (5.23)$$

where $p(x)$ is the probability of observing k-tuple x in a coding region, and $q(x)$ the corresponding probability in noncoding regions. The measure d takes values between 0 and 1, and will be near 0.5 for k-tuples with low discriminative power. Given a novel sequence $Y_1^T = (Y_1, \ldots, Y_T)$, this measure can then be computed for each position in the sequence, generating a series of k-tuple scores

$$\mathbf{d} = \{d_t\}_{t=1}^{T-k+1}, \qquad (5.24)$$

where $d_t = d(Y_t^{t+k-1})$ is the d-score in (5.23) for the k-tuple starting in position t. To obtain a more smooth score profile over the sequence, the d_t values are averaged over a window of size $2w + 1$, generating a sequence profile

$$P = \left\{ P_t = \sum_{i=t-w}^{t+w} d_t \right\}_{t=w+1}^{T-w-k+1} \qquad (5.25)$$

The choice of k and w is empirical and varies with the application. Typically, when comparing coding and noncoding nucleotide sequences, the optimal settings are $k = 6$ and $w = 20$ [21].

To determine the reading frame in a potential coding region, *phase-dependent* frequency tables may be useful. In that case the reference coding sequences give rise to three different frequency tables, one for each reading frame. That is, each frequency table consists of all k-tuple counts in phase $i = 0, 1, 2$, where i indicates the codon position of the first base in the k-tuple. Three different discriminant profiles are then constructed, one for each combination of pairs of phases. However, due to the third base redundancy in codons, comparisons between the first $i = 0$ and the second $i = 1$ phases tend to be the most informative [21].

The k-tuple frequency analysis method has several other uses than those mentioned. For instance, by comparing the test sequence toward its own k-tuple distribution can reveal repeat structures in the sequence. Moreover, comparing the test sequence to the entire genome of the corresponding specie will provide insights of the abundance of certain k-tuples in the genome, and the variation in information content between different sequence positions.

5.3.5 Markov Chain Content Sensors

Rather than just counting word frequencies, such as in k-tuple frequency analysis, a more flexible and adaptable model is to use a Markov chain that incorporates the interdependencies between nucleotides. In Example 2.2 we showed how a first-order Markov chain models dinucleotide frequencies when applied to DNA sequence. However, since coding sequences are organized into triplets, a better model would be a second-order Markov chain, such that the probability of the current base depends on the two preceding bases.

The transition probabilities in a second-order Markov chain can be written as

$$a_{ij}^{(2)} = \mathbf{P}(X_t = j | X_{t-1} = i, X_{t-2} = i_2), \tag{5.26}$$

for some states i, j, i_2 in a state space S. This model represents a homogeneous Markov chain. An even more accurate model, that makes better use of the codon structure of coding sequence, is to use an *inhomogeneous* Markov chain. More specifically, a 3-periodic Markov chain that takes into account the fact that the distribution of nucleotides, given the previous two (or more), varies depending on the position in the codon. This is incorporated into the model by training four different sets of initial and transitional probabilities, one for each of the three coding frames and one for noncoding sequences.

Example 5.2 Markov chain classification of E. coli (cont.)
Recall from Example 2.2 that if we were to classify a DNA sequence $Y = (Y_1, \ldots, Y_T)$, a common decision rule is to use the log-odds ratio

$$\log \frac{\mathbf{P}(Y|\text{coding})}{\mathbf{P}(Y|\text{noncoding})} \begin{cases} > \eta \Rightarrow \text{coding} \\ < \eta \Rightarrow \text{noncoding} \end{cases} \tag{5.27}$$

where η is some empirically determined threshold. We want to model the sequence using a second-order 3-periodic Markov chain, and for this we need three separate parameter sets, one for each codon position. We let π^1, π^2 and π^3 denote the three initial probability vectors, where, for instance $\pi^1 = \{\pi_A^1, \pi_C^1, \pi_G^1, \pi_T^1\}$ gives initial probabilities on the form

$$\pi_A^1 = \mathbf{P}(Y_1 = A | \text{codon position} = 1). \tag{5.28}$$

Fig. 5.13 Distribution of log-odds ratio scores of length-normalized coding (*dark gray*) and noncoding (*light gray*) sequences in *E. coli*

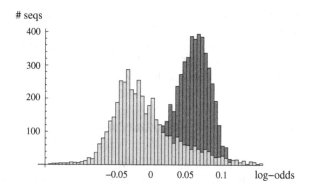

Furthermore, let $\mathbf{A}^1$, $\mathbf{A}^2$ and $\mathbf{A}^3$ denote the transition probability matrices for each respective reading frame, with transition probabilities on the form

$$a_{ijk}^r = \mathbf{P}_r(Y_t = k | Y_{t-1} = j, Y_{t-2} = i) \tag{5.29}$$

where $\mathbf{P}_r$ denotes the probability when Y_t is in codon position $r = 1, 2, 3$. The decision rule above could then be adjusted to

$$\log \frac{\max_k \mathbf{P}(Y | \text{coding}_k)}{\mathbf{P}(Y | \text{noncoding})} \begin{cases} > \eta & \Rightarrow \text{coding (in frame } k) \\ < \eta & \Rightarrow \text{noncoding} \end{cases} \tag{5.30}$$

where, for instance

$$\mathbf{P}(Y | \text{coding}_1) = \pi_{Y_1}^1 a_{Y_1 Y_2}^2 a_{Y_1 Y_2 Y_3}^3 a_{Y_2 Y_3 Y_4}^1 \cdots a_{Y_{T-2} Y_{T_1} Y_T}^3. \tag{5.31}$$

Naturally, the superscript of the last probability depends on the length of the sequence. Note that the first transition probability, $a_{Y_1 Y_2}^2$, is of first-order, since we only have one previous base to condition on. This probability also depends on codon position, however. The length-normalized log-odds scores of coding versus noncoding sequences in *E. coli* are illustrated in Fig. 5.13. Note that the peaks are more distinctly separated compared to the dinucleotide model in Example 2.2, indicating that this is a better model than the first-order model to discriminate between coding and noncoding sequences. □

We see in the example above, that using a second-order, 3-periodic Markov model improves the separation between coding and noncoding sequences compared to a first-order model. There is still considerable overlap between the coding and noncoding measures, however, which would result in prediction errors if we were using this model as our sole indicator of coding sequences. Extending the second-order model to a 5th order, a *hexamer* model, and thereby incorporating dependencies between adjacent codons as well, may improve the prediction accuracy further. One problem with higher order models is the increased computational complexity, but in eukaryotic DNA, where the exons are of moderate length compared to prokaryotic genes, this price may be well worth to pay. A more serious difficulty is the requirements on the training set in order to achieve reasonable frequency estimates for all parameters. The use of *interpolated* Markov models, presented next, is one possible solution to the training problem.

5.3.6 Interpolated Markov Models

Interpolated Markov models (IMMs), described in more detail in Chap. 2, have been successfully applied to various sequence analysis problems, where the object is to determine the likelihood and, ultimately, the class membership of certain subsequences.

Recall from Sect. 2.3 that the likelihood of a sequence $Y_1^T = Y_1, \ldots, Y_T$ can be decomposed into

$$\mathbf{P}(Y_1^T) = \prod_{t=1}^{T} \mathbf{P}(Y_t | Y_1^{t-1}). \tag{5.32}$$

The subsequence Y_1^{t-1} that the probability of Y_t is conditioned on, is often called the *context* of Y_t. Since we cannot handle contexts of arbitrary lengths, it is common to limit the length of the context to some constant k,

$$\mathbf{P}(Y_1^T) \approx \prod_{t=1}^{T} \mathbf{P}(Y_t | Y_{t-k}^{t-1}). \tag{5.33}$$

The resulting model is thus a kth-order Markov chain. For the discrimination between coding and noncoding sequences, for instance, we would like to use contexts of length $k = 5$ or more in order to incorporate dependencies between adjacent codons, but since the training set rarely is large enough to support such high orders, it is common to resort to within-codon dependencies only, and use second-order models.

The idea of IMMs is that instead of using a context of a rather low order, we use as long contexts as the training set allows for each specific context, and only fall back on shorter contexts when the frequency of the current context cannot be reliably estimated. We exemplify the use of IMMs as content sensors by recalling from Sect. 2.3 that a *linear interpolation* of the conditional probabilities in (5.33) is a weighted sum of the maximum likelihood estimates of all contexts up to length k [57]

$$\tilde{P}(Y_t | Y_{t-k}^{t-1}) = \rho_0 \frac{1}{L} + \rho_1 \hat{P}(Y_t) + \rho_2 \hat{P}(Y_t | Y_{t-1}) + \cdots + \rho_k \hat{P}(Y_t | Y_{t-k}^{t-1}) \tag{5.34}$$

where ρ_k are interpolation weights that sum up to one, L is a factor that ensures that no contexts are given a zero probability, and $\hat{P}$ denotes the ML estimate of the corresponding conditional probability. An alternative to linear interpolation is *rational interpolation*, in which the weights are allowed to depend on the context in question. Rational interpolation is described in Sect. 2.3.

Once we have estimated the parameters of the interpolation model, we can use it as a content sensor of novel sequences. It can be incorporated as a submodel in a gene finding algorithm, or we can use it as a stand-alone indicator of coding and noncoding sequences, for instance by using the same type of likelihood ratio test as in Example 2.2.

An IMM is then trained for each of the two models, "coding" and "noncoding," and a common form of the decision rule is a log-odds ratio

$$\log \frac{\mathbf{P}(Y_1^T | \text{coding})}{\mathbf{P}(Y_1^T | \text{noncoding})} \begin{cases} > \eta & \text{coding,} \\ < \eta & \text{noncoding,} \end{cases} \tag{5.35}$$

for some empirically determined threshold η. A similar application is promoter recognition [47], in which the two models now correspond to "promoter" and "background" sequences.

5.4 Splice Site Detection

The major problem in gene finding is to determine the correct gene structure. While coding sensors provide strong indicators of potential protein-coding regions, it still remains a difficult task to determine the exact borders of these regions, and to correctly predict the final protein product. Splice site sequences exhibit characteristics that clearly deviate from that of other sequences, and provide strong indicators of the internal boundaries between exons and introns in multi-exon genes. The characteristics involve both sequence composition as well as strong internal dependencies between signal positions. The task of splice site detection is to use these characteristics to distinguish true signal sequences from background sequences. Clearly, just as any other gene finding submodel, splice site predictors perform best when used as part of a larger gene finding machinery, which incorporates other pieces of evidence as well. In this section, however, we describe various splice site models as stand-alone discriminators, using only the sequence of the signal itself as evidence.

5.4.1 Weight Matrices and Weight Array Models

The absolutely simplest model of a biological signal is to assume that signal positions are independent and follow the same base composition. This is a poor model for biological signals such as splice sites, however, as both the independence and the identical distribution assumptions are clearly violated. A slightly more sensitive model, introduced by Staden [63], is the *weight matrix model* (WMM), also known as a *position-specific scoring matrix* (PSSM). In this model we still assume independence between sequence positions, but apply separate base compositions for each position. A weight matrix W for a sequence signal of length λ is an $(N \times \lambda)$-matrix, where the columns correspond to the λ signal positions, the rows to the N possible residues in the alphabet ($N = 4$ for nucleotides and $N = 20$ for proteins). An entry W_{ij}, $i = 1, \ldots, N$, $j = 1, \ldots, \lambda$ gives the frequency of residue i at position j. Still fairly simple, a WMM is a great improvement over having equal base compositions for all positions and manages, for instance, to capture the invariant dinucleotide consensus 'GT' and 'AG' in donors and acceptors, respectively.

Given a new sequence $Y = (Y_1, \ldots, Y_\lambda)$, we can use the weight matrix to calculate the probability, or the *likelihood*, that this sequence is a member of that type of signal

$$\mathbf{P}_{\text{WMM}}(Y) = \prod_{t=1}^{\lambda} P_t(Y_t = y), \qquad (5.36)$$

where P_t is the probability frequency over the sequence alphabet in position t, and $P_t(Y_t = y)$ is the weight matrix element W_{yt} in row y and column t. To discriminate between signals and non-signals, we can construct a similar weight matrix for a set of aligned non-signals, typically corresponding to the background distribution of the genome. The decision rule can then be a log-odds ratio, where the numerator is calculated using the weight matrix for signals, and the denominator using the weight matrix for non-signals.

$$\log \frac{\mathbf{P}_{\text{WMM}}^{S}(Y)}{\mathbf{P}_{\text{WMM}}^{N}(Y)} \begin{cases} > \eta & \text{signal,} \\ < \eta & \text{non-signal,} \end{cases} \qquad (5.37)$$

for some threshold η, where $\mathbf{P}^S$ and $\mathbf{P}^N$ correspond to the weight matrices for signal (S) and non-signal (N) respectively.

Example 5.3 Generation of a weight matrix
Assume that we want to generate a weight matrix for a set of human donor sites, based on the aligned sequences covering the last three exonic bases and the first six intronic bases surrounding the donor junction (the bar signifies the exon–intron junction in the table below).

```
C T G G T A A G G
C G G G T G A G C
A A A G T A A G T
A A G G T A C T T
C A G G T A A G G
C A G G T G C G G
G A C G T A T G T
C A A G T A G G T
G A G G T A A G C
C A G G T T T G T
```

The weight matrix is then obtained by simply counting the occurrences of each residue in each position, and dividing it by the number of sequences $n = 10$. That is, the entries are given by

$$W_{bj} = n_{bj}/n \qquad (5.38)$$

where n_{bj} is the observed frequency of base b at position j.

	Position								
	−3	−2	−1	+1	+2	+3	+4	+5	+6
A	0.2	0.8	0.2	0.0	0.0	0.7	0.5	0.0	0.0
C	0.6	0.0	0.1	0.0	0.0	0.0	0.2	0.0	0.2
G	0.2	0.1	0.7	1.0	0.0	0.2	0.1	0.9	0.3
T	0.0	0.1	0.0	0.0	1.0	0.1	0.2	0.1	0.5

The frequency matrix is commonly transformed into a log-odds matrix, that weighs the observed base frequencies either against the frequencies of non-signal sequences, or against the background composition of the input sequences. Moreover, we want to avoid zero frequencies in the matrix, since we usually do not know if these reflect the reality, or are simply the result of a too sparse training set. Therefore, it is common to insert *pseudocounts* in some manner. The simplest pseudocount model just adds 1 to all entries, and divide the frequencies by $n + 4$ instead (since four pseudocounts are added to each column). The entries of the log-odds matrix W is then given by

$$W_{bj} = \log \frac{(n_{bj} + 1)/(n + 4)}{e_b} \qquad (5.39)$$

where e_b is the expected frequency of base b. A uniform distribution over the bases would yield $e_b = 0.25$, while using the overall base composition of the input sequences in this case yields $\{e_A, e_C, e_G, e_T\} = \{0.27, 0.12, 0.39, 0.22\}$. The resulting log-odds matrix (using background base composition) becomes

	Position								
	−3	−2	−1	+1	+2	+3	+4	+5	+6
A	−0.22	0.88	−0.22	−1.32	−1.32	0.76	0.47	−1.32	−1.32
C	1.41	−0.54	0.16	−0.54	−0.54	−0.54	0.56	−0.54	0.56
G	−0.60	−1.00	0.38	0.70	−1.69	−0.60	−1.00	0.61	−0.31
T	−1.13	−0.44	−1.13	−1.13	1.26	−0.44	−0.04	−0.44	0.66

More sophisticated pseudocount models are discussed in Sect. 6.2. □

A major drawback with weight matrix models is that the independence assumption between positions is clearly violated in reality. An alternative approach, that models dependencies between adjacent residues, is to use an inhomogeneous Markov model, also known as a *weight array model* (WAM) [71]. Recall from Chap. 2 that in an inhomogeneous Markov model the model parameters are allowed to vary over the sequence. Thus, a WAM consists of a set of transition matrices, one for each position in the signal sequence, and the probability of a new sequence Y now takes the form

$$\mathbf{P}_{\text{WAM}}(Y) = P_1(Y_1) \prod_{t=2}^{\lambda} P_t(Y_t|Y_{t-1}) \tag{5.40}$$

where P_1 is an initial probability over the residues, and P_t is a position-dependent transition probability that comes from the transition matrix for position t.

A natural extension of the WAM is to allow for higher order dependencies. However, this immediately puts extra requirements on the size of the training set. In a comparison between WMMs and WAMs for splice site prediction, a second-order WAM in fact performed worse than the first-order model in [14]. The sole explanation was that the training set was too small to provide reliable parameter estimates for the higher order model. In an attempt to circumvent this problem, the *windowed weight array model* (WWAM) was introduced [14]. In a WWAM the second-order transition probabilities are estimated by taking the average of the same second-order conditional probabilities of the positions surrounding position t in a small window centered at t. That is, a second-order transition probability $P_t(Y_t = A|Y_{t-2} = C, Y_{t-1} = A)$ is estimated using all counts of triplet CAA within this window, rather than just the counts in the specific position. The idea is that nearby positions are expected to have a similar trinucleotide compositions, and can thereby provide more information to the parameter estimates than using just the observed frequencies at a single position. As it turns out, the WWAM appears to have better discriminative power than both the WMM and the WAM, as it manages to capture some significant triplet biases present in certain signal positions [14].

The application of *alignment profiles* to splice site detection, is essentially equivalent to using weight matrices. As described in Sect. 3.2.9, a profile is a motif description using a position-specific scoring matrix (PSSM). A main difference from a WMM is that the underlying alignment of a profile is allowed to include gaps. However, since splice site alignments are gap-free the difference vanishes. The *profile HMMs* described in Sect. 3.2.9 can naturally be applied to splice sites as well, but since they incorporate position dependencies they are closer to WAMs than to WMMs.

5.4.2 Variable-Length Markov Models (VLMMs)

As discussed in the previous section, sequence signals in general, and splice sites in particular, tend to exhibit dependencies between positions. While higher order Markov models, such as weight array models (WAMs) may be able to capture dependencies between adjacent positions, the requirements on the size of the training set increases exponentially with the order of the model, and with a sparse training set we often have to resort to smaller model orders. As we have seen before, one possible solution is the use of interpolated Markov models (IMMs), which interpolate lower order models to estimate higher order models.

Another possibility is the use of *variable-length Markov models* (VLMMs), also called *variable-order Markov models* (VOMs) or *context trees* [16, 54]. There are many different variants of VLMMs, most notably in the field of lossless data compression, but the common ingredient is that the order of the transition probabilities is allowed to vary depending on the availability in the training set. The model includes longer contexts, when these are available, and shorter contexts when the estimates of longer contexts become unreliable.

We introduce a *context function* $c(Y_1^{t-1})$ that maps the entire previous sequence of Y_t to a "relevant" context, typically resulting in a considerably shorter memory string [16]. The probability of a sequence then takes the form

$$\mathbf{P}(Y_1^T) = P_1(Y_1) \prod_{t=2}^{T} \mathbf{P}(Y_t | c(Y_1^{t-1})), \tag{5.41}$$

where, again, P_1 is the initial probability of the signal. The length of the context is given by

$$l(Y_t) = |c(Y_1^{t-1})| = \min \left\{ k : \mathbf{P}(Y_t | Y_1^{t-1}) = \mathbf{P}(Y_t | Y_{t-k}^{t-1}) \right\}. \tag{5.42}$$

If $l(Y_t) = 0$ the residues of the sequence are independent. Furthermore, let k be the smallest positive integer so that

$$l(Y_t) \le k \text{ for all } t \in [2, T]. \tag{5.43}$$

Then k signifies the maximum order for each position t, and the context function c is said to be of order k. If k is finite we say that we have a homogeneous VLMM of order k, and if the context length is constant $l(Y_t) = k$ (for $t > k$ naturally), the VLMM is equivalent to a full kth-order Markov model. Moreover, if the context function does not depend on the sequence position, the VLMM is homogeneous. However, the structure of a splice signal is typically inhomogeneous. In an inhomogeneous VLMM, instead of estimating a single VLMM for the entire sequence, we would estimate a separate VLMM for each position of the sequence signal. That is, for a sequence of length T we would need $T - 1$ different context functions c_t, where $t = 2, \dots, T$ [19].

Figure 5.14 illustrates an example from [19] of a tree representation of a second-order context function for a given splice site position. The probability of the observed residue at the given position can be obtained directly from the tree. Each node at level k represents one of the 4^k possible contexts of the current sequence position, and contains the probabilities for that context. The top tree in Fig. 5.14 illustrates the full

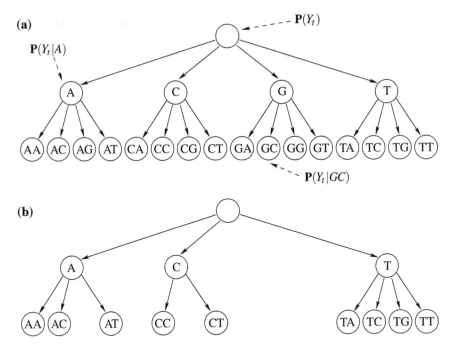

Fig. 5.14 A tree representation of a context function for a splice site position [19]. The *top* figure illustrates the full second-order Markov model, while the pruned version represents the context function for a VLMM. The nodes contain the conditional distribution of $\mathbf{P}(Y_t|c)$ for the corresponding context c

tree corresponding to a full second-order Markov model, while the bottom figure illustrates a pruned VLMM representation.

Splice sites exhibit strong dependencies not only between adjacent positions, but between positions enough far apart to require a too high order of a standard VLMM. A further generalization of the VLMM structure, that is able to capture such dependencies, is to use a *permuted* context. That is, instead of conditioning only on the adjacent previous residues in order of appearance in the sequence, the order of the context may be arbitrary. For instance, the context may be presented in the order of influence on the current position. Such a model was first mentioned in [54], but more recent references include [26, 70]. Another model that attempts to capture the most significant of the nonadjacent dependencies is the *Maximal Dependence Decomposition* algorithm, which is presented next.

5.4.3 Maximal Dependence Decomposition (MDD)

Weight array matrices or other kinds of Markov models can successfully capture dependencies between adjacent positions in a sequence signal. Eukaryotic splice sites, however, display significant dependencies between both adjacent and nonadjacent sequence positions. The *Maximal Dependence Decomposition* (MDD) algorithm [13, 14] attempts to capture at least the most informative of such dependencies by using a combination of weight matrices and decision trees.

Suppose we want to model a signal $Y = (Y_1, \ldots, Y_\lambda)$ of length λ based on an aligned set $\mathscr{A}$ of n such signal sequences. The MDD procedure can be summarized as follows.

1. Determine the consensus sequence of $\mathscr{A}$. If there are "ties" in the nucleotide frequencies, the consensus at that position consists of all tied residues.
2. Determine the sequence position that has the strongest influence on the others, using a χ^2-test on pairs of positions (see below for details). Denote this position t^*.
3. Calculate the base composition $P_{t^*} = \{p_A, p_C, p_G, p_T\}$ in position t^*, using $\mathscr{A}$.
4. Split the sequence set $\mathscr{A}$ into two subsets, $\mathscr{A}^+$ and $\mathscr{A}^-$, where $\mathscr{A}^+$ consists of signal sequences having the consensus nucleotide(s) in position t^*, and where $\mathscr{A}^-$ consists of the remaining sequences.
5. Calculate the base composition $P_{t^*} = \{p_A, p_C, p_G, p_T\}$ in position t^*, using the entire sequence set $\mathscr{A}$.
6. Calculate a weight matrix W for each subset $\mathscr{A}^+$ and $\mathscr{A}^-$ for all signal positions $t \neq t^*$.

Repeat the procedure in steps 2–5 recursively on the subsets, resulting in further subsets of the subsets and so on. The result can be viewed in a binary tree (see Fig. 5.15), where each split of the tree represents a split of the sequence set. The procedure is repeated until either of the following conditions occur:

(i) The $(\lambda - 1)$th level of the tree is reached, so no further splits are possible.
(ii) No position stands out as more influential than the others in the current subset.
(iii) The current subset has to few sequences to provide reliable estimates of the base composition. A rule of thumb is to stop when the number of sequences falls below 100 in a subset.

The Position with the Strongest Influence

The position having the strongest influence on the other positions in the sequence is determined by using a χ^2-*test of independence* on pairs of sequence positions. Suppose we want to test if positions t and u are independent in the signal, $1 \leq t, u \leq \lambda$. In a standard χ^2-test we would construct a (4×4) *contingency table*, where the rows correspond to the four possible residues $\{A, C, G, T\}$ in position t, and the columns to the same residues in position u. We would then insert the counts of the number of sequences in the current sequence set that have the corresponding residues in their respective positions. That is, in cell (i, j) in the table we insert the number of sequences having residue i in position t and residue j in position u. The χ^2-test then compares these observed counts to the expected number of counts, if the sequence positions were independent. The test used by the MDD method in [14] is also a

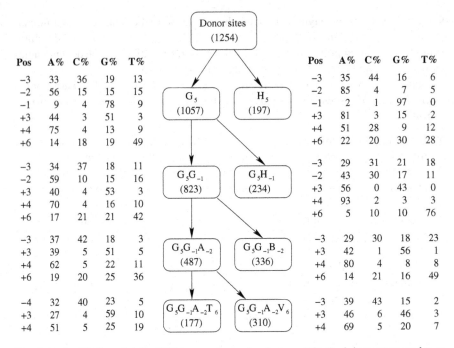

Pos	A%	C%	G%	T%
-3	33	36	19	13
-2	56	15	15	15
-1	9	4	78	9
+3	44	3	51	3
+4	75	4	13	9
+6	14	18	19	49
-3	34	37	18	11
-2	59	10	15	16
+3	40	4	53	3
+4	70	4	16	10
+6	17	21	21	42
-3	37	42	18	3
+3	39	5	51	5
+4	62	5	22	11
+6	19	20	25	36
-4	32	40	23	5
+3	27	4	59	10
+4	51	5	25	19

Pos	A%	C%	G%	T%
-3	35	44	16	6
-2	85	4	7	5
-1	2	1	97	0
+3	81	3	15	2
+4	51	28	9	12
+6	22	20	30	28
-3	29	31	21	18
-2	43	30	17	11
+3	56	0	43	0
+4	93	2	3	3
+6	5	10	10	76
-3	29	30	18	23
+3	42	1	56	1
+4	80	4	8	8
+6	14	21	16	49
-3	39	43	15	2
+3	46	6	46	3
+4	69	5	20	7

Fig. 5.15 An MDD model for 1257 human donor signals from [13]. Each box correspond to a subset of the donor sequence set. For instance G_5G_{-1} corresponds to donor signals with consensus nucleotide G in position 5 and -1, while H indicates a nonconsensus "not G" (A, C, or T), B indicates "not A" (C, G or T), and V indicates "not T" (A, C or G)

χ^2-test, but instead of counting all residues separately, the residues in position t are grouped into two classes, "consensus" and "nonconsensus" residue, while position u remains ungrouped. The resulting contingency table is now a (2×4)-table, where the rows correspond to the two classes at position t, and the columns to the four nucleotides in position u. The position t is then tested against all other positions u in the sequence signal. To illustrate the procedure, we define a new random variable, called a *consensus indicator variable* K_t, where

$$K_t = \begin{cases} 1 & \text{if } Y_t \text{ matches the consensus at position } t \\ 0 & \text{otherwise.} \end{cases} \tag{5.44}$$

Thus, instead of performing a χ^2-test between variables Y_t and Y_u, we test K_t against Y_u instead. The position with the strongest influence is then extracted as follows.

1. For each pair of positions (t, u), $t \neq u$, construct a (2×4) contingency table with counts of K_t and Y_u.
2. Compute the χ^2-statistic, denote it X^2_{tu}, of the contingency table above of positions t and u, $t \neq u$ (see below).

3. Collect all the X^2_{tu}-values for all positions t and u in the signal, resulting in a $(\lambda \times \lambda)$-table of χ^2-statistic values, where the rows correspond to position t and the columns to position u.
4. Calculate the row sums $S_t = \sum_{u:u \neq t} X^2_{tu}$ in the table above. The largest row sum, denote it S^*_t, indicates the position t^* with the strongest influence.

The χ^2-statistic for K_t and Y_u is calculated by constructing a (2×4) contingency table, where the rows correspond to the two K_t values "consensus" or "nonconsensus", and the columns correspond to the four possible nucleotides in position u. The χ^2-statistic is given by

$$X^2_{tu} = \sum_i \sum_j \frac{(O_{ij} - E_{ij})^2}{E_{ij}} \tag{5.45}$$

where O_{ij} is the observed number of sequences in $\mathscr{A}$ having residues i and j in positions t and u respectively. E_{ij} is the expected count if the positions were independent, given by

$$E_{ij} = \frac{n_i . n_{.j}}{n} \tag{5.46}$$

where $n_i.$ is the number of sequences having value i in position t, $n_{.j}$ the corresponding count for residue j in position u, and n the total number of sequences. If sequence position t is independent of all other positions, the row sum S_t will be approximately χ^2-distributed with $(4 - 1)(\lambda - 1)$ degrees of freedom. If the largest row sum S_{t^*} is significantly larger than zero, position t^* can be used to split the sequence set as described above.

Example 5.4 MDD of donor splice sites
We borrow the example in [14] to illustrate the use of the MDD algorithm. The sequence set consists of 1254 human donor sequences, and we want to create an MDD model that can be used for splice site detection in novel sequences. Each training sequence is 9 bases long, including the last three exon nucleotides with indices -3 to -1, and the first six intron nucleotides with indices $+1$ to $+6$.

1. Determine the consensus sequence
First, the positions $+1$ and $+2$ are excluded from the MDD analysis, as they represent the invariant donor sequence GT and do not depend on anything else. Next, the base frequencies and consensus residues of the remaining positions are determined (see Table 5.4).

2. Determine the sequence position with the strongest influence
Table 5.5 illustrates the contingency tables between positions -2 and $+6$, where the -2 position has been grouped into sequences having the consensus nucleotide 'A' in position -2, and sequences with a nonconsensus nucleotide in position, -2. The rightmost table with expected counts is computed according to (5.46). For instance,

Table 5.4 Base composition and consensus sequence of the 1254 donor sites analyzed in [14]

Base	Position								
	−3	−2	−1	+1	+2	+3	+4	+5	+6
A%	33	60	8	0	0	49	71	6	15
C%	37	13	4	0	0	3	7	5	19
G%	18	14	81	100	0	45	12	84	20
T%	12	13	7	0	100	3	9	5	46
Cons	a/c	A	G	G	T	a/g	A	G	T

Table 5.5 Contingency table of position −2 and +6 in a set of donor splice sites, where the consensus in position −2 is A

K_{-2}	Y_{+6}				
	A	C	G	T	Total
Observed					
A	136	144	182	292	754
Not A	55	88	75	282	500
Total	191	232	257	574	1254
Expected					
A	114.84	139.50	154.53	345.13	754
Not A	76.16	92.50	102.47	228.87	500
Total	191	232	257	574	1254

the expected count of observing an A in position −2 and an A in position +6, if the positions were independent is given by

$$E_{AA} = (754 \cdot 191)/1254 = 114.84. \qquad (5.47)$$

The χ^2-statistic $X^2_{-2,+6}$ for K_{-2} and Y_{+6} becomes

$$X^2_{-2,+6} = \frac{(136 - 114.84)^2}{114.84} + \cdots + \frac{(282 - 228.87)^2}{228.87} \approx 42.9. \qquad (5.48)$$

Table 5.6 shows the χ^2-statistics for all (t, u) pairs in the signal sequence. The bold values are significant when using a confidence level of $\alpha = 0.001$. The individual pair values were tested using a χ^2-distribution with 3° of freedom $((2 - 1)(4 - 1) = 3)$, such that values larger than $\chi^2_3(0.001) = 16.27$ are deemed significant. Similarly, the row sums were compared to a χ^2-distribution with 18° of freedom $((4 - 1)(7 - 1) = 18)$. The table value of this distribution is $\chi^2_{18}(0.001) = 42.31$, resulting in significance for all rows. That is, the hypothesis of independence could be rejected for all positions. Position +5 with consensus 'G' achieves the largest row sum, and, thus, becomes the first split point.

Table 5.6 The χ^2-statistic values of pairwise positions between consensus indicator K_t and nucleotide Y_u

Position i	Consensus	Position j							
		-3	-2	-1	$+3$	$+4$	$+5$	$+6$	S_i
-3	c/a	–	**61.8**	14.9	5.8	**20.2**	11.2	**18.0**	131.8
-2	A	**115.6**	–	**40.5**	**20.3**	**57.5**	**59.7**	**42.9**	336.5
-1	G	15.4	**82.8**	–	13.0	**61.5**	**41.4**	**96.6**	310.8
$+3$	a/g	8.6	**17.5**	13.1	–	**19.3**	1.8	0.1	60.5
$+4$	A	**21.8**	**56.0**	**62.1**	**64.1**	–	**56.8**	0.2	260.9
$+5$	G	11.6	**60.1**	**41.9**	**93.6**	**146.6**	–	**33.6**	387.3*
$+6$	T	**22.2**	**40.7**	**103.8**	**26.5**	17.8	**32.6**	–	243.6

The bold values are significant with $p < 0.0001$

3. Split the sequence set into consensus and nonconsensus subsets

Next we split the sequence set into two subsets according to whether the sequences had the consensus 'G' in position $+5$ or not. Let G_5 denote the subset of consensus sequences, and H_5 the remaining sequences.

4-5. Calculate base composition and weight matrices

The base composition at $+5$ is given in Table 5.4 as

$$P_{+5} = \{0.06, 0.05, 0.84, 0.05\} \tag{5.49}$$

and weight matrices of G_5 and H_5 are calculated for the remaining positions. The procedure in steps 2–5 is repeated recursively for each subset until a stop condition is reached. The resulting MDD is given in Fig. 5.15. A minimum number of sequences to allow for new splits was set to 175 in this example. Thus, H_5 was not divided further. G_5 was split in position -1, with consensus 'G', resulting in subgroups $G_5 G_{-1}$ and $G_5 H_{-1}$ both having consensus residue 'G' in position $+5$, but with consensus and nonconsensus in position -1, respectively. The consensus group $G_5 G_{-1}$ was split yet again at position -2, and the consensus group $G_5 G_{-1} A_{-2}$ was split one last time in position $+6$.

Score a New Sequence

The resulting MDD model M can be used to score a new potential signal sequence $Z = (Z_{-3}, \ldots, Z_{+6})$, and thus can be used for splice site prediction. The score is achieved as follows. Let $p_{-3}, p_{-2}, p_{-1}, p_1, \ldots, p_6$ denote the base frequencies of the consensus sequence given in Table 5.4, and let q_t denote the weight matrix scores of the current residue and the current subset of sequences. Sequence Z can then be scored by following the decision tree illustrated in Fig. 5.16. For instance, if $Z_1 Z_2$ coincides with the consensus dinucleotide 'GT' the process moves to the next

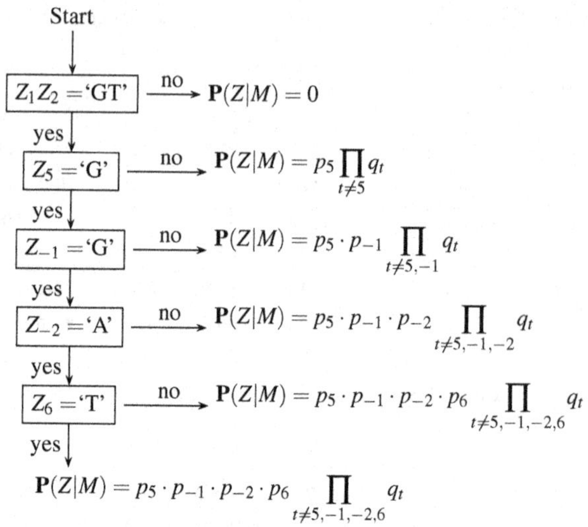

Fig. 5.16 A decision tree illustration of how to use the MDD for scoring new sequences

Start

$Z_1 Z_2 =$ 'GT' $\xrightarrow{\text{no}}$ $\mathbf{P}(Z|M) = 0$

yes $\downarrow$

$Z_5 =$ 'G' $\xrightarrow{\text{no}}$ $\mathbf{P}(Z|M) = p_5 \prod_{t \neq 5} q_t$

yes $\downarrow$

$Z_{-1} =$ 'G' $\xrightarrow{\text{no}}$ $\mathbf{P}(Z|M) = p_5 \cdot p_{-1} \prod_{t \neq 5, -1} q_t$

yes $\downarrow$

$Z_{-2} =$ 'A' $\xrightarrow{\text{no}}$ $\mathbf{P}(Z|M) = p_5 \cdot p_{-1} \cdot p_{-2} \prod_{t \neq 5, -1, -2} q_t$

yes $\downarrow$

$Z_6 =$ 'T' $\xrightarrow{\text{no}}$ $\mathbf{P}(Z|M) = p_5 \cdot p_{-1} \cdot p_{-2} \cdot p_6 \prod_{t \neq 5, -1, -2, 6} q_t$

yes $\downarrow$

$\mathbf{P}(Z|M) = p_5 \cdot p_{-1} \cdot p_{-2} \cdot p_6 \prod_{t \neq 5, -1, -2, 6} q_t$

level. Otherwise the candidate donor gets score $P(Z|M) = 0$. At the next level, if $Z_5 \neq$ 'G', the sequence is scored using the weight matrix entries of the subset H_5. The q_t at the next level comes from the $G_5 H_{-1}$ weight matrix, and so on. □

5.4.4 Neural Networks

Neural networks has been successfully applied to various sequence analysis problems, and the basic theory of these models are described in Sect. 2.4. The application of neural networks to splice site detection was first introduced by Brunak et al. [12], and has also been integrated in gene finders such as GRAIL [67, 68] described in Sect. 2.4.4, and Genie [38, 53]. Here we give a brief overview of the neural networks used for splice site detection in [12].

The neural network in question is a two-layer feed-forward network, such as described in Sect. 2.4.3, consisting of an input layer processing the sequence, a hidden layer of nonlinear processing elements, and an output layer, typically consisting of a single element that assigns a score to the input sequence. We recall that the output unit is a function on the form

$$y = \phi \left(\sum_{j=0}^{M} w_{jk}^{(2)} \cdot \phi \left(\sum_{i=0}^{N} w_{ij}^{(1)} x_i \right) \right) \qquad (5.50)$$

where x_i are the input units, z_j the hidden units, and y the output unit. The weights $w_{ij}^{(1)}$ correspond to the connections between the input layer and the hidden layer, and $w_{jk}^{(2)}$ to the connections between the hidden layer and the output layer. The same sigmoid activation function ϕ is used for both the input and the hidden layer

$$\phi(a) = \frac{1}{1 + e^{-a}}. \tag{5.51}$$

The output unit takes values between 0 and 1, where true splice sites are assumed to receive scores close to 1 and non-signals receive scores close to 0. Typically, the cutoff value used to separate signal sequences from non-signals is set to 0.5. The network is trained using the *backpropagation algorithm* described in Sect. 6.7.

Example 5.5 Splice site detection in GRAIL
The neural network-based gene finder GRAIL, described in Sect. 2.4.4, uses a standard neural network approach for splice site prediction. The splice site detector slides a window across the sequence and assigns a score to the middle nucleotide of each window sequence. For instance, the network for acceptor sites slides a window of 95 bp across the sequence, searching for signals sequences $Y_{-60}, \ldots, Y_{35}$, where $Y_{-60}, \ldots, Y_2$ is the intron part with the consensus dinucleotide 'AG' located at $Y_1 Y_2$, and the exon part is in $Y_3, \ldots, Y_{35}$. The network score is a combination of seven different frequency measures of nucleotide "words" that are characteristic to splice sites.
The seven frequency measures are defined as follows:

1. Position-specific 5-tuple frequencies:

$$\sum_{i=-23}^{-4} \log \frac{f_i^+(Y_i^{i+4})}{f_i^-(Y_i^{i+4})}, \tag{5.52}$$

 where f_i^+ and f_i^- are the position-specific frequency counts of the 5-tuple Y_i^{i+4} for true and false splice sites, respectively.
2. Position independent 5-tuple frequencies:

$$\sum_{i=-27}^{0} \log \frac{f^+(Y_i^{i+4})}{f^-(Y_i^{i+4})}, \tag{5.53}$$

 where f^+ and f^- now correspond to the overall background frequencies of 5-tuples in true and false splice sites, respectively.
3. Pyrimidine count in the intron region:

$$\sum_{i=-27}^{0} \mathbb{I}(Y_i = \text{'C' or 'T'})\sqrt{i + 28}, \tag{5.54}$$

 where $\mathbb{I}$ is the indicator function.

4. The normalized distance between Y_0 and the nearest upstream 'YAG' consensus, where 'Y' signifies a pyrimidine ('C' or 'T').
5. Nonadjacent pairwise counts:

$$\sum_{i=-27}^{4} \sum_{j \geq i}^{4} \log \frac{f_i^+(Y_i Y_j)}{f_i^-(Y_i Y_j)}. \tag{5.55}$$

6. Coding potential in the intron region $Y_{-60}, \ldots, Y_{-1}$.
7. Coding potential in the exon region $Y_3, \ldots, Y_{35}$.

Coding potential is measured using a periodic hexamer model, and is used in addition to the consensus dinucleotide 'AG' to confirm a transition from noncoding to coding sequence. The seven measures are fed into the network, which is a two-layer feedforward network, consisting of seven input nodes, a hidden layer of three nodes, and one output node. The network is trained using the backpropagation algorithm described in Sect. 6.7.   □

5.4.5 Linear Discriminant Analysis

Linear discriminant analysis (LDA) is a method in *multivariate analysis* concerned with finding the best separation of different classes of objects, based on a set of features that characterize the objects. There are two main areas of application: dimensionality reduction and data classification. Dimensionality reduction involves reducing the set of features to a set that best *explains* the data in the sense of separating the classes. Data classification involves determining a decision rule that can be used to allocate a new object to one of the existing classes, based on its feature values. We focus on data classification here, and give a brief overview of linear discriminant analysis for the purpose of splice site detection.

Suppose that we want to assign an object Y to one of two possible classes, C_1 or C_2 say, based on a set of features $\mathbf{x} = (x_1, \ldots, x_p)$ that characterize Y. In terms of splice site detection, the classes are "signal" or "non-signal," and the features typically include various compositional measures of the signal sequence. The idea of linear discriminant analysis is to find a model of the features that best separates the classes. When only dealing with two classes, a discriminative model can be written as a linear combination of the features

$$z = \sum_{i=1}^{p} \alpha_i x_i, \tag{5.56}$$

and we can use a decision rule on the form

$$\text{class}(Y) = \begin{cases} C_1 & \text{if } z \geq c, \\ C_2 & \text{if } z < c. \end{cases} \tag{5.57}$$

for some empirically determined threshold c. In what follows we focus on the case with two classes, but the theory can be extended to an arbitrary number of classes. For more details confer for instance [43].

Quadratic Discriminant Analysis (QDA)

Formally, the decision rule given in (5.57) assigns object Y to the class that obtains the highest probability for the given set of features. That is,

$$\text{class}(Y) = \begin{cases} C_1 & \text{if } \mathbf{P}(C_1|\mathbf{x}) > \mathbf{P}(C_2|\mathbf{x}), \\ C_2 & \text{otherwise.} \end{cases} \qquad (5.58)$$

The conditional probability of each class $\mathbf{P}(C_i|\mathbf{x})$ can be obtained using Bayes' rule

$$\mathbf{P}(C_i|\mathbf{x}) = \frac{\mathbf{P}(\mathbf{x}|C_i)\mathbf{P}(C_i)}{\sum_{j=1}^{2} \mathbf{P}(\mathbf{x}|C_j)\mathbf{P}(C_j)} \qquad (5.59)$$

where $\mathbf{P}(C_i)$ are called the *prior probabilities* of the classes, $\mathbf{P}(C_i|\mathbf{x})$ the *posterior probabilities*, and $\mathbf{P}(\mathbf{x}|C_i)$ are sometimes called *likelihood functions* of C_i. The prior probabilities are assumed to be known, and if they are not known typically the uniform distribution is used, assigning equal probabilities to all classes. We will assume uniformly distributed priors throughout the rest of this section. If we apply (5.59) to (5.58) we get that the decision on how to classify Y is based on whether

$$\mathbf{P}(\mathbf{x}|C_1) > \mathbf{P}(\mathbf{x}|C_2). \qquad (5.60)$$

The probabilities $\mathbf{P}(\mathbf{x}|C_i)$ can be estimated from data, but this often requires a very large training set. Here we instead assume that these conditional distributions are *multivariate normal* $MN(\boldsymbol{\mu}_i, \Sigma_i)$. That is,

$$\mathbf{P}(\mathbf{x}|C_i) = \frac{1}{(2\pi)^{p/2}|\Sigma_i|^{1/2}} \exp\left(-\frac{1}{2}(\mathbf{x} - \boldsymbol{\mu}_i)^T \Sigma_i^{-1}(\mathbf{x} - \boldsymbol{\mu}_i)\right) \qquad (5.61)$$

where $\boldsymbol{\mu}_i$ is the mean vector and Σ_i the covariance matrix of the p features in class C_i. Inserting (5.61) into (5.60) and manipulating the formulas somewhat leads us to the *quadratic discriminant function*, which assigns Y to class C_1 if

$$-\log|\Sigma_1| - (\mathbf{x} - \boldsymbol{\mu}_1)^T \Sigma_1^{-1}(\mathbf{x} - \boldsymbol{\mu}_1) \geq \log|\Sigma_2| + (\mathbf{x} - \boldsymbol{\mu}_2)^T \Sigma_2^{-1}(\mathbf{x} - \boldsymbol{\mu}_2). \qquad (5.62)$$

The term $(\mathbf{x} - \boldsymbol{\mu}_i)^T \Sigma_i^{-1}(\mathbf{x} - \boldsymbol{\mu}_i)$ is called the *Mahalanobis distance* and measures the dissimilarity between the classes. The term "quadratic" in QDA comes from the fact that the surface that separates the classes is quadratic.

Linear Discriminant Analysis (LDA)

The linear form in (5.56) occurs when, in addition to the multivariate normal assumption, we assume that the covariance matrices of the two classes are equal, $\Sigma_1 = \Sigma_2 = \Sigma$. This simplifies the expression in (5.62) further,

$$(\mu_1 - \mu_2)^T \Sigma^{-1} \mathbf{x} \geq \frac{1}{2}(\mu_1 - \mu_2)^T \Sigma^{-1}(\mu_1 + \mu_2). \tag{5.63}$$

The left hand side of the inequality is in fact a linear combination of the features,

$$z = \alpha^T \mathbf{x} = \sum_{i=1}^{p} \alpha_i x_i, \tag{5.64}$$

where the coefficients vector is given by

$$\alpha = \Sigma^{-1}(\mu_1 - \mu_2). \tag{5.65}$$

The threshold c in (5.57) is then given by

$$c = \frac{\alpha^T (\mu_1 + \mu_2)}{2}. \tag{5.66}$$

In practice, the mean vectors μ_i and the covariance matrices Σ_i of the feature vectors are typically unknown and need to be estimated from a training set with known class labels and feature vectors. Let $Y_1, \ldots, Y_n$ denote such a training set with corresponding feature vectors $\mathbf{x}_1, \ldots, \mathbf{x}_n$. Let $n_i, i = 1, 2$ denote the number of objects in class C_i, with $n = n_1 + n_2$, and $\mathbf{x}_1^{(i)}, \ldots, \mathbf{x}_{n_i}^{(i)}$ the corresponding feature vectors of that class. Then the sample mean vectors and sample covariance matrices for each class are then computed as

$$\bar{\mathbf{x}}_i = \frac{1}{n_i} \sum_{j=1}^{n_i} \mathbf{x}_j^{(i)}, \quad \mathbf{S}_i = \frac{1}{n_i - 1} \sum_{j=1}^{n_i} (\mathbf{x}_j^{(i)} - \bar{\mathbf{x}}_i)(\mathbf{x}_j^{(i)} - \bar{\mathbf{x}}_i)^T. \tag{5.67}$$

The mean vectors μ_1 and μ_2 are estimated by the sample means $\bar{\mathbf{x}}_1$ and $\bar{\mathbf{x}}_2$, respectively, and the covariance matrix Σ is estimated using the *pooled* covariance matrix given by

$$\mathbf{S}_p = \frac{(n_1 - 1)\mathbf{S}_1 + (n_2 - 1)\mathbf{S}_2}{n_1 + n_2 - 2}. \tag{5.68}$$

Faced with a new, unknown object Y_0 with feature vector $\mathbf{x}_0$, we classify it to class C_1 if

$$(\bar{\mathbf{x}}_1 - \bar{\mathbf{x}}_2)^T \mathbf{S}_p^{-1} \mathbf{x}_0 \geq \frac{1}{2}(\bar{\mathbf{x}}_1 - \bar{\mathbf{x}}_2)^T \mathbf{S}_p^{-1}(\bar{\mathbf{x}}_1 + \bar{\mathbf{x}}_2), \tag{5.69}$$

and to C_2 otherwise. This formulation is in fact equal to the *Fisher's discrimi-nant* [28], which is often used interchangeably with LDA, but is slightly different. Although arriving at the same statistic, Fisher's approach does not assume multi-variate normal distributions of the classes. It does, however, implicitly assume equal covariance matrices of the classes, since a pooled estimate of the sample covariance matrices is used.

Example 5.6 Splice site detection in FGENEH
FGENEH [61, 62] is a gene finder that uses linear discriminant analysis both for the prediction of exons and for prediction of splice sites. The method searches for open reading frames (ORFs) surrounded by potential splice sites, and combines measures of coding potential and intronic sequence composition with signal sensors for initial, terminal, and internal exon boundaries. The candidate exons are passed on to a dynamic programming algorithm that produces the final gene model.

Initially, the sequence is scanned for all occurrence of the consensus dinucleotides GT and AG representing candidate donors and acceptors, respectively. The splice site predictor then combines a set of measures (features) for each candidate site, and classifies the site into "signal" or "pseudosignal" using linear discriminant analysis. In LDA terms, the *object* is thus the candidate splice site, and the features are the set of measures that characterize a splice site.

The characteristics used for donor site classification, where negative indices sig-nify positions in the preceding exon and positive indices positions in the succeeding intron, are

1. The average triplet preferences in the potential coding region (-30 to -5 bp).
2. The average triplet preferences in the conserved consensus region (-4 to $+6$ bp).
3. The average triplet preferences in the G-rich region ($+7$ to $+50$ bp).
4. The number of significant triplets in the conserved consensus region (-4 to $+6$ bp).
5. The octanucleotide preference for being coding in the (-60 to -1 bp) region.
6. The octanucleotide preference for being noncoding in ($+1$ to $+54$ bp).
7. The number of G-bases, GG-doublets, and GGG-triplets in the ($+6$ to $+50$ bp) region.

The characteristics of acceptor sites, where now negative indices indicate positions in the preceding intron and positive indices positions in the succeeding exon, are

1. The average triplet preferences in the branch point region (-48 to -34 bp).
2. The average triplet preferences in the Poly(T/C)-tract region (-33 to -7 bp).
3. The average triplet preferences in the conserved consensus region (-6 to $+5$ bp).
4. The average triplet preferences in the coding region ($+6$ to $+30$ bp).
5. The octanucelotide preferences for being coding in the ($+1$ to $+54$ bp) region.
6. The octanucelotide preferences for being intron in the (-1 to -54 bp) region.
7. The number of Ts and Cs in the poly(T/C)-tract region (-33 to -7 bp).

The triplet preferences are determined in a sequence window (L, R) where L is the number of positions upstream (5' side) and R the number of positions downstream (3' side) of the exon–intron junction of the specific splice site. For donors, $L = 30$ and $R = 50$, and for acceptors, $L = 80$ and $R = 30$. The training set consists of a set of true splice sites, as well as a set of pseudosites containing the consensus dinucleotides without being a real splice site. The preference of a triplet at position i in the (L, R) window is given by

$$P(i) = \frac{f_i^+(k)}{f_i^+(k) + f_i^-(k)}, \tag{5.70}$$

where $f_i^+(k)$ and $f_i^-(k)$ are the frequencies of triplet $k = 1, \ldots, 64$ at position i for splice sites and pseudosites, respectively. The discrimination function used is a *mean preference index* and is calculated for each candidate splice site using

$$P_{sp}(j) = \frac{1}{m}\left(\sum_{i=L}^{R} P(i)\right), \tag{5.71}$$

where j is the position of the G in the AG or GT dinucleotide, and m is the number of triplets in the (L, R) window. The discrimination function is calculated only for those triplets that differ significantly in frequencies between splice sites and pseudosites. That is, only triplets rendering $P(i) - 0.5 > \eta$, for some threshold η, are included in (5.71), and thus, m is now the number of significant triplets in (L, R).

Internal exons are predicted by considering all open reading frames (OFRs) flanked by the consensus dinucleotides for donors and acceptors. A candidate ORF is scored using a discriminant function that combines the splice site scores with sequence composition scores of the potential exon and the flanking introns. The sequence composition scores are computed using a sliding window of length λ across the sequence, starting 70 bp upstream of the candidate donor site and ending 70 bp downstream of the candidate acceptor site.

The probability that the window oligonucelotide $x = (x_1, \ldots, x_\lambda)$ is coding is estimated using Bayes' rule

$$\begin{aligned} \mathbf{P}(C|x) &= \frac{\mathbf{P}(x|C)\mathbf{P}(C)}{\mathbf{P}(x|C)\mathbf{P}(C) + \mathbf{P}(x|N)\mathbf{P}(N)} \\ &= \frac{f_C(x)}{f_C(x) + f_N(x)}, \end{aligned} \tag{5.72}$$

where C and N denotes "coding" and "noncoding", and $f_C(x)$ and $f_N(x)$ are the frequencies of oligonucleotide y in coding and noncoding sequences, respectively. The prior probabilities of the two classes are set equal, $\mathbf{P}(C) = \mathbf{P}(N) = \frac{1}{2}$. The simplest discriminant function for classifying a potential coding region $Y = (Y_1, \ldots, Y_T)$

of length T say, is then the average oligonucleotide probability in a sliding window across the sequence

$$\mathbf{P}(C|Y) = \frac{1}{n-\lambda} \left(\sum_{t=1}^{n-\lambda} \mathbf{P}(C|Y_t^{t+\lambda-1}) \right), \qquad (5.73)$$

where $Y_t^{t+\lambda-1}$ is the oligonucleotide starting at position t. □

5.4.6 Maximum Entropy

Suppose we have a random process $Y_1, Y_2, \ldots$ that we want to build a model for. Suppose further that we have a set of *constraints* that we want to impose on our model. We may for instance have some knowledge of the process, such as marginal distributions, expected values, or at least know the bounds on such values. In the case of splice sites, we would typically want to impose constraints that capture base compositions and position dependencies of the signal. The idea of maximum entropy is that, among all possible models that satisfy our constraints, choose the one that is the most random. That is to say, we want to choose the model with the highest entropy. By doing so, we assure that we do not inadvertently include more assumptions or biases than we really have information on. The idea behind the *principle of maximum entropy*, first proposed by Jaynes [34, 35], is to stipulate that among all possible distributions that satisfy the given constraints, the one that "best" approximates the true distribution, is the one with the largest (Shannon) entropy. This needs to be applied with some caution, however, in particular since in the unconstrained situation the distribution with the largest entropy is the uniform, something that we have shown to be a poor model for splice site sequences in earlier sections of this chapter.

Entropy is a measure of the level of uncertainty in a random variable or process. Or, on the other side of the coin, the entropy tells us the level of information contained in the variable.

The entropy measure can be written as

$$H(p) = \sum_y p(y) \log \frac{1}{p(y)} = -\sum_x p(y) \log p(y) \qquad (5.74)$$

where the sum runs over all possible states (or events) y of the process, and $p(y)$ denotes the probability of state y. In information theory it is common to use log base 2, to enable an interpretation in "bits," but any base will do as long as it is used consistently. The entropy $H(p)$ runs between 0 and infinity, and if no constraints are placed on the model, the distribution with the highest entropy is the uniform distribution, which places equal probabilities to all possible states of the process.

The Maximum Entropy Method

The setup of maximum entropy modeling is to determine the possible states of the process and decide on a set of constraints. The constraints need to be *consistent*, which means that they are not allowed to contradict one another. We borrow the following example from [35].

Example 5.7 Maximum entropy motivation
Assume for instance that we roll a die a large number of times, N say, and observe an average of 4.5 (note that a fair die would give an average of 3.5). Given this *constraint*, we want to estimate the outcome probabilities

$$p(y) = \mathbf{P}(Y = y), \quad y = 1, \ldots, 6 \tag{5.75}$$

of the die. We can formalize our constraints as

$$\sum_{y=1}^{6} p(y) = 1, \tag{5.76a}$$

$$E[Y] = \sum_{y=1}^{6} y \cdot p(y) = 4.5. \tag{5.76b}$$

There are infinitely many distributions that satisfy these two constraints, and the question is which one to choose. For instance $\{p(1), p(2), p(3), p(4), p(5), p(6)\} = \{0, 0, 0, 0.5, 0.5, 0\}$ is a member of the class, but without any added information we have no reason to believe that the die only shows four or five. By choosing that particular distribution we have implicitly included extra constraints that we might not have call for. A reasonable assignment must not only satisfy the given constraints, it may not include any extra assumptions or limitations.   □

The principle of maximum entropy basically states that the distribution that is the most fair, or most conservative given what we know, is the one that spreads the probability mass as evenly as possible. To formalize, assume that we have a random variable Y, that can take values in a *sample space* $\Omega = \{\omega_1, \ldots, \omega_M\}$. Assume further that we have a set of *feature functions* $f_1, \ldots, f_K$, $K < M$, where each function maps the outcome of Y to a real number, $f_k : \Omega \to \mathbb{R}, k = 1, \ldots, K$. The feature functions correspond to our desired constraints, and are often indicators of specific events or sets of events, such that

$$f_k(y) = \begin{cases} 1 & \text{if } Y = y, \\ 0 & \text{otherwise.} \end{cases} \tag{5.77}$$

Now assume that the state probabilities $p(y) = \mathbf{P}(Y = y), y \in \Omega$, are unknown. Instead we are given the expected values of the feature functions

$$F_k = E[f_k], \quad k = 1, \ldots, K. \tag{5.78}$$

The problem we want to solve is to find the probabilities $p(y)$, $y \in \Omega$, that maximize the entropy

$$H(p) = -\sum_y p(y) \log p(y), \tag{5.79}$$

under the constraints

$$\sum_y p(y) = 1,$$

$$F_k = \sum_y p(y) f_k(y), \quad k = 1, \ldots, K. \tag{5.80}$$

This is an optimization problem that is commonly solved using *Lagrange multipliers*, which is a method for optimizing a function under a set of given constraints. In short, we introduce the *Lagrangian function*

$$L = H(p) + \sum_{k=1}^{K} \lambda_k \left(E[f_k] - \sum_y p(y) f_k(y) \right) + \gamma \left(\sum_y p(y) - 1 \right), \tag{5.81}$$

where $\lambda_1, \ldots, \lambda_K$ and γ are the Lagrange multipliers, and solve the equation system

$$\frac{\partial L}{\partial p(y)} = 0, \quad y \in \Omega. \tag{5.82}$$

As a result we achieve the expression

$$p(y) = \frac{1}{Z(\lambda_1, \ldots, \lambda_K)} \exp\left(\sum_k \lambda_k f_k(y) \right), \tag{5.83}$$

which is on the form of a *Boltzmann distribution*. $Z(\lambda_1, \ldots, \lambda_K)$ is a normalizing factor, sometimes called a *partition function*, that makes the $p(y)$ probabilities sum to one

$$Z(\lambda_1, \ldots, \lambda_K) = \sum_y \exp\left(\sum_k \lambda_k f_k(y) \right). \tag{5.84}$$

The coefficients $\lambda_1, \ldots, \lambda_K$ are then determined using the constraints in (5.80) and by solving the equation system

$$E[f_k] = \frac{\partial}{\partial \lambda_k} Z(\lambda_1, \ldots, \lambda_K). \tag{5.85}$$

Example 5.8 Maximum entropy motivation (cont.)
We apply the Lagrange multipliers method to Example 5.7 in order to find the probabilities $p(y)$, $y = 1, \ldots, 6$ that maximizes the entropy. We begin by formulating the Lagrangian function based on our constraints (5.76a) and (5.76b),

$$
L = -\sum_{y=1}^{6} p(y) \log p(y) + \lambda \left(4.5 - \sum_{y=1}^{6} y \cdot p(y) \right) + \gamma \left(\sum_{y=1}^{6} p(y) - 1 \right). \quad (5.86)
$$

By differentiating L with respect to the frequencies $p(y)$ and normalizing in order to make the frequencies sum to one, according to (5.83) and (5.84) we achieve

$$
p(y) = \frac{e^{\lambda y}}{\sum_{y=1}^{6} e^{\lambda y}}. \quad (5.87)
$$

The coefficient γ cancels out in the normalization, and the coefficient λ is given by solving the equation

$$
4.5 = \frac{d}{d\lambda} \sum_{y=1}^{6} e^{\lambda y} = \sum_{y=1}^{6} y e^{\lambda y}. \quad (5.88)
$$

This equation can be solved numerically by any standard computational software, giving $\lambda \approx -0.400$ and the corresponding maximum entropy distribution

$$
\{p(1), p(2), p(3), p(4), p(5), p(6)\} = \{0.363, 0.243, 0.163, 0.109, 0.073, 0.049\}.
$$
$\square$

An generalization of the maximum entropy method, called the *relative entropy model*, uses the Kullback–Leibler distance to relate the probabilities $p(Y)$ to a more general distribution than the uniform. For instance, under the given constraints we may want to choose the model that is the closest to the background distribution rather than to the uniform distribution. The Kullback–Leibler distance can be written as

$$
KL(p||q) = \sum_{y} p(y) \log \frac{p(y)}{q(y)} \quad (5.89)
$$

where q is now the background distribution. The *principle of minimum relative entropy* is to choose the distribution p^* with the smallest Kullback–Leibler distance. That is, among all distributions satisfying the given constraints p^* is the distribution closest to the background distribution q. If q is the uniform distribution, minimizing the Kullback–Leibler distance is equivalent to maximizing the Shannon entropy.

While Lagrange multipliers are a common choice for solving the maximum entropy optimization problem, it becomes impractical in the application to splice site detection. Another approach is to solve the problem numerically, using a strategy called *iterative scaling*, which will be briefly introduced below.

Application to Splice Site Detection

In what follows we describe the application of maximum entropy to splice site detection presented in [69]. We assume as in previous sections that we have a splice signal $Y = (Y_1, \ldots, Y_\lambda)$ of length λ, where $Y_t \in \{A, C, G, T\}$, and we want to determine a model for the joint probability $p(Y) = \mathbf{P}(Y_1 = y_1, \ldots, Y_\lambda = y_\lambda)$. A DNA signal of length λ can take on 4^λ values, or we can say that the signal has 4^λ possible "states." Based on a training set $\mathscr{A}$ of known signal sequences, we would like to determine a set of constraints and find the maximum entropy approximation of $p(Y)$. When applied to splice site detection we would like the constraints to incorporate base compositions and position dependencies that are characteristic for such signals.

In [69] the constraints are expressed in terms of subsets of the marginal distributions of $p(Y)$, estimated from the training set. Two types of constraints are defined: *complete* constraints and *specific* constraints. The complete constraints involve all lower order marginal distributions of $p(Y)$. For instance, if $\lambda = 3$, the set of all lower order marginal distributions consist of all first- and second-order margins

$$S_Y = \{p(Y_1), p(Y_2), p(Y_3), p(Y_1, Y_2), p(Y_1, Y_3), p(Y_2, Y_3)\}. \tag{5.90}$$

Different subsets of S_Y then specify different models of $p(Y)$. If, for instance, only the first-order marginals $p(Y_t)$ are used as constraints, the resulting model is an ordinary weight matrix. If second-order, nearest-neighbor constraints are used such as $p(Y_1, Y_2)$ and $p(Y_2, Y_3)$ but not $p(Y_1, Y_3)$), the maximum entropy model is an inhomogeneous first-order Markov model.

The specific constraints are observed frequencies for certain complete constraints. For instance, the specific constraints of $p(Y_1, Y_3)$ are the 16 frequencies of the trinucleotides

$$\{ANA, ANC, ANG, ANT, \ldots, TNC, TNG, TNT\}, \quad N \in \{A, C, G, T\}.$$

Once we have settled on the set of constraints we want to find the model with the largest entropy under these constraints. A common approach is to use Lagrange multipliers [34], but due to the sheer size of our set of constraints, this tends to be impractical. Instead we use a method called *iterative scaling*, that starts out with the uniform distribution and iteratively improves it toward the maximum entropy distribution.

Iterative Scaling

We would like to determine the distribution $p(Y)$ that has the maximum entropy under our set of specific constraints. The *iterative scaling* algorithm, first introduced by [10], is a method for computing approximate probability distributions. It can be shown that the following procedure converges to the distribution with the largest entropy among those consistent with the given marginal distributions [69].

We begin by specifying a set of complete constraints S_Y and a corresponding list of specific constraints f_k, $k = 1, \ldots, K$, where the subscript k now signifies the order in the list. The iteration is initialized with a uniform distribution over all λ-lengthed sequences

$$\hat{p}^{(0)}(Y) = 4^{-\lambda}, \tag{5.91}$$

where we let $\hat{p}^{(j)}(Y)$ denote the approximation of the joint probability $p(Y)$ after the jth iteration. In iteration j, we apply each of the constraints, one at a time. The approximated probability is then updated using the formula

$$\hat{p}^{(j)}(Y) = \hat{p}^{(j-1)}(Y)\frac{f_k(Y)}{\hat{f}_k^{(j-1)}(Y)}, \tag{5.92}$$

where $\hat{f}_k^{(j-1)}(Y)$ is the value of the marginal corresponding to the kth constraint in the $(j-1)$th step of the iteration. To illustrate, consider the example with a signal length of $\lambda = 3$ above, and consider the specific constraint $f_k = ANA$. In the jth step of the iteration we calculate

$$\hat{p}^{(j)}(ANA) = \hat{p}^{(j-1)}(ANA)\frac{f_k(ANA)}{\hat{f}_k^{(j-1)}(ANA)}, \tag{5.93}$$

where

$$\hat{f}_k^{(j-1)}(ANA) = \sum_{N \in \{A,C,G,T\}} \hat{p}^{(j-1)}(ANA). \tag{5.94}$$

The probability of all other triplets (not matching ANA), is derived as follows

$$\hat{p}^{(j)}(VNW) = \hat{p}^{(j-1)}(VNW)\frac{1 - f_k(VNW)}{1 - \hat{f}_k^{(j-1)}(VNW)}, \tag{5.95}$$

where $V \neq A$, $W \neq A$ and $N \in \{A, C, G, T\}$. In each iteration, each constraint is imposed on the approximated probability in the order of the list. The process is iterated until convergence. To speed things up it is possible to rank the constraints and impose them in order of importance. This can be done roughly as follows.

1. Determine the maximum entropy distribution that satisfies the so far ranked constraints (uniform in the first step).
2. For each unranked constraint, determine the increase in entropy achieved by applying that constraint.
3. Place the constraint that causes the largest increase in entropy next in rank among the ranked constraints.

Continue these steps until all constraints have been ranked. Naturally, the constraints can be ranked individually like this, or in groups if that is more sufficient for the application. As mentioned earlier, the Kullback–Leibler distance can be used instead of the Shannon entropy, if we instead want to minimize the distance to the background distribution rather than to the uniform. In the ranking step, the highest ranking constraint is then the one that causes the largest reduction in Kullback–Leibler distance.

5.4.7 Bayesian Networks

Bayesian networks, first coined by J. Pearl [51], can be seen as graphical representations of the joint probability distribution of a set of random variables. The graph is a *directed acyclic graph* (DAG), where the nodes in the graph correspond to the random variables, and the edges to the conditional independence structure between the nodes. The causal relationships are represented by the conditional probability distributions that are connected with each node. Given the state of the parent nodes, each node contains the possible values (or states) of the corresponding random variable, as well as the conditional probability distribution of these states.

Preliminaries

Any joint distribution of a set of random variables $Y_1, \ldots, Y_T$ can be factorized as

$$\mathbf{P}(Y_1, \ldots, Y_T) = \prod_{t=1}^{T} \mathbf{P}(Y_t | Y_1, \ldots, Y_{t-1}). \tag{5.96}$$

Moreover, any such joint distribution could be represented by a directed acyclic graph $G = (V, E)$ (see Fig. 5.17), where the nodes V correspond to the random variables, and the directed edges E to the conditional distributions in the factorization.
Now assume that for every variable Y_t, $t = 1, \ldots, T$ there is a nonempty subset $V_t \subseteq \{Y_1, \ldots, Y_{t-1}\}$ such that

$$\mathbf{P}(Y_t | Y_1, \ldots, Y_{t-1}) = \mathbf{P}(Y_t | V_t). \tag{5.97}$$

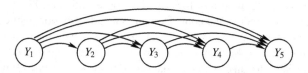

Fig. 5.17 A complete Bayesian network of the joint distribution $\mathbf{P}(Y_1, \ldots, Y_5)$

That is, Y_t is conditionally independent of $\{Y_1, \ldots, Y_{t-1}\} \backslash V_t$ given V_t. The joint probability thus becomes

$$\mathbf{P}(Y_1, \ldots, Y_T) = \prod_{t=1}^{T} \mathbf{P}(Y_t | V_t). \tag{5.98}$$

A Bayesian network is such a graph G, where the set V_t is simply the parent nodes $\mathbf{Y}_{\mathrm{pa}(t)}$ of Y_t. Thus, the joint probability can be factorized as

$$\mathbf{P}(Y_1, \ldots, Y_T) = \prod_{t=1}^{T} \mathbf{P}(Y_t | \mathbf{Y}_{\mathrm{pa}(t)}). \tag{5.99}$$

This conditioning is known as the *local*, or *causal* Markov property: given the parent nodes $\mathbf{Y}_{\mathrm{pa}(t)}$, the node Y_t is conditionally independent of all other previous nodes (or, in fact, all other *non-descendant* nodes). The process of constructing a Bayesian network can be summarized as follows:

1. Specify the set of variables of the network.
2. Order the variables, and determine the causal relationships between them.
3. Determine the different states of each variable.
4. Determine the conditional probability distributions of each variable, given its parents.

It is the causal relationships that determine the parent-descendant structure in (5.98). A key issue is to order the variables properly, since a "bad" ordering may miss some of the conditional dependencies present among the variables. In the worst case scenario we need to search through all $T!$ orderings of the variables in order to determine the best one. Fortunately, there are efficient methods for learning causal relationships from data. It is the last step above, however, that we will treat in this section: determining the conditional probabilities of each variable.

Some Bayesian Theory

Let X denote a discrete random variable with probability distribution $\mathbf{P}(X = x | \theta)$ characterized by some parameter θ. Suppose we want to determine the value of θ using a random sample $\mathbf{X} = (X_1, \ldots, X_n)$ of independent, identically distributed (i.i.d.) variables drawn from the distribution of X. Note the difference between the sample $X_1, \ldots, X_n$ here and the set of random variables $Y_1, \ldots, Y_T$ above; $X_1, \ldots, X_n$ are independent and come from the same distribution, while $Y_1, \ldots, Y_T$ are not assumed independent and not (necessarily) identically distributed.

In classical inference theory we would treat θ as a fixed, but unknown constant, and we would search for an estimate $\hat{\theta}$, that is optimal in some sense. Typical characteristics of interest are the *bias* and *variance* of the estimator, and the most common approach to the optimization is maximum likelihood (ML) estimation (see Sect. 6.3), resulting in an estimator that is both unbiased and has a minimal variance among all unbiased estimators. In Bayesian theory, however, we treat the random sample $\mathbf{X}$ as fixed, and introduce a level of uncertainty into the parameter θ instead. That is, the value θ is considered to be an observation of some random variable Θ, which has a corresponding probability distribution $\mathbf{P}(\Theta = \theta)$. Moreover, in Bayesian theory we use not only the observations $\mathbf{X}$ in the estimation of θ, but include additional background knowledge, or beliefs, denote it ξ, to infer and update the distribution of the underlying random variable Θ. That is, we first formulate a *prior* distribution of the model parameter Θ that captures our background knowledge ξ. After observing the data, we then apply Bayes' rule to obtain the *posterior* distribution of Θ, representing the updated beliefs of Θ after the new evidence $\mathbf{X}$ has been presented. The posterior distribution of Θ is then given by

$$\mathbf{P}(\Theta|\mathbf{X}, \xi) = \frac{\mathbf{P}(\Theta|\xi)\mathbf{P}(\mathbf{X}|\Theta, \xi)}{\mathbf{P}(\mathbf{X}|\xi)}, \qquad (5.100)$$

where $\mathbf{P}(\Theta|\xi)$ is the prior probability, and $\mathbf{P}(\Theta|\mathbf{X}, \xi)$ is the posterior probability. The term $\mathbf{P}(\mathbf{X}|\Theta, \xi)$, which is the probability distribution of $\mathbf{X}$, is called the *likelihood* when Θ is fixed, and gives the probability of observing sample $\mathbf{X}$ for a given Θ. The term in the denominator, $\mathbf{P}(\mathbf{X}|\xi)$, is a normalizing factor that can be achieved by summing over all possible values of Θ

$$\mathbf{P}(\mathbf{X}|\xi) = \int_{\Theta} \mathbf{P}(\mathbf{X}|\theta, \xi)\mathbf{P}(\theta|\xi)d\theta. \qquad (5.101)$$

From the posterior distribution we can calculate *predictive* distributions of future observation X_{n+1}, given our training sample and our background information,

$$\mathbf{P}(X_{n+1}|\mathbf{X}, \xi) = \int_{\Theta} \mathbf{P}(X_{n+1}|\theta, \xi)\mathbf{P}(\theta|\mathbf{X}, \xi)d\theta. \qquad (5.102)$$

If the likelihood function in (5.102) belongs to the *exponential family* this probability exists in closed form and is fairly easy to compute. Examples of distributions that belong to the exponential family are the normal (Gaussian), the exponential, the Poisson, the Gamma, the Beta, the binomial, the geometric and the multinomial distributions. In terms of Bayesian inference, a convenient property of the exponential family is that if the likelihood function $\mathbf{P}(\mathbf{X}|\Theta, \xi)$ belongs to the exponential family it has a simple *conjugate prior* that often belongs to the exponential family as well. Choosing the prior $\mathbf{P}(\Theta|\xi)$ is often a compromise between invoking background knowledge of Θ and choosing a mathematically convenient form of the distribution. A useful strategy is to choose the prior distribution in such a way that the posterior distribution $\mathbf{P}(\Theta|\mathbf{X}, \xi)$ follows the same distribution. The distribution depends on the likelihood function, and a prior chosen this way is said to be *conjugate* to the likelihood.

Now assume that the distribution of X is the outcome of a *multinomial* experiment. That is, while a *binomial* experiment results in an outcome belonging to one of two categories (e.g., success or failure), the multinomial distribution is the generalization to when each experiment has m possible outcomes $(c_1, \ldots, c_m)$. We denote the individual probabilities of falling into a specific outcome category

$$\theta_i = \mathbf{P}(X = c_i | \boldsymbol{\theta}, \xi), \quad i = 1, \ldots, m, \tag{5.103}$$

where $\boldsymbol{\theta} = (\theta_1, \ldots, \theta_m)$ is the vector of all cell probabilities such that $\sum_{i=1}^{m} \theta_i = 1$. Thus, if $\mathbf{X} = (X_1, \ldots, X_n)$ is a set of independent multinomial experiments, $\mathbf{X}$ has a multinomial distribution with cell probabilities $\boldsymbol{\theta} = (\theta_1, \ldots, \theta_m)$. The probability distribution of $\mathbf{X}$ is given by

$$\mathbf{P}(X_1 = x_1, \ldots, X_n = x_n | \boldsymbol{\theta}, \xi) = \frac{n!}{n_1! \cdots n_m!} \prod_{i=1}^{m} \theta_i^{n_i}, \tag{5.104}$$

where $n_1 + \cdots + n_m = n$ and $n_1, \ldots, n_m$ are the number of X_i's in each cell category $c_1, \ldots, c_m$, respectively. The conjugate prior of the multinomial distribution is the *Dirichlet distribution*,

$$\mathbf{P}(\Theta | \xi) = \mathrm{Dir}(\alpha_1, \ldots, \alpha_m) = \frac{1}{B(\boldsymbol{\alpha})} \prod_{i=1}^{m} \theta_i^{\alpha_i - 1}. \tag{5.105}$$

The term $B(\boldsymbol{\theta})$ is a normalizing factor that can be expressed in terms of the *Gamma distribution* Γ

$$B(\boldsymbol{\alpha}) = \frac{\prod_{i=1}^{m} \Gamma(\alpha_i)}{\Gamma(\boldsymbol{\alpha})}, \tag{5.106}$$

where $\Gamma(x) = (x-1)!$ and $\Gamma(1) = 1$ for an integer $x > 0$. The α parameters, with $\alpha_i > 0$ and $\boldsymbol{\alpha} = \sum_{i=1}^{m} \alpha_i$, are sometimes referred to as *hyperparameters*. The posterior distribution of Θ is also a Dirichlet distribution given by

$$\mathbf{P}(\Theta | \mathbf{X}, \xi) = \mathrm{Dir}(\alpha_1 + n_1, \ldots, \alpha_m + n_m) = \frac{1}{B(\boldsymbol{\alpha} + n)} \prod_{i=1}^{m} \theta_i^{\alpha_i + n_i - 1}. \tag{5.107}$$

Now, the probability of a new observation X_{n+1} falling into category c_i say, is given by (5.102). That is,

$$\mathbf{P}(X_{n+1} = c_i | \mathbf{X}, \boldsymbol{\theta}, \xi) = \int_{\Theta} \theta_i \, \mathbf{P}(\Theta | \mathbf{X}, \xi) \, \mathrm{d}\theta = \frac{\alpha_i + n_i}{\alpha + n} \tag{5.108}$$

In other words, the prediction probability of X_{n+1} is the expectation of Θ_i, under the posterior distribution.

Training a Bayesian Network

There are many methods for training a Bayesian network, involving various levels of modification to the initial network structure and estimation of the conditional probability distributions. Here we only give a brief overview of the simplest case, in which we assume that the network structure is given, and we only need to train the probabilities. For a more thorough treatment of Bayesian networks, see for instance [18].

Assume that we have decided on a structure of the Bayesian network, $G = (V, E)$, where the nodes V correspond to a set of random variables $\mathbf{Y} = (Y_1, \ldots, Y_T)$, with distributions parametrized by vectors $\boldsymbol{\theta}_G = (\boldsymbol{\theta}_1, \ldots, \boldsymbol{\theta}_T)$. Now we are back to the general case where the Y_t's may be both dependent and have differing distributions. We want to train the parameters using a training set $\mathscr{A} = \{\mathbf{Y}_1, \ldots, \mathbf{Y}_p\}$ of *examples*, or *configurations*, of the network. In accordance with Bayesian inference we assume that $\boldsymbol{\theta}_G$ is an observation of a vector of random variables Θ_G, and that the training problem involves determining the posterior probability distribution $\mathbf{P}(\Theta_G|\mathscr{A}, G)$, given the training set $\mathscr{A}$ and the background information, which is now the network structure G. Recall that a Bayesian network is just a graphical representation of a joint distribution over the variables $Y_1, \ldots, Y_T$. There are many methods available to train such a distribution, coming from a variety of fields including regression theory, neural networks and decision trees. The most commonly used approach in Bayesian networks, however, is to use the multinomial sampling methodology described in the previous section, which is also the case that we treat here. For more details, see for instance [22].

Hence, we let each variable Y_t in the network represent a multinomial experiment, each with m possible outcomes $c_1, \ldots, c_m$. Furthermore, if $\mathbf{Y}_{\mathrm{pa}(t)}$ denotes the set of parent nodes of Y_t, we let $\mathbf{y}^1_{\mathrm{pa}(t)}, \ldots, \mathbf{y}^{r_t}_{\mathrm{pa}(t)}$ denote the set of possible configurations of $\mathbf{Y}_{\mathrm{pa}(t)}$, where r_t is the number of possible configurations ($r_t = m^{a_t}$ for a_t number of nodes in $\mathbf{Y}_{\mathrm{pa}(t)}$). There is, thus, mr_t conditional probabilities associated with each node Y_t; one for each combination of m outcomes of Y_t and r_t possible configurations of $\mathbf{Y}_{\mathrm{pa}(t)}$. We denote the probability of Y_t being in category c_j, given that the parents are in configuration $\mathbf{y}^i_{\mathrm{pa}(t)}$, as

$$\theta_{tij} = \mathbf{P}(Y_t = c_j|\mathbf{y}^i_{\mathrm{pa}(t)}, \boldsymbol{\theta}_t, G) \tag{5.109}$$

where $\boldsymbol{\theta}_t = (\boldsymbol{\theta}_{t1}, \ldots, \boldsymbol{\theta}_{tr_t})$ and $\boldsymbol{\theta}_{ti} = (\theta_{ti1}, \ldots, \theta_{tir_t})$. Assuming independence between different configurations of the nodes $\{Y_t, \mathbf{Y}_{\mathrm{pa}(t)}\}$, we can write the posterior probability of Θ_G as

$$\mathbf{P}(\Theta_G|E, G) = \prod_{t=1}^{T} \prod_{i=1}^{r_t} \mathbf{P}(\Theta_{ti}|E, G). \tag{5.110}$$

In other words, with this independence assumption, we can update each parameter vector $\boldsymbol{\theta}_{ti}$ separately.

Since the conjugate prior of the multinomial distribution is the Dirichlet distribution, the posterior distribution of the parameter vector Θ_{ti} is given by

$$\mathbf{P}(\Theta_{ti}|E, G) = \mathrm{Dir}(\alpha_{ti1} + n_{ti1}, \ldots, \alpha_{tir_t} + n_{tir_t}), \tag{5.111}$$

where n_{tij} is the number of examples in the training set $\mathscr{A}$ with $Y_t = c_j$ and $\mathbf{Y}_{\mathrm{pa}(t)} = \mathbf{y}^i_{\mathrm{pa}(t)}$. Recall that the prediction probability of a new configuration $\mathbf{Y}_{p+1}$ is the expectation of Θ_G under the posterior distribution. Thus, the probability of observing category c_j in node Y_t in the new configuration, given some parent configuration $\mathbf{y}^i_{\mathrm{pa}(t)}$ is given by

$$\mathbf{P}(Y_t = c_j | \mathbf{y}^i_{\mathrm{pa}(t)}, G) = \frac{\alpha_{tij} + n_{tij}}{\alpha_{ti} + n_{ti}} \qquad (5.112)$$

where $\alpha_{ti} = \sum_{j=1}^{r_t} \alpha_{tij}$ and $n_{ti} \sum_{j=1}^{r_t} n_{tij}$. Because of the independence assumption, the probability of the entire configuration $\mathbf{Y}_{p+1}$ is then just the product of all individual node probabilities

$$\mathbf{P}(\mathbf{Y}_{p+1} | E, G) = \prod_{t=1}^{T} \mathbf{P}(Y_t | \mathbf{y}_{\mathrm{pa}(t)}, G). \qquad (5.113)$$

Application to Splice Site Detection

Since a Bayesian network can model any joint distribution it has the ability to capture both adjacent and nonadjacent dependencies between positions in a sequence signal, and has been successfully applied to splice site detection for instance in [17]. The set of variables $Y_1, \ldots, Y_\lambda$ is now the splice signal of length λ, and the multinomial distributions used for the nodes run over the four possible nucleotide "categories" $\{A, C, G, T\}$.

In the previous section we assumed that the network structure $G = (V, E)$ was given, but this can be determined from the training set as well. In the Bayesian mindset this means that we introduce uncertainty into the structure as well, such that the network G is thought to be drawn from some distribution of networks. The training set $\mathscr{A}$ is then used to compute the posterior distributions of both the parameter vector $\mathbf{P}(\Theta_G | \mathscr{A}, G)$ and the network structure $\mathbf{P}(G | \mathscr{A})$. The probability of a new configuration is given as before by averaging the expectation in (5.113) over all possible structures

$$\mathbf{P}(\mathbf{Y}_{p+1} | \mathscr{A}) = \sum_G \mathbf{P}(G | \mathscr{A}) \mathbf{P}(\mathbf{Y}_{p+1} | \mathscr{A}, G). \qquad (5.114)$$

The conditional probability on the right hand side is as in the previous section, and the posterior distribution of the network is given using Bayes' rule

$$\mathbf{P}(G | \mathscr{A}) = \frac{\mathbf{P}(G)\mathbf{P}(\mathscr{A} | G)}{\mathbf{P}(\mathscr{A})}, \qquad (5.115)$$

where $\mathbf{P}(\mathscr{A})$ is a normalizing factor and does not depend on the network structure. In the general case, the likelihood $\mathbf{P}(\mathscr{A} | G)$ can be complicated to compute, but with multinomial sampling and independence between configurations, the likelihood is simply the product of likelihoods for each configuration of $\{Y_t, \mathbf{Y}_{\text{pa}(t)}\}$ [22]

$$\mathbf{P}(\mathscr{A}|G) = \prod_{t=1}^{T} \prod_{i=1}^{r_t} \frac{\Gamma(\alpha_{ti})}{\Gamma(\alpha_{ti}) + n_{ti}} \prod_{j=1}^{m} \frac{\Gamma(\alpha_{tij} + n_{tij})}{\Gamma(\alpha_{tij})}. \tag{5.116}$$

One problem, however, is that the summation over all possible structures in (5.114) becomes infeasible already for fairly small networks. Therefore, in order to automatically train both structure and parameters of the network from training data, we need a good search method and a scoring metric that ranks the candidate structures. Typically the search is initiated by a graph G that has no edges, and then a set of candidate DAGs are created successively using the search method. The algorithm used in [17] is an *inclusion-driven structure learning algorithm* first introduced in [36]. The algorithm is optimal in the sense that if the graph is sampled from the correct distribution and if the scoring metric is consistent, the limiting structure will be the correct one [18]. The resulting inclusion-driven Bayesian network, or *idl*BN, is then used to score new configurations as in the previous section.

For the purpose of splice site detection [17] two networks are trained, one for true signals and one for false. The decision for a new candidate splice site $\mathbf{Y} = (Y_1, \ldots, Y_T)$ is then made using the usual likelihood ratio test

$$\frac{\mathbf{P}(\mathbf{Y}|E_S, G)}{\mathbf{P}(\mathbf{Y}|E_N, G)} \begin{cases} > \eta & \Rightarrow \text{signal}, \\ < \eta & \Rightarrow \text{non-signal}, \end{cases} \tag{5.117}$$

where E_S and E_N are the training sets of signals and non-signals respectively.

5.4.8 Support Vector Machines

Support vector machines (SVMs), first introduced by Cortes and Vapnik [23], constitute a machine learning technique typically used in regression and binary classification problems. As we shall see, SVMs share large similarities with linear discriminant analysis (LDA) described in Sect. 5.4.5. In both methods the goal is to find the best separation of two classes of objects, based on a set of features defined for each object. The main difference is that while LDA uses a probabilistic approach, assuming multivariate normal distributions of the feature vectors given, SVMs are purely deterministic.

The idea of SVMs is to identify the boundary that best separates the classes. If the classes are linearly separable, this boundary is a *hyperplane* (i.e., a linear combination), in feature space. In the nonlinear case we instead define a *kernel function* that transforms the nonlinear problem into a higher dimensional space, in which the classes are linearly separable again.

Linearly Separable Classes

When the two classes are linearly separable, the SVM approach is to construct two parallel *supporting* hyperplanes, one for each class, at a maximal distance from each other. The model that best separates the classes is then the hyperplane that lies right in between these two. A plane is said to *support* the class, if all the points of the class lie on the same side of that plane, and the procedure is to construct two parallel hyperplanes that lie between the two classes of points, and then push them apart until we bump into the first point(s) of the respective class. These touching points are then called the *support vectors* of the respective class.

Just as in linear discriminant analysis we are given a training set of objects $Y_1, \ldots, Y_n$, each with a known class label $C_i \in \{-1, +1\}$, and each represented by a feature vector $\mathbf{x}_i = (x_{i1}, \ldots, x_{ip})^T$, $i = 1, \ldots, n$, corresponding to a point in a p-dimensional real space $\mathbb{R}^p$. The assumption that the classes are linearly separable means that they can be separated by a hyperplane given by

$$\boldsymbol{\alpha}^T \mathbf{x} + b = 0, \tag{5.118}$$

where $\boldsymbol{\alpha}$ is a *normal* vector (orthogonal to the hyperplane), $b/\|\boldsymbol{\alpha}\|$ is the orthogonal distance from the hyperplane to the origin (i.e., along the vector $\boldsymbol{\alpha}$), and $\|\boldsymbol{\alpha}\|$ is the *norm* giving the length of the vector $\boldsymbol{\alpha}$. We want to determine the supporting planes of the two classes, which is to say that we want to find the vector $\boldsymbol{\alpha}$ and a constant b such that, for $i = 1, \ldots, n$

$$\boldsymbol{\alpha}^T \mathbf{x}_i + b \leq -1 \quad \text{for } C_i = -1, \tag{5.119}$$
$$\boldsymbol{\alpha}^T \mathbf{x}_i + b \geq +1 \quad \text{for } C_i = +1. \tag{5.120}$$

Due to our class representation, using -1 and $+1$ instead of for instance 0 and 1, the two inequalities can be combined into a single expression as

$$C_i(\boldsymbol{\alpha}^T \mathbf{x}_i + b) \geq 1, \quad i = 1, \ldots, n. \tag{5.121}$$

The support vectors, specifying the two supporting hyperplanes, call them H_{-1} and H_{+1}, would then be the points satisfying

$$H_{-1} = \{\mathbf{x}_i : \boldsymbol{\alpha}^T \mathbf{x}_i + b = -1, \ C_i = -1\},$$
$$H_{+1} = \{\mathbf{x}_i : \boldsymbol{\alpha}^T \mathbf{x}_i + b = +1, \ C_i = +1\}.$$

We want to place a separating hyperplane H in between H_{-1} and H_{+1} such that the distances d_{-1} and d_{+1} between the respective class hyperplanes and H are equal, that is $d_{-1} = d_{+1}$. This distance is called the *margin* of the SVM, and the hyperplane that best separates the two classes is the one that maximizes the margin.

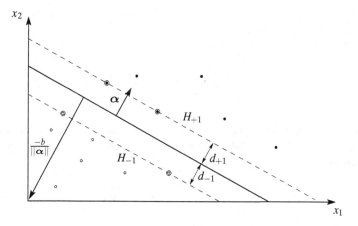

Fig. 5.18 An SVM in two dimensions

Example 5.9 An illustration SVMs in 2D
Suppose we are given a set of training objects $Y_1, \ldots, Y_n$ with two-dimensional feature vectors $\mathbf{x}_i = (x_{i1}, x_{i2})^T$ and class labels $C_i \in \{-1, +1\}$, $i = 1, \ldots, n$. Assuming that the two classes are linearly separable, a separating "hyperplane" is now a line given by

$$\boldsymbol{\alpha}^T \mathbf{x} + b = \alpha_1 x_1 + \alpha_2 x_2 + b = 0. \tag{5.122}$$

Figure 5.18 illustrates the situation. The feature vectors are represented as points in a two-dimensional space, where the white dots belong to class -1, and the black dots to class $+1$. The two supporting hyperplanes, represented by dashed lines, are determined by the support vectors, which are the circled points coinciding with the dashed lines, and the separating hyperplane is the solid line drawn right in between. A new object Y_{n+1} is then classified to class -1 or $+1$ depending on which side of the separating hyperplane its feature vector falls. $\square$

Simple geometry gives that the margin is in fact, $d_{-1} = d_{+1} = 1/\|\boldsymbol{\alpha}\|$, so in order to maximize this distance, we need to minimize $\|\boldsymbol{\alpha}\|$ under the constraints $C_i(\boldsymbol{\alpha}^T \mathbf{x}_i + b) \geq 1$, $i = 1, \ldots, n$. Thus, the optimization problem we want to solve can be stated as follows:

$$\boldsymbol{\alpha}^* = \underset{\boldsymbol{\alpha}}{\operatorname{argmin}} \{ \|\boldsymbol{\alpha}\| : C_i(\boldsymbol{\alpha}^T \mathbf{x}_i + b) \geq 1, \ \forall i \}. \tag{5.123}$$

This is a rather difficult problem, however, since the norm $\|\boldsymbol{\alpha}\|$ involves the square root of $\boldsymbol{\alpha}$. A simpler, but equivalent problem is to minimize $\frac{1}{2}\|\boldsymbol{\alpha}\|^2$ instead,

$$\boldsymbol{\alpha}^* = \underset{\boldsymbol{\alpha}}{\operatorname{argmin}} \left\{ \frac{1}{2}\|\boldsymbol{\alpha}\|^2 : C_i(\boldsymbol{\alpha}^T \mathbf{x}_i + b) \geq 1, \ \forall i \right\}. \tag{5.124}$$

As mentioned in Sect. 5.4.6, optimization under a set of given constraints is typically done using *Lagrange multipliers*.

We therefore introduce the Lagrangian function

$$L = \frac{1}{2}||\boldsymbol{\alpha}||^2 - \sum_{i=1}^{n} \lambda_i \left(C_i (\boldsymbol{\alpha}^T \mathbf{x}_i + b) - 1 \right), \tag{5.125}$$

where $\lambda_1, \ldots, \lambda_n$ are the Lagrange multipliers. We want to find the $\boldsymbol{\alpha}$ and the b that minimize L, which involves differentiation of L and setting the derivatives to zero. This results in

$$\frac{\partial L}{\partial \boldsymbol{\alpha}} = 0 \Leftrightarrow \boldsymbol{\alpha} = \sum_{i=1}^{n} \lambda_i C_i \mathbf{x}_i, \tag{5.126a}$$

$$\frac{\partial L}{\partial b} = 0 \Leftrightarrow \sum_{i=1}^{n} \lambda_i C_i = 0. \tag{5.126b}$$

Substituting these equalities into L gives us the *dual form* of L,

$$L_D = \sum_{i=1}^{n} \lambda_i - \frac{1}{2} \sum_{i,j} \lambda_i \lambda_j C_i C_j \mathbf{x}_i^T \mathbf{x}_i, \tag{5.127}$$

which we now need to *maximize* over $\boldsymbol{\alpha}$ subject to $\lambda_i \geq 0$ and $\sum_{i=1}^{n} \lambda_i C_i = 0$. This is a convex quadratic problem which allows us to use something called *quadratic programming* (QP) to determine the λ_i's. While *linear programming* involves optimizing a linear objective function, quadratic programming includes methods for optimizing a quadratic objective function subject to linear constraints. Then, inserting the solution achieved by QP for the λ_i's in (5.126a) gives us the solution for $\boldsymbol{\alpha}$, and what remains is to solve for the parameter b.

First we note that the support vectors of the two supporting hyperplanes are those points $\mathbf{x}_i$ where the Lagrange multiplier is positive, $\lambda_i > 0$. Thus, let V be this set of support vectors

$$V = \{\mathbf{x}_i : \lambda_i > 0\}. \tag{5.128}$$

We also note that for each such support vector $\mathbf{x}_v \in V$ it holds that

$$C_v (\boldsymbol{\alpha}^T \mathbf{x}_v + b) = 1. \tag{5.129}$$

Thus, using (5.126a) again gives us the result for b

$$b = \frac{1}{|V|} \sum_{v \in V} \left(C_v - \sum_{u \in V} \lambda_u C_u \mathbf{x}_u^T \mathbf{x}_v \right) \tag{5.130}$$

where $|V|$ is the number of elements in V.

Once we have determined $\boldsymbol{\alpha}$ and b we are done, and the model can be used for classification. A new object Y with feature vector $\mathbf{x}$ is classified using the decision function

$$\text{class}(Y) = \text{sign}(\boldsymbol{\alpha}^T \mathbf{x} + b), \tag{5.131}$$

such that Y is classified as -1 if $\text{class}(Y) < 0$ and as $+1$ otherwise.

Nearly Linear SVMs

The assumption of linearly separable classes is often violated in practice, and classification of DNA sequence signals certainly constitute such an example. When the classes are no longer linearly separable, the classification task is slightly more complicated.

If the classes are only *almost* linearly separable, we can deal with this by relaxing the constraints in (5.119) and (5.120) slightly and construct a *soft margin* SVM:

$$\boldsymbol{\alpha}^T \mathbf{x}_i + b \leq -1 - z_i \quad \text{for } C_i = -1, \tag{5.132a}$$
$$\boldsymbol{\alpha}^T \mathbf{x}_i + b \geq +1 + z_i \quad \text{for } C_i = +1, \tag{5.132b}$$

where $z_i \geq 0$ for all $i = 1, \ldots, n$. These constraints can be combined as before into a single constraint

$$C_i(\boldsymbol{\alpha}^T \mathbf{x}_i + b) \geq 1 - z_i. \tag{5.133}$$

In this model objects that fall within the margin between the separating hyperplane and the support vectors are penalized with a penalty that decreases with the distance to the separating hyperplane. The optimization problem becomes

$$(\boldsymbol{\alpha}^*, \mathbf{z}^*) = \underset{\boldsymbol{\alpha}, \mathbf{z}}{\text{argmin}} \left\{ \frac{1}{2}||\boldsymbol{\alpha}||^2 + C \sum_{i=1}^{n} z_i : C_i(\boldsymbol{\alpha}^T \mathbf{x}_i + b) \geq 1 - z_i \right\}, \tag{5.134}$$

where C is a parameter that represents a trade-off between minimizing the training error and maximizing the margin between the supporting hyperplanes. The dual form in (5.127) remains unchanged, but the maximization is now subject to the modified constraints $0 \leq \lambda_i \leq C$ and $\sum_{i=1}^{n} \lambda_i C_i = 0$. The variable b is calculated as before. Note however, that the set of support vectors is now given by $V = \{\mathbf{x}_i : 0 < \lambda_i < C\}$.

Nonlinear SVMs

When the classes are not linearly separable a useful trick, commonly referred to as the *kernel trick* [1], is to map the original data into a higher dimensional space where the classes are linearly separable. The classification problem in the higher dimensional space can then be solved using linear methods, and the corresponding solution is equivalent to the nonlinear classification in the original space. For a given mapping $\phi : \mathbb{R}^p \to \mathbb{R}^m$, $m > p$, a *kernel function* is defined as the dot product

$$K(\mathbf{x}_i, \mathbf{x}_j) = \phi(\mathbf{x}_i)^T \phi(\mathbf{x}_j) \qquad (5.135)$$

for real-valued p-dimensional feature vectors $\mathbf{x}_i$ and $\mathbf{x}_j$. Note that in the linearly separable case, we work with dot products on the form $\mathbf{x}_i^T \mathbf{x}_j$, which are in fact examples of *linear kernels*. To exploit the kernel trick, we thus need to define a suitable map ϕ. An alternative would be to define the kernel function directly, without explicitly stating the mapping.

Example 5.10 Mapping to higher dimensions
An example of a mapping from $\mathbb{R}^2$ to $\mathbb{R}^3$ for vector $\mathbf{x} = (x_1, x_2)^T$ is

$$\phi(\mathbf{x}) = (x_1^2, \sqrt{2}x_1 x_2, x_2^2)^T. \qquad (5.136)$$

Thus, we could for instance define the corresponding kernel as

$$K(\boldsymbol{\alpha}, \mathbf{x}) = \phi(\boldsymbol{\alpha})^T \phi(\mathbf{x}). \qquad (5.137)$$

This is just one example of many $\mathbb{R}^2 \to \mathbb{R}^3$ mappings. However, one useful feature is that we get

$$\begin{aligned} \phi(\boldsymbol{\alpha})^T \phi(\mathbf{x}) &= \alpha_1^2 x_1^2 + 2\alpha_1 \alpha_2 x_1 x_2 + \alpha_2^2 x_2^2 \\ &= (\alpha_1 x_1 + \alpha_2 x_2)^2 \\ &= (\boldsymbol{\alpha}^T \mathbf{x})^2. \end{aligned}$$

Thus, an equivalent formulation would be to define the kernel function directly

$$K(\boldsymbol{\alpha}, \mathbf{x}) = (\boldsymbol{\alpha}^T \mathbf{x})^2, \qquad (5.138)$$

without having to bother about what the underlying mapping looks like. The question is how this makes life easier.   □

Now, *Mercer's theorem* [44] gives that a continuous, symmetric, positive semi-definite kernel function can be written as the dot product of vectors in a higher dimension

$$K(\mathbf{x}_i, \mathbf{x}_j) = \phi(\mathbf{x}_i)^T \phi(\mathbf{x}_j). \qquad (5.139)$$

The kernel trick is that instead of trying to find the map ϕ, we determine a kernel directly that maps our nonlinear feature data into a space where the classes are linearly separable. By inserting the kernel in place of the dot products in our linear formulation above, we can apply the linear classification methods to our new space. The resulting solution is then a solution for the original data as well.

More specifically, instead of defining a separating hyperplane, we seek a separating boundary given by

$$\alpha^T \phi(\mathbf{x}) + b = 0, \tag{5.140}$$

and the optimization problem we want to solve is given by

$$\alpha^* = \underset{\alpha}{\text{argmin}} \left\{ \frac{1}{2} ||\alpha||^2 : C_i(\alpha^T \phi(\mathbf{x}_i) + b) \geq 1 \right\}. \tag{5.141}$$

This is solved using Lagrange multipliers as before, by simply putting $K(\mathbf{x}_i, \mathbf{x}_j)$ in place of $\mathbf{x}_i^T \mathbf{x}_j$ in (5.127) and (5.130). The dual Lagrangian L_D is maximized using quadratic programming subject to the constraints $0 \leq \lambda_i \leq C$ and $\sum_{i=1}^{n} \lambda_i C_i = 0$, where C is a parameter that controls the penalty for misclassification. Similarly, the parameter b is summarized over the support vector set given by $V = \{\mathbf{x}_i : 0 < \lambda_i < C\}$. The SVM method can be summarized as follows:

1. Select the parameter C, which determines how much misclassification should be penalized.
2. Choose a kernel function $K(\mathbf{x}_i, \mathbf{x}_i)$ and the corresponding parameters.
3. Determine the λ_i's that maximize

$$L_D = \sum_{i=1}^{n} \lambda_i - \frac{1}{2} \sum_{i,j} \lambda_i \lambda_j C_i C_j K(\mathbf{x}_i, \mathbf{x}_i) \tag{5.142}$$

   subject to the constraints $0 \leq \lambda_i \leq C$ and $\sum_i \lambda_i C_i = 0$, using a QP algorithm.
4. Determine the set of support vectors

$$V = \{\mathbf{x}_i : 0 < \lambda_i < C\}. \tag{5.143}$$

5. Calculate the threshold parameter

$$b = \frac{1}{|V|} \sum_{v \in V} \left(C_v - \sum_{u \in V} \lambda_u C_u K(\mathbf{x}_u, \mathbf{x}_v) \right). \tag{5.144}$$

6. Classify a new object Y with feature vector $\mathbf{x}$ using the nonlinear decision function

$$\text{class}(Y) = \text{sign} \left(\sum_{i=1}^{n} C_i \lambda_i K(\mathbf{x}, \mathbf{x}_i) + b \right). \tag{5.145}$$

The tricky thing is to choose a suitable kernel function. Common examples are:

- The *Gaussian* or *radial basis* kernel:

$$K(\mathbf{x}_i, \mathbf{x}_j) = e^{-\left(\frac{||\mathbf{x}_i - \mathbf{x}_j||^2}{2\sigma^2}\right)}. \tag{5.146}$$

In this case the map ϕ is infinite dimensional, resulting in an infinite summation, if we were to calculate the dot product of the map.

- The *polynomial* kernel:

$$K(\mathbf{x}_i, \mathbf{x}_j) = (\mathbf{x}_i^T \mathbf{x}_j + a)^d \tag{5.147}$$

for some parameters a and d. The kernel is said to be *homogeneous* if $a_0 = 0$, and *inhomogeneous* otherwise. The parameter d signifies the *degree* of the kernel.

- The *sigmoidal* kernel:

$$K(\mathbf{x}_i, \mathbf{x}_j) = \tanh(a_0 \mathbf{x}_i^T \mathbf{x}_j - a_1) \tag{5.148}$$

for some parameters a_0 and a_1.

There are many choices of kernel functions, with a variety of applications. However, the conditions and requirements to be placed upon such kernels are often domain specific, and beyond the scope of this book. For a more thorough treatment of SVMs and kernel-based methods, see for instance [20]. The near linear approach presented above can naturally be applied to the nonlinear situation, if transformation into the higher dimensional space is still resulting in slightly noisy data.

SVMs in Splice Site Detection

Support vector machines have been applied to various bioinformatics problems (see [46] for a review), such as translation initiation site prediction [72], gene classification [50], gene expression analysis [11, 29, 31], protein homology detection [33, 41], protein classification [40], and protein fold recognition [25]. In terms of splice site detection, a number of different kernels have been developed. Here we briefly present the ones reviewed in [55].

A feature vector $\mathbf{x} = x_1, \ldots, x_T$ is now the DNA sequence of a potential splice site, and the kernels are thus functions of pairs of such sequences. The decision function for classification is written as before

$$\text{class}(Y) = \text{sign}\left(\sum_{i=1}^n C_i \lambda_i K(\mathbf{x}, \mathbf{x}_i) + b\right). \tag{5.149}$$

The *polynomial kernel* in (5.147) is defined on real-valued objects. To accommodate this, we can for instance transform the sequence $\mathbf{x}$ into binary code. For instance, $\mathbf{x} = AACG$ could be transformed as

$$\tilde{\mathbf{x}} = \big[\mathbb{I}(x_1 = A), \mathbb{I}(x_1 = C), \mathbb{I}(x_1 = G), \mathbb{I}(x_1 = T),$$
$$\mathbb{I}(x_2 = A), \mathbb{I}(x_2 = C), \mathbb{I}(x_2 = G), \mathbb{I}(x_2 = T),$$

$$\ldots$$

$$\mathbb{I}(x_T = A), \mathbb{I}(x_T = C), \mathbb{I}(x_T = G), \mathbb{I}(x_T = T)\big]^T$$
$$= (1, 0, 0, 0, 1, 0, 0, 0, 0, 1, 0, 0, 0, 0, 1, 0)^T,$$

where $\mathbb{I}$ is the indicator function.
The polynomial kernel

$$K(\tilde{\mathbf{x}}, \tilde{\mathbf{x}}') = (\tilde{\mathbf{x}}^T \tilde{\mathbf{x}}')^d \tag{5.150}$$

takes into account all correlations of matches $\mathbb{I}(\tilde{x}_t = \tilde{x}'_t)$ up to order d. The features are position-dependent, and any position can be combined with any other position to form a feature.

The *locality improved* (LI) kernel has previously been applied to translation initiation site prediction [72]. It is similar to the polynomial kernel, as it also considers correlations of matches up to order d. The main difference, however, is that the dependencies between positions are not formed over the entire sequence, but the sequence positions are compared within a small window of length $2l + 1$. The LI kernel is presented in [55] as

$$K(\mathbf{x}, \mathbf{x}') = \sum_{p=l+1}^{T-l} w_p \mathrm{win}_p(\mathbf{x}, \mathbf{x}') \tag{5.151}$$

where $p = l + 1, \ldots, T - l$ and

$$\mathrm{win}_p(\mathbf{x}, \mathbf{x}') = \left(\frac{1}{2l+1} \sum_{j=-l}^{+l} \mathbb{I}(x_{p+j} = x'_{p+j}) \right)^d. \tag{5.152}$$

The weight w_p is a window score used to assign higher weights to more important regions. If we assume that the splice site is located in the center of the sequence, we could for instance use

$$w_p = \begin{cases} p - l & p \leq T/2 \\ N - p - l + 1 & p > T/2 \end{cases} \tag{5.153}$$

which assigns the highest weight to the center position, and then decreases linearly toward the ends.

The *weighted degree* (WD) kernel [55] is similar to both the previous kernels, except that here we count matches between "words" $\mathbf{x}_{t,k} = x_t, x_{t+1}, \ldots, x_{t+k-1}$ of length k, starting in position t. The WD kernel is given by

$$K(\mathbf{x}, \mathbf{x}') = \sum_{k=1}^{d} w_k \sum_{t=1}^{T-d} \mathbb{I}(\mathbf{x}_{t,k} = \mathbf{x}'_{t,k}) \qquad (5.154)$$

where the weights are chosen as $w_k = d - k + 1$, assigning lower weights to higher order matches.

The *TOP* kernel [65] is similar to the more common *Fisher kernel* [33] in that it introduces prior knowledge into a probabilistic model.

The TOP kernel is given by

$$K(\mathbf{x}, \mathbf{x}') = \mathbf{f}_\theta(\mathbf{x})^T \mathbf{f}_\theta(\mathbf{x}'), \qquad (5.155)$$

where $\boldsymbol{\theta} = (\theta_1, \ldots, \theta_p)^T$ is a parameter vector and

$$\mathbf{f}_\theta(\mathbf{x}) = \big(v(\mathbf{x}, \boldsymbol{\theta}), \partial_{\theta_1} v(\mathbf{x}, \boldsymbol{\theta}), \ldots, \partial_{\theta_p} v(\mathbf{x}, \boldsymbol{\theta})\big)^T. \qquad (5.156)$$

The terms $\partial_{\theta_i} v(\mathbf{x}, \boldsymbol{\theta})$ denote the partial derivatives of v with respect to θ_i, $i = 1, \ldots, p$, and

$$v(\mathbf{x}, \boldsymbol{\theta}) = \log \mathbf{P}(c = +1|\mathbf{x}, \boldsymbol{\theta}) - \log \mathbf{P}(c = -1|\mathbf{x}, \boldsymbol{\theta}). \qquad (5.157)$$

A comparison in [55] showed that the more specialized kernels, the LI and WD kernels, are best suited for single site prediction, but that the standard polynomial kernel did surprisingly well. Combining the splice site prediction into a larger, gene finding framework improved the performances considerably, as expected.

References

1. Aizerman, M., Braverman, E., Rozonoer, L.: Theoretical foundations of the potential function method in pattern recognition learning. Autom. Remote Control **25**, 821–837 (1964)
2. Alexandersson, M., Cawley, S., Pachter, L.: SLAM: cross-species gene finding and alignment with a generalized pair hidden Markov model. Genome Res. **13**, 496–502 (2003)
3. Axelson-Fisk, M., Sunnerhagen, P.: Gene finding in fungal genomes. In: Sunnerhagen, P., Piskur, J. (eds.) Topics in Current Genetics: Comparative Genomics Using Fungi as Models, pp. 1–29. Springer, Berlin (2005)
4. Bennetzen, J.L., Hall, B.D.: Codon selection in yeast. J. Biol. Chem. **257**, 3026–3031 (1982)
5. Bernardi, G.: Isochores and the evolutionary genomics of vertebrates. Gene **241**, 3–7 (2000)
6. Bernardi, G., Olofsson, B., Filipski, J., Zerial, M., Salinas, J., Cuny, G., Menier-Rotival, M., Rodier, F.: The mosaic genome of warm-blooded vertebrates. Science **228**, 953–958 (1985)
7. Biémont, C., Vieira, C.: Junk DNA as an evolutionary force. Nature **443**, 521–524 (2006)
8. Bobbio, A., Horvath, A., Telek, M.: PhFit: a general phase-type fitting tool. Proc. Dep. Syst. Netw. (DSN-02) **1**, 1 (2002)
9. Bobbio, A., Horvath, A., Scarpa, M., Telek, M.: Acyclic discrete phase type distributions: properties and a parameter estimation algorithm. Perform. Eval. **54**, 1–32 (2003)
10. Brown, D.: A note on approximations to probability distributions. Inf. Control **2**, 386–392 (1959)

11. Brown, M.P.S., Grundy, W.N., Lin, D., Cristianini, N., Sugnet, C.W., Furey, T.S., Ares, M., Haussler, D.: Knowledge-based analysis of microarray gene expression data by using support vector machines. Proc. Natl. Acad. Sci. USA **97**, 262–267 (2000)
12. Brunak, S., Engelbrecht, J., Knudsen, S.: Prediction of human mRNA donor and acceptor sites from the DNA sequence. J. Mol. Biol. **220**, 49–65 (1991)
13. Burge, C.: Identification of genes in human genomic DNA. Ph.D. thesis, Stanford University, Stanford (1997)
14. Burge, C.B.: Modeling dependencies in pre-mRNA splicing signals. In: Salzberg, S.L., Searls, D.B., Kasif, S. (eds.) Computational Methods in Molecular Biology, pp. 109–128. Elsevier, Amsterdam (1998)
15. Burge, C., Karlin, S.: Prediction of complete gene structures in human genomic DNA. J. Mol. Biol. **268**, 78–94 (1997)
16. Bühlmann, P., Wyner, A.J.: Variable length Markov chains. Ann. Stat. **27**, 480–513 (1999)
17. Castelo, R., Guigó, R.: Splice site identification with idlBNs. Bioinformatics **20**, 169–171 (2004)
18. Castelo, R., Kočka, T.: On inclusion-driven learning of Bayesian networks. J. Mach. Learn. Res. **4**, 527–574 (2003)
19. Cawley, S.: Statistical models for DNA sequencing and analysis. Ph.D. thesis, University of California, Berkeley (2000)
20. Cristianini, N., Shawe-Taylor, J.: An Introduction to Support Vector Machines and Other Kernel-Based Learning Methods. Cambridge University Press, Cambridge (2000)
21. Claverie, J.-M., Sauvaget, I., Bougueleret, L.: k-Tuple frequency analysis: from intron/exon discrimination to T-cell epitope mapping. Methods Enzymol. **183**, 237–252 (1990)
22. Cooper, G.F., Herskovits, E.: A Bayesian method for the induction of probabilistic networks from data. Mach. Learn. **9**, 309–347 (1992)
23. Cortes, C., Vapnik, V.: Support-vector networks. Mach. Learn. **20**, 273–297 (1995)
24. Crooks, G.E., Hon, G., Chandonia, J.-M., Brenner, S.E.: WebLogo: a sequence logo generator. Genome Res. **14**, 1188–1190 (2004)
25. Ding, C.H.Q., Dubchak, I.: Multi-class protein fold recognition using support vector machines and neural networks. Bioinformatics **17**, 349–358 (2001)
26. Ellrott, K., Yang, C., Sladek, F.M., Jiang, T.: Identifying transcription factor binding sites through Markov chain optimization. Bioinformatics **18**, S100–S109 (2002)
27. Fickett, J.W., Tung, C.-S.: Assessment of protein coding measures. Nucleic Acids Res. **20**, 6441–6450 (1992)
28. Fisher, R.A.: The use of multiple measurements in taxonomic problems. Ann. Eugen. **7**, 179–188 (1936)
29. Furey, T.S., Cristianini, N., Duffy, N., Bednarski, D.W., Schummer, M., Haussler, D.: Support vector machine classification and validation of cancer tissue samples using microarray expression data. Bioinformatics **16**, 906–914 (2000)
30. Gregory, T.R.: Coincidence, coevolution, or causation? DNA content, cell size, and the C-value enigma. Biol. Rev. **76**, 65–101 (2001)
31. Guyon, I., Weston, J., Barnhill, S., Vapnik, V.: Gene selection for cancer classification using support vector machines. Mach. Learn. **46**, 389–422 (2002)
32. Ikemura, T.: Correlation between the abundance of *Escherichia coli* transfer RNAs and the occurence of the respective codons in its protein genes: a proposal for a synonymous codon choice that is optimal for the *E. coli* translational system. J. Mol. Biol. **151**, 389–409 (1981)
33. Jaakola, T.S., Diekhans, M., Haussler, D.: Using the Fisher kernel method to detect remote protein homologies. Proc. Int. Conf. Intell. Syst. Mol. Biol. **7**, 149–158 (1999)
34. Jaynes, E.T.: Information theory and statistical mechanics. Phys. Rev. **106**, 620–630 (1957)
35. Jaynes, E.T.: Information theory and statistical mechanics II. In: Ford, K. (ed.) Statistical Physics, pp. 181–218. Benjamin, New York (1963)
36. Kočka, T., Castelo, R.: Improved learning of Bayesian networks. In: Proceedings of Uncertainty in Artificial Intelligence, pp. 269–276 (2001)

37. Kozak, M.: Point mutations define a sequence flanking the AUG initiator codon that modulates translation by eukaryotic ribosomes. Cell **44**, 283–292 (1986)
38. Kulp, D., Haussler, D., Reese, M.G., Eeckman, F.H.: A generalized hidden Markov model for the recognition of human genes in DNA. Proc. Int. Conf. Intell. Syst. Mol. Biol. **4**, 134–142 (1996)
39. Lander, E.S., Linton, L.M., Birren, B., Nusbaum, C., Zody, M.C., Baldwin, J., Devon, K., Dewar, K., Doyle, M., FitzHugh, W., et al.: Initial sequencing and analysis of the human genome. Nature **409**, 860–921 (2001)
40. Leslie, C.S., Eskin, E., Cohen, A., Weston, J., Noble, W.S.: Mismatch string kernels for discriminative protein classification. Bioinformatics **20**, 467–476 (2004)
41. Liao, L., Noble, W.S.: Combining pairwise sequence similarity and support vector machines for detecting remote protein evolutionary and structural relationships. J. Comput. Biol. **10**, 857–868 (2003)
42. Lukashin, A.V., Borodvsky, M.: GeneMark.hmm: new solutions for gene finding. Nucleic Acids Res. **26**, 1107–1115 (1998)
43. McLachlan, G.J.: Discriminant Analysis and Statistical Pattern Recognition. Wiley, New York (2004)
44. Mercer, J.: Functions of positive and negative type and their connection with the theory of integral equations. Philos. Trans. R. Soc. Lond. A **209**, 415–446 (1909)
45. Munch, K., Krogh, A.: Automatic generation of gene finders for euakryotic species. BMC Bioinform. **7**, 263–274 (2006)
46. Noble, W.S.: Support vector machine applications in computational biology. In: Schölkopf, B., Tsuda, K., Vert, J.-P. (eds.) Kernel Methods in Computational Biology, pp. 1–31. MIT Press, London (2004)
47. Ohler, U., Harbeck, S., Niemann, H., Nöth, E., Reese, M.G.: Interpolated Markov chains for eukaryotic promoter recognition. Bioinformatics **15**, 362–369 (1999)
48. Ohno, S.: So much "junk" DNA in our genome. Brookhaven Symp. Biol. **23**, 366–370 (1972)
49. Oliver, J.L., Bernaola-Galván, P., Carpena, P., Román-Roldán, R.: Isochore chromosome maps of eukaryotic genomes. Gene **276**, 47–56 (2001)
50. Pavlidis, P., Furey, T.S., Liberto, M., Haussler, D., Grundy, W.N.: Promoter region-based classification of genes. In: Altman, R.B., Dunker, A.K., Hunter, L., Lauderdale, K., Kelin, T.E. (eds.) Pacific Symposium of Biocomputing, pp. 151–163. World Scientific, Singapore (2001)
51. Pearl, J.: Probabilistic Reasoning in Intelligent Systems: Networks of Plausible Inference. Morgan Kaufmann, San Francisco (1988)
52. Perna, N.T., Plunkett, G., Burland, V., Mau, B., Glasner, J.D., Rose, D.J., Mayhew, G.F., Evans, P.S., Gregor, J., Kirkpatrick, H.A., Pósfai, G., Hackett, J., Klink, S., Boutin, A., Shao, Y., Miller, L., Grotbeck, E.J., Davis, N.W., Lim, A., Dimalanta, E.T., Potamousis, K.D., Apodaca, J., Anantharaman, T.S., Lin, J., Yen, G., Schwartz, D.C., Welch, R.A., Blattner, F.R.: Genome sequence of enterohaemorrhagic *Escherichia coli* O157:H7. Nature **409**, 529–533 (2001)
53. Reese, M.G., Eeckman, F.H., Kulp, D., Haussler, D.: Improved splice site detection in genie. J. Comput. Biol. **4**, 311–323 (1997)
54. Rissanen, J.: A universal data compression system. IEEE Trans. Inf. Theory **29**, 656–664 (1983)
55. Rätsch, G., Sonnenburg, S.: Accurate splice site detection for *Caenorhabditis elegans*. In: Schölkopf, B., Tsuda, K., Vert, J.-P. (eds.) Kernel Methods in Computational Biology, pp. 277–298. MIT Press, London (2004)
56. Schneider, T.D., Stephens, R.M.: Sequence logos: a new way to display consensus sequences. Nucleic Acids Res. **18**, 6097–6100 (1990)
57. Schukat-Talamazzini, E.G., Gallwitz, F., Harbeck, S., Warnke, V.: Rational interpolation of maximum likelihood predictors in stochastic language modeling. In: Proceedings of Eurospeech'97, pp. 2731–2734. Rhodes, Greece (1997)
58. Sharp, P.M., Li, W.H.: The codon adaptation index—a measure of directional synonymous codon usage bias, and its potential applications. Nucleic Acids Res. **15**, 1281–1295 (1987)
59. Shine, J., Dalgarno, L.: Determinant of cistron specificity in bacterial ribosomes. Nature **254**, 34–38 (1975)

60. Snyder, E.E., Stormo, G.D.: Identification of protein coding regions in genomic DNA. J. Mol. Biol. **248**, 1–18 (1995)
61. Solovyev, V.V., Salamov, A.A., Lawrence, C.B.: Predicting internal exons by oligonucleotide composition and discriminant analysis of spliceable open reading frames. Nucleic Acids Res. **22**, 5156–5163 (1994)
62. Solovyev, V.V., Salamov, A.A., Lawrence, C.B.: 82: identification of human gene structure using linear discriminant functions and dynamic programming. Proc. Int. Conf. Intell. Syst. Mol. Biol. **3**, 367–375 (1995)
63. Staden, R.: Computer methods to locate signals in nucleic acid sequences. Nucleic Acids Res. **12**, 505–519 (1984)
64. Staden, R., McLachlan, A.D.: Codon preference and its use in identifying protein coding regions in long DNA sequences. Nucleic Acids Res. **10**, 141–156 (1982)
65. Tsuda, K., Kawanabe, M., Rätsch, G., Sonnenburg, S., Müller, K.-R.: A new discriminative kernel from probabilistic models. Neural Comput. **14**, 2397–2414 (2002)
66. Wright, F.: The 'effective number of codons' used in a gene. Gene **87**, 23–29 (1990)
67. Xu, Y., Mural, R.J., Einstein, J.R., Shah, M.B., Uberbacher, E.C.: GRAIL: a multi-agent neural network system for gene identification. Proc. IEEE **84**, 1544–1552 (1996)
68. Xu, Y., Uberbacher, E.C.: Computational gene prediction using neural networks and similarity search. In: Salzberg, S.L., Searls, D.B., Kasif, S. (eds.) Computational Methods in Molecular Biology, pp. 109–128. Elsevier, Amsterdam (1998)
69. Yeo, G., Burge, C.B.: Maximum entropy modeling of short sequence motifs with applications to RNA splicing signals. J. Comput. Biol. **11**, 377–394 (2004)
70. Zhao, X., Huang, H., Speed, T.P.: Finding short DNA motifs using permuted Markov models. J. Comput. Biol. **12**, 894–906 (2005)
71. Zhang, M.Q., Marr, T.G.: Weight array methods for splicing signal analysis. Comput. Appl. Biosci. **9**, 499–509 (1993)
72. Zien, A., Rätsch, G., Mika, S., Schölkopf, B., Lengauer, T., Müller, K.-R.: Engineering support vector machine kernels that recognize translation initiation sites. Bioinformatics **16**, 799–807 (2000)

Chapter 6
Parameter Training

The training problem is the most difficult problem of the three hidden Markov model (HMM) problems mentioned in Sect. 2.1. It involves finding the estimates of the model parameters that maximizes the probability of the observed sequence under the given model. If we are given an observed sequence where the state labels are known for each observed residue, we could estimate the parameters using regular maximum likelihood methods. Sometimes, however, the likelihood equations may be intractable to solve analytically, and we need to apply more elaborate techniques. In addition, when the state labels are unknown for the training data as well, the situation is even more tricky. In fact, in such a situation the optimization problem is NP-complete, so that in particular there is no known algorithm that can guarantee an optimal solution in reasonable time. Given an observed sequence with an unknown state path, the best we can do is to find a *local maximum* under the given model. The method commonly used is an iterative procedure called the *Baum–Welch algorithm* [4], which is a version of the more general *EM-algorithm* [6].

In this chapter, we describe some of the most common techniques used for training in gene finding. The methods are typically well established, such that they even may be considered outdated to researchers within optimization theory. While the methods covered here for the most part are sufficient for the purpose of training gene finding algorithms, there may be interesting room for improvement by applying more up-to-date optimization approaches.

6.1 Introduction

Parameter estimation, or parameter training, is an important area in mathematical statistics, in which one attempts to fit a certain model to observed data. The model can then be used to analyze a system and test certain hypotheses about the reality the model attempts to capture. A model typically consists of a set of variables and a set of mathematical formulations that describe the relationships between these variables. The variables are, in their turn, often dependent on a number of parameters, that

M. Axelson-Fisk, *Comparative Gene Finding*, Computational Biology 20, DOI 10.1007/978-1-4471-6693-1_6

characterize the system. As the values of these parameters are usually unknown, we use a training set of known observations to estimate these parameters. We say that we *train* the model parameters. For instance, in an HMM the model is the state space and their possible connections, and the parameters to be estimated are the initial, transition, and emission probabilities of the model. In neural networks, it is mainly the weights and the biases of the nodes that need training.

There exist many different methods to estimate parameters, and a number of different features that we may want such estimates to fulfill. A central theme, however, is that we want the estimates to be *optimal*, which basically means that we want to be able to extract as much information as possible from the training data. For instance, we may want to find the estimates that maximize the fit of the model to the data, such as in *maximum likelihood*, or the estimates that optimize the classification ability of the model, such as in *discriminative training*. Either way, the optimality requirement turns the parameter estimation problem into one of optimization.

In a general setting, we first define an *objective function* that maps the training data measurements onto the space of possible parameter values, and then we search for the parameter settings that maximize (or minimize) this function. That is, for a parameter set θ and an objective function f, we want to solve the optimization problem

$$\hat{\theta} = \underset{\theta}{\operatorname{argmax}} \; f(\theta). \tag{6.1}$$

The objective function f can also be referred to as a *cost function* or an *energy function*. In what follows, we describe a number of objective functions and optimization methods commonly used in the training of gene finding algorithms.

6.2 Pseudocounts

One problem when using a limited training set is that only events or subsequences appearing in the data will get a positive probability. While some of the nonobserved events may be rare, setting the probability to zero may be too strict and affect the model accuracy negatively. A common solution is to insert extra counts, called *pseudocounts*, in some manner to ensure that all events are possible. The simplest approach is to just add a constant value to each of the counts. When this constant equals 1 this is called the *Laplace's rule*,

$$\mathbf{P}(a) = \frac{c_a + 1}{\sum_b (c_b + 1)}, \tag{6.2}$$

where, c_a is the observed count of residue a. This method is rather crude, however, and a slightly more sophisticated approach is to take the background frequencies of the residues into account when determining on a pseudocount. The advantage of this is that resulting counts resemble the prior distribution of residues when the dataset is small, while the pseudocounts have less and less effect as the training set grows. The *Bayesian prediction method* [22] uses this strategy and adds a residue-specific number to each count,

$$\tilde{c}_a = c_a + z_a, \tag{6.3}$$

where z_a is the pseudocount for residue a. For instance, a log-odds weight matrix is typically on the form

$$W_{ak} = \log \frac{p_{ak}}{p_a} \tag{6.4}$$

where p_{ak} is the probability that residue a occurs in position k in the sequence motif, and p_a is the background probability of a. Using the Bayesian prediction method to insert pseudocounts, we would estimate p_{ak} by

$$\hat{p}_{ak} = \frac{c_{ak} + B\, p_a}{N + B} \tag{6.5}$$

where N is the number of aligned sequences and B is a pseudocount distributed according to the background distribution of the residues. The method is referred to as the Bayesian prediction method because the estimate uses a *Dirichlet prior distribution* (see Sect. 5.4.7). Empirical studies have shown $B \approx \sqrt{N}$ to be an efficient approximation of the pseudocount [22].

One problem with the Bayesian prediction method is that it does not take into account the similarities between residues, such as is done by the PAM and BLOSUM matrices for amino acids. A simple solution to this is implemented in the *data-dependent pseudocount method* [32], which utilizes the statistical significance of alignments [17]. The probability p_{ak} is then estimated by

$$\hat{p}_{ak} = \frac{c_{ak} + B\, p_a \sum_{b=1}^{20} \frac{c_{bk}}{N} e^{\lambda s(a,b)}}{N + B} \tag{6.6}$$

where λ is the same as in BLAST (see Sect. 3.1.9).

Even more sophisticated is the use of *Dirichlet mixtures*, which are composed of weighted sums of *Dirichlet distributions*. We describe this method next, in the setting of the multiple alignment program SAM [12].

The SAM Regularizer

The multiple alignment program SAM [12] uses *regularization* as a method to avoid overfitting of parameters. When using Bayesian statistics, regularization is closely connected to the use of prior distributions (see Sect. 5.4.7). The use of regularizers is a way of adding random amounts of pseudocounts to the observed counts, according to some distribution. The regularizer used in SAM is the *Dirichlet distribution*, which is the *conjugate prior* of the *multinomial distribution*. While the *binomial distribution* is the distribution of the number of successes in n Bernoulli trials (success/failure), the multinomial distribution is the generalization to when each trial can have one of m possible outcomes (instead of just two). The parameters of the multinomial distribution are the cell probabilities $p_1, \ldots, p_m$ of choosing each the respective m outcomes in each trial. The Dirichlet distribution is the conjugate prior of this, meaning that it models the outcome probabilities $p_1, \ldots, p_m$ as random variables $P_1, \ldots, P_m$. That is, the Dirichlet probability density function with parameters $\boldsymbol{\alpha} = (\alpha_1, \ldots, \alpha_m)$ of variables $P_1, \ldots, P_m$ is given by

$$\mathbf{P}(P_1 = p_1, \ldots, P_1 = p_1; \boldsymbol{\alpha}) = \frac{1}{B(\boldsymbol{\alpha})} \prod_{i=1}^{m} p_i^{\alpha_i - 1}, \tag{6.7}$$

where $\alpha_0 = \sum_{i=1}^{m} \alpha_i$. In other words, the Dirichlet distribution is the probability distribution of the prior beliefs that the individual outcome probabilities in each trial are $p_1, \ldots, p_m$.

In the case of HMMs, the Dirichlet regularizer is the probability distribution over the prior beliefs of the model parameters, and these prior beliefs are the standard parameter estimates such as the proportional counts of the events occurring in the training set. That is, for transition probability a_{ij} between states $i, j \in S$, the prior belief is given by

$$\hat{a}_{ij} = \frac{n_{ij}}{\sum_{k \in S} n_{ik}}, \tag{6.8}$$

where n_{ij} is the number of observed transitions from state i to state j in the training set. The Dirichlet regularizer assumes that this proportional count is the outcome of a random variable that is distributed according to a Dirichlet distribution with parameter α_{ij}. The reestimation formula used is then given by

$$\hat{a}_{ij} = \frac{n_{ij} + \alpha_{ij}}{\sum_{k \in S} (n_{ik} + \alpha_{ik})}. \tag{6.9}$$

The emission probabilities of the HMM are estimated in the same way, by counting the occurrences in the training set and adding a Dirichlet distributed random amount to each count. The set of all the α's are the corresponding regularizers. The advantage of this method over the standard, in which we add a constant pseudocount to each observed count, is that while the regularizers become important for small training sets, their influence decreases as the training set increases. Moreover, in sequence alignment one often has prior knowledge of the alignment problem at hand, and it only makes sense to include this knowledge into the model.

6.3 Maximum Likelihood Estimation

The method of maximum likelihood (ML) is a very useful technique in statistics, and is the most widely used method for training HMMs on labeled sequences. The idea of maximum likelihood is to find the parameter settings of the given model that maximize the probability, or the *likelihood*, of the observed data. In what follows, we only consider discrete distributions, but the theory is readily expandable to continuous distributions.

Let Y be a discrete random variable with a probability mass function $f_Y(y|\theta)$ characterized by a parameter θ. When the parameter value is known, the probability density function can be used directly to make inferences about the random variable. If we instead reverse the roles of Y and θ and view θ as the unknown variable and Y as the given information, we get for an observed value $Y = y$ the *likelihood function* of θ

$$L(\theta|Y) = f_Y(y|\theta). \tag{6.10}$$

Note, however, that the likelihood is not a probability density in itself, but only a function of the model parameter. The method of maximum likelihood uses this likelihood function as its objective function, and attempts to find the value $\hat{\theta}$ of θ that solves the equation

$$\hat{\theta} = \operatorname*{argmax}_{\theta} L(\theta|Y). \tag{6.11}$$

The resulting maximum likelihood estimate is the value of θ that makes the observed data $Y = y$ "most probable" or "most likely" under the model.

Suppose now that we have a sample of random variables $Y_1, \ldots, Y_n$, drawn from the distribution of Y, with observed values $Y_i = y_i, i = 1, \ldots, n$. Since the random variables in a sample are independent, the likelihood function for the entire sample becomes

$$L(\theta|Y_1, \ldots, Y_n) = \prod_{i=1}^{n} f_Y(y_i|\theta). \tag{6.12}$$

To simplify the calculations, we transform the product in (6.12) to a sum by taking the logarithm of the likelihood. Since the logarithm is a strictly monotone function, the maximum of $L(\theta|Y)$ is reached in the same point as its logarithm. The *log-likelihood* function is then defined as

$$l(\theta|Y_1, \ldots, Y_n) = \log L(\theta|Y_1, \ldots, Y_n) = \sum_{i=1}^{n} \log f_Y(y_i|\theta). \tag{6.13}$$

The maximum likelihood estimate of θ is then obtained by maximizing the log-likelihood function over θ

$$\hat{\theta} = \operatorname*{argmax}_{\theta} l(\theta|Y). \tag{6.14}$$

Example 6.1   Coin flips
Suppose that we are given a coin that shows "heads" (H) with probability p and "tails" (T) with probability $1 - p$. The random variable Y can thus take values in $\{H, T\}$. Now suppose that we would like to estimate the parameter p based on the following sequence of flips:

$$HHTHHHTTHH$$

The likelihood function becomes (Fig. 6.1)

$$L(p|Y_1, \ldots, Y_{10}) = \prod_{i=1}^{10} \mathbf{P}(Y_i = y_i) = p^7 (1 - p)^3,$$

and the log-likelihood

$$l(p|Y_1, \ldots, Y_{10}) = 7 \log p + 3 \log(1 - p).$$

The maximum likelihood estimate is given by the equation

$$\hat{p} = \underset{p}{\mathrm{argmax}}\; l(p|Y_1, \ldots, Y_{10}),$$

which can be obtained by solving the equation

$$\frac{dl(p|Y_1, \ldots, Y_{10})}{dp} = 0.$$

As a solution, we get

$$\frac{dl(p|Y_1, \ldots, Y_{10})}{dp} = \frac{7}{p} - \frac{3}{1 - p} = 0 \iff p = \frac{7}{10}.$$

Fig. 6.1 The likelihood function of 10 coin flips. The maximum is achieved at $\hat{p} = 7/10$

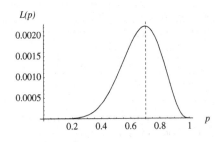

More generally, for n coin flips showing heads c times, the maximum likelihood estimate of p is given by

$$\hat{p} = \frac{c}{n}.$$

□

The extension to the case when the probability distribution of Y depends on more than one parameter, $\Theta = \{\theta_1, \ldots, \theta_k\}$ is straightforward. The log-likelihood function is formulated in the same way, and the maximum likelihood estimate of Θ is found by solving the equation system

$$\frac{\partial l(\Theta|Y)}{\partial \theta_i} = 0, \quad i = 1, \ldots, k. \tag{6.15}$$

Dempster [6] uses an example to illustrate the use of the EM-algorithm on a multinomial distribution, and we will use the same in Sect. 6.4. But first, we show how to determine the maximum likelihood estimate of the multinomial parameters directly.

Example 6.2 A multinomial distribution
In the example by Dempster in [6], 197 animals have been categorized into four "cells" as $\mathbf{y} = (y_1, y_2, y_3, y_4) = (125, 18, 20, 34)$. The underlying genetic model depends on a single parameter $0 \leq \pi \leq 1$, with cell probabilities given by

$$(p_1, p_2, p_3, p_4) = \left(\frac{1}{2} + \frac{1}{4}\pi, \frac{1}{4}(1 - \pi), \frac{1}{4}(1 - \pi), \frac{1}{4}\pi\right). \tag{6.16}$$

Here we illustrate how to estimate the parameter π by using the maximum likelihood method. The probability mass function of the multinomial distribution is given by

$$f_Y(y|\pi) = \frac{n!}{y_1! y_2! y_3! y_4!}\left(\frac{1}{2} + \frac{1}{4}\pi\right)^{y_1}\left(\frac{1}{4}(1 - \pi)\right)^{y_2}\left(\frac{1}{4}(1 - \pi)\right)^{y_3}\left(\frac{1}{4}\pi\right)^{y_4}. \tag{6.17}$$

The log-likelihood function of π becomes

$$l(\pi|Y) = c + y_1 \log\left(\frac{1}{2} + \frac{1}{4}\pi\right) - (y_2 + y_3) \log\left(\frac{1}{4}(1 - \pi)\right) + y_4\left(\frac{1}{4}\pi\right) \tag{6.18}$$

where $c = \log n! + \sum_{i=1}^{4} \log(y_i!)$ is a constant not depending on π. Taking the derivative of l with respect to π yields

$$\frac{dl}{d\pi} = \frac{y_1}{2 + \pi} + \frac{y_2 + y_3}{1 - \pi} + \frac{y_4}{\pi}, \tag{6.19}$$

and setting the derivative to zero yields a quadratic equation with roots

$$\hat{\pi} = \frac{1}{394}\left(15 \pm \sqrt{53809}\right). \tag{6.20}$$

Since we require $0 \leq \pi \leq 1$ there is only one valid solution,

$$\hat{\pi} \approx 0.6268. \tag{6.21}$$

□

HMM Training on Labeled Sequences

Turning to the case of training an HMM on labeled sequences, we recall from Sect. 2.1 that an HMM is composed of an observed process and a hidden process, where the hidden process is a Markov chain jumping between states in a state space, and the observed process is a function of the underlying state sequence. We assume that we are given an observed sequence Y_1^T with the corresponding known state path X_1^L. The HMM is characterized by the model parameters $\theta = \{\pi_i, a_{ij}, b_j(\cdot)\}$ corresponding to the initial distribution of the first hidden state, the transition probabilities of the hidden process, and the emission probabilities of the observed process, respectively. In this case, the likelihood function takes the form

$$L(\theta|X, Y) = \mathbf{P}(X_1^L, Y_1^T|\theta) \tag{6.22}$$

and the maximum likelihood estimate becomes

$$\hat{\theta} = \underset{\theta}{\operatorname{argmax}} \, L(\theta|X_1^L, Y_1^T). \tag{6.23}$$

The following example illustrates how to derive the exact expression of these estimates in a simple HMM where the state space consists of two dice.

Example 6.3 Two dice model
Consider Example 2.3 where we had two dice, A and B, where A had six sides, generating numbers between 1 and 6, and B had four sides, generating numbers between 1 and 4 (see Fig. 6.2). The hidden Markov model constitutes of a hidden

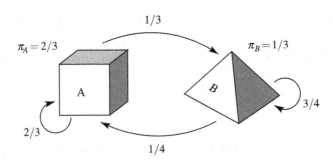

Fig. 6.2 A two-state HMM, where the hidden states are the dice, and the observed outputs are the roll outcomes

state sequence, $X_1^T = X_1, \ldots, X_T$ taking values in the state space $S = \{s_A, s_B\}$, and an observed sequence Y_1^T taking values in $V = \{1, 2, 3, 4, 5, 6\}$. Note that since this is a standard HMM, the indices of the observed and the hidden process will remain the same, and thus both sequences have the same length T. The model parameters to be estimated are thus the initial probabilities $\pi = \{\pi_A, \pi_B\}$ and the transition probabilities $A = \{a_{ij}\}_{i,j \in S}$ of the hidden process, and the emission probabilities $B = \{b_i(v)\}_{i \in S, v \in V}$ of the observed process.

Given a training set of labeled sequences, where we know the underlying state sequence of each observed sequence, we can use maximum likelihood to estimate the model parameters.

If we let $\theta = \{\pi, A, B\}$ denote the model, the likelihood function is given by

$$L(\theta | X, Y) = \mathbf{P}(X_1^T, Y_1^T | \theta) = \pi_{X_1} \prod_{t=2}^{T} a_{X_{t-1}, X_t} b_{X_t}(Y_t), \tag{6.24}$$

and the log-likelihood function

$$l(\theta | X, Y) = \log L(\theta) = \log \pi_{X_1} + \sum_{t=2}^{T} \left(\log a_{X_{t-1}, X_t} + \log b_{X_t}(Y_t) \right). \tag{6.25}$$

As before, the maximum likelihood estimate is given by

$$\hat{\theta} = \operatorname*{argmax}_{\theta} l(\theta). \tag{6.26}$$

To illustrate the ML procedure, we focus on the transition probabilities a_{ij} and note that $a_{AB} = 1 - a_{AA}$ and $a_{BB} = 1 - a_{BA}$. Furthermore, if c_{ij} denotes the number of times the transition a_{ij} occurs in the training set, the transition probability part of the likelihood function becomes

$$\prod_{t=2}^{T} a_{X_{t-1}, X_t} = (a_{AA})^{c_{AA}} (1 - a_{AA})^{c_{AB}} (a_{BA})^{c_{BA}} (1 - a_{BA})^{c_{BB}}. \tag{6.27}$$

Taking the logarithm of (6.27), the maximum likelihood estimates can be achieved as before by setting the partial derivatives to zero. For instance, since

$$\frac{\partial l}{\partial a_{aa}} = \frac{c_{AA}}{a_{AA}} - \frac{c_{AB}}{1 - a_{AA}} \tag{6.28}$$

the maximum likelihood estimate becomes

$$\hat{a}_{AA} = \frac{c_{AA}}{c_{AA} + c_{AB}}. \tag{6.29}$$

That is, the transition probability is estimated by the observed frequency among all transitions out of the s_A state. The remaining model parameters are estimated in a similar fashion.   □

The result in the example above applies directly to the general case of training an HMM on labeled training sequences. For an HMM where the hidden states X_1^T take values in some state space $S = \{s_1, \ldots, s_N\}$, and the observed sequence Y_1^T takes values in some symbol set $V = \{v_1, \ldots, v_M\}$, the maximum likelihood parameter estimates are given by their frequencies of occurrence in the training set. That is, for $i, j \in S$ and $y \in V$

$$\hat{\pi}_i = \frac{c_i}{T}, \tag{6.30}$$

$$\hat{a}_{ij} = \frac{c_{ij}}{c_i}, \tag{6.31}$$

$$\hat{b}_i(y) = \frac{c_i(y)}{c_i}, \tag{6.32}$$

where c_{ij} is the number of transitions from state i to j, $c_i = \sum_j c_{ij}$ is the counts of state i in X_1^T, and $c_i(y)$ is the number of times state i has emitted symbol y.

So far we have only considered one single training sequence. The extension to the case of $p > 1$ training sequences $(\mathbf{Y}_1, \mathbf{X}_1), \ldots, (\mathbf{Y}_p, \mathbf{X}_p)$ is simply obtained by taking the product of the likelihoods of the individual sequences

$$\hat{\theta} = \underset{\theta}{\mathrm{argmax}} \prod_{i=1}^{p} L(\theta | \mathbf{Y}_i, \mathbf{X}_i). \tag{6.33}$$

Maximum likelihood estimation and the likelihood function has several desirable asymptotic properties. Although the estimates are typically not unbiased for finite samples, the bias tends to zero as the sample size increases. Moreover, the estimator is asymptotically "efficient," meaning that it obtains the smallest variance of all unbiased estimators. Another useful property is that the probability distribution of the maximum likelihood estimate tends to the normal distribution as the sample size grows. As a result, likelihood functions can be used when constructing confidence bounds or conducting hypothesis tests of the parameters. One disadvantage is that although tending toward an unbiased estimate in the limit, if the sample size is small, maximum likelihood estimates can be heavily biased. Another disadvantage is that the likelihood equations may be difficult to solve analytically, and we may have to use numerical techniques. One such technique is the *expectation–maximization (EM) algorithm* described in the next section. The special case, when training HMMs on unlabeled sequences, is more known as the *Baum–Welch algorithm* described in Sect. 6.5. The Baum–Welch algorithm utilizes a combination of the forward and the backward variables in Sects. 2.1.4 and 2.1.5, often referred to as the *forward–backward algorithm*.

Gene finding models have been greatly inspired by the models used in speech recognition. As a consequence, the problem of parameter training is very similar. In speech recognition, however, an alternative approach to maximum likelihood estimation has emerged during the last decade, namely that of *discriminative training* or *conditional maximum likelihood* (CML) [24]. The argument for this approach is that while the maximum likelihood method aims at estimating parameters at a rather local level, by estimating the parameters for each submodel individually, discriminative training optimizes the ability to discriminate between model states at a more global level. The discriminative training approach is described in more detail in Sect. 6.8.

6.4 The Expectation–Maximization (EM) Algorithm

The *expectation–maximization (EM) algorithm* [6] is an iterative procedure that is used to determine maximum likelihood estimates of parameters in situations when data is incomplete or missing. There are a number of methods in statistics that are in fact special cases of the EM-algorithm, including the Baum–Welch algorithm for HMMs described in Sect. 6.5. The Baum–Welch algorithm is used to train HMMs on unlabeled sequences where, thus, the underlying state sequence is "missing."

In general, there are two main situations when the EM-algorithm is suitable. One is when data really is incomplete or missing, perhaps due to limitations of the observation process. As an example, consider the problem of estimating the life time of the light bulbs in a series of street lights. Instead of changing each bulb at the time of failure, usually all light bulbs are exchanged at the same time, and usually before all of them fails. Thus, in terms of the lifetimes of the bulbs, some are observed and some are unobserved, also called censored, data points. The only lifetimes we know, are those of the light bulbs that already failed. However, instead of just ignoring the still functioning bulbs, their lifetimes *so far* may still provide additional information to the failure time distribution and improve the estimates of the model parameters. The other main application of the EM-algorithm is a situation which is common in, for instance, computational pattern recognition. In this case, there is not really any missing data, but the likelihood equations are too hard to solve analytically, and can sometimes be made tractable by assuming the occurrence of additional hidden parameters. We give a brief overview of the EM-algorithm here. For a nice (and longer) presentation, with a large number of examples, see for instance [25].

We assume we have a random variable on the form $W = (X, Y)$, where only Y is observed and X is missing or hidden. We call W the *complete* data, and Y the *incomplete* data. We have a joint density function $f(w|\theta) = f(x, y|\theta)$ for the complete data, which is characterized by some parameter θ, and a marginal density function $f_Y(y|\theta)$ for the incomplete data,

$$f_Y(y|\theta) = \int_X f(x, y|\theta)\,dx. \tag{6.34}$$

We want to estimate the parameter θ using the information of both the observed data and the knowledge of the complete data distribution. In EM-algorithm applications there is typically some sort of constraint describing the dependence between the observed and the hidden variables, for instance $Y = h(X)$ for some function h, but this is not necessary. The only difference from the independent situation is that instead of considering the entire range of X, the integral in (6.34) runs over all x that fulfills the constraint $y = h(x)$.

Now, if we had access to all the data, we could estimate the parameter directly by formulating the log-likelihood function called the *complete-data likelihood*

$$l(\theta|X, Y) = \log f(X, Y|\theta). \tag{6.35}$$

If X is missing, however, this computation becomes impossible.

Example 6.4 The three coin game
Assume that we have three possibly biased coins, showing "heads" (H) or "tails" (T) as we flip them. Coin c_0 shows heads with probability λ, coin c_1 shows heads with probability p_1, and coin c_2 shows heads with probability p_2. The game goes on as follows:

1. Flip coin c_0.
2. If $c_0 = H$, flip coin c_1 three times. Otherwise, flip coin c_2 three times.
3. Repeat from 1.

The observed outcome is thus series of triplets, each coming from either c_1 or c_2. What is hidden from the observer is which coin gave rise to which triple. Assume that we observe the following series

$$\langle HHH \rangle, \langle TTT \rangle, \langle HHH \rangle, \langle TTT \rangle, \langle HHH \rangle.$$

How do we estimate the coin probabilities λ, p_1, and p_2?
Let X denote outcome of c_0, and Y the observed triplet using the coin determined by X. That is,

- $X \in \{H, T\}$,
- $Y \in \{HHH, HHT, HTH, THH, HTT, THT, TTH, TTT\}$,
- $\Theta = \{\lambda, p_1, p_2\}$.

If we had access to the complete data, that is including the information of which coin that was used, we could estimate the parameters directly using maximum likelihood. Assume, for instance, that the complete data is

$$(H : \langle HHH \rangle), (T : \langle TTT \rangle), (H : \langle HHH \rangle), (T : \langle TTT \rangle), (H : \langle HHH \rangle)$$

where the first letter (H or T) shows the outcome of coin c_0. The corresponding maximum likelihood estimates are simply the relative frequencies of each coin. The estimates become

$$\hat{\Theta} = \{\hat{\lambda}, \hat{p}_1, \hat{p}_2\} = \left\{\frac{3}{5}, \frac{9}{9}, \frac{0}{6}\right\}.$$

The problem is how to estimate the parameters when the outcomes of X are hidden. This is typically where the EM-algorithm becomes useful. □

Since X is an unknown random variable, the log-likelihood $l(\theta|X, Y)$ becomes a random variable as well. That is, we can think of $l(\theta|X, y)$ as a function of X, where y and θ are constants or parameters. The idea behind the EM-algorithm is that since we cannot maximize the complete-data likelihood $l(\theta|X, Y)$ directly, we maximize the conditional expectation $E[l(\theta|X, Y)|Y]$ instead, using the observations of Y and a provisional estimate of θ. We define an *auxiliary function*

$$Q(\theta'|\theta) = E[l(\theta'|X, Y)|Y, \theta] = E[\log f(X, Y|\theta')|Y, \theta], \qquad (6.36)$$

which is used iteratively to produce an improved estimate θ' of the current estimate θ. Let $\theta^{(k)}$ denote the estimate of θ after k iteration cycles, and initiate the algorithm with some value $\theta = \theta^{(0)}$. The EM procedure then iterates over the following expectation (E) and maximization (M) steps until convergence:

1. *E-step*: calculate $Q(\theta|\theta^{(k)})$.
2. *M-step*: determine $\theta^{(k+1)} = \text{argmax}_\theta \, Q(\theta|\theta^{(k)})$.

That is, the E-step calculates the expected value of the complete-data likelihood as a function of θ, given the observed data Y and the current parameter estimate $\theta^{(k)}$. In the M-step, we take the conditional expectation from the E-step and maximize it over θ. The algorithm is initiated by choosing an initial value of θ. The initial value can be chosen at random, although making a "reasonable" guess based on prior knowledge, if available, may both reduce the number of iterations and improve the results.

Example 6.5 The three coin game (cont.)
We continue Example 6.4 by assuming that the information about coin c_0 is unknown, and use the EM-algorithm to estimate the coin probabilities λ, p_1 and p_2. For the complete data, we have the density function

$$f(x, y|\Theta) = \mathbf{P}(X = x, Y = y|\Theta) = \mathbf{P}(X = x|\Theta)\mathbf{P}(Y = y|X = x, \Theta), \quad (6.37)$$

where

$$\mathbf{P}(X = x|\Theta) = \begin{cases} \lambda, & \text{for } x = H, \\ 1 - \lambda, & \text{for } x = T, \end{cases} \qquad (6.38)$$

$$\mathbf{P}(Y_i = y|X = x, \Theta) = \begin{cases} p_1^{h_i}(1 - p_1)^{3-h_i}, & \text{if } x = H, \\ p_2^{h_i}(1 - p_2)^{3-h_i}, & \text{if } x = T. \end{cases} \qquad (6.39)$$

Y_i denotes the observed triplet number i, and h_i is the number of heads in that triplet. The E-step of the EM-algorithm involves calculating the conditional expectation of the complete-data log-likelihood, given the observed data and the current parameter estimates. To simplify the notation, we introduce the notation

$$\tilde{p}_i = \mathbf{P}(X = H | Y = y_i, \Theta) \tag{6.40}$$

corresponding to the *posterior* probability of having heads of coin c_0 in the ith round. The conditional expectation of the complete-data log-likelihood becomes

$$E[\log f(X, Y | \Theta') | Y, \Theta] = \tag{6.41}$$
$$= \sum_i E[\log f(X, Y_i | \Theta') | Y_i, \Theta]$$
$$= \sum_i \sum_{x \in \{H,T\}} f(x | y_i, \Theta) \log f(x, y_i | \Theta')$$
$$= \sum_i \sum_x \mathbf{P}(X = x | Y = y_i, \Theta) \log \left(\mathbf{P}(X = x | \Theta') \mathbf{P}(Y = y_i | X = x, \Theta') \right)$$
$$= \sum_i \tilde{p}_i \log \left(\lambda' {p_1'}^{h_i} (1 - p_1')^{3-h_i} \right) + (1 - \tilde{p}_i) \log \left((1 - \lambda') {p_2'}^{h_i} (1 - p_2')^{3-h_i} \right).$$

The update formula used in the M-step is obtained by setting the partial derivatives of Θ' to 0 and solving the following equation system:

$$\lambda' = \frac{\sum_i \tilde{p}_i}{n}, \tag{6.42a}$$

$$p_1' = \frac{\sum_i (h_i/3) \tilde{p}_i}{\sum_i \tilde{p}_i}, \tag{6.42b}$$

$$p_2' = \frac{\sum_i (h_i/3)(1 - \tilde{p}_i)}{\sum_i (1 - \tilde{p}_i)}, \tag{6.42c}$$

where n is the number of tossing rounds. Using Bayes' theorem, we get that

$$p_H = \mathbf{P}(X = H | Y = \langle HHH \rangle, \Theta) = \frac{\lambda p_1^3}{\lambda p_1^3 + (1 - \lambda) p_2^3}, \tag{6.43a}$$

$$p_T = \mathbf{P}(X = H | Y = \langle TTT \rangle, \Theta) = \frac{\lambda (1 - p_1)^3}{\lambda (1 - p_1)^3 + (1 - \lambda)(1 - p_2)^3}. \tag{6.43b}$$

As an example, if we initiate the algorithm with $\lambda^{(0)} = 0.5$, $p_1^{(0)} = 0.4$ and $p_2^{(0)} = 0.5$ we get in the E-step

$$p_H^{(0)} \approx 0.3386,$$
$$p_T^{(0)} \approx 0.6334.$$

Using this estimates in the M-step yields the updated parameter estimates

$$\lambda^{(1)} = \frac{3}{5}0.3386 + \frac{2}{5}0.6334 \approx 0.4565, \tag{6.44a}$$

$$p_1^{(1)} = \frac{3 \cdot 0.3386}{3 \cdot 0.3386 + 2 \cdot 0.6334} \approx 0.4450, \tag{6.44b}$$

$$p_2^{(1)} = \frac{3(1 - 0.3386)}{3(1 - 0.3386) + 2(1 - 0.6334)} \approx 0.7302. \tag{6.44c}$$

Iterations of the E- and the M-step progress as in Table 6.1. It is interesting to note that if we had started in $\{\lambda^{(0)}, p_1^{(0)}, p_2^{(0)}\} = \{0.5, 0.5, 0.5\}$ we would have ended up in a local maximum $\{0.5, 0.6, 0.6\}$ almost immediately, but by moving p_1 just slightly away from 0.5, to 0.49 say, the algorithm pretty soon ends up in the global maximum. This shows the importance of choosing initial estimates not too close to a local optimum. □

The EM-algorithm can sometimes be used even if a problem does not appear on the form of incomplete/complete data. By reformulating it as such, for instance by introducing artificial "missing" random variables, the maximum likelihood estimation can sometimes be greatly simplified. To illustrate this, we borrow an example presented in [6], which involves estimating the parameters in a multinomial distribution. In Example 6.2, we showed how to produce maximum likelihood estimates for that example directly, and now we continue by introducing additional hidden data.

Example 6.6 A multinomial distribution (cont.)
Recall from Example 6.2 that we have a multinomial distribution of 4 cells, with observations

$$\mathbf{y} = (y_1, y_2, y_3, y_4) = (125, 18, 20, 34). \tag{6.45}$$

The corresponding density function is given by

$$f_{\mathbf{Y}}(\mathbf{y}|\pi) = \frac{n!}{y_1!y_2!y_3!y_4!}\left(\frac{1}{2} + \frac{1}{4}\pi\right)^{y_1}\left(\frac{1}{4}(1 - \pi)\right)^{y_2}\left(\frac{1}{4}(1 - \pi)\right)^{y_3}\left(\frac{1}{4}\pi\right)^{y_4} \tag{6.46}$$

where $n = 197$ is the total number of observations.

Table 6.1 Iterations in the EM-algorithm

Cycle	p_H	p_T	λ	p_1	p_2
0			0.5000	0.4000	0.5000
1	0.3386	0.6334	0.4565	0.4450	0.7302
2	0.1598	0.8797	0.4477	0.2141	0.9129
3	0.0104	0.9983	0.4055	0.0153	0.9989
4	0.0000	1.0000	0.4000	0.0000	1.0000

Moreover, the cell probabilities depend on a single parameter $0 \le \pi \le 1$, where

$$(p_1, p_2, p_3, p_4) = \left(\frac{1}{2} + \frac{1}{4}\pi, \frac{1}{4}(1 - \pi), \frac{1}{4}(1 - \pi), \frac{1}{4}\pi\right). \qquad (6.47)$$

Now we assume that this is the *observed* data, and that the *complete* data actually consists of 5 cells with cell probabilities

$$(q_1, q_2, q_3, q_4, q_5) = \left(\frac{1}{2}, \frac{1}{4}\pi, \frac{1}{4}(1 - \pi), \frac{1}{4}(1 - \pi), \frac{1}{4}\pi\right). \qquad (6.48)$$

That is, we assume that the observation of the first cell $y_1 = 125$ is a combination of two hidden cells $y_1 = x_1 + x_2$. The complete data is thus on the form

$$(\mathbf{x}, \mathbf{y}) = (x_1, x_2, y_2, y_3, y_4) \qquad (6.49)$$

and the density function of the complete data can be written as

$$f(\mathbf{x}, \mathbf{y}|\pi) = \frac{n!}{x_1! x_2! y_2! y_3! y_4!} \left(\frac{1}{2}\right)^{x_1} \left(\frac{1}{4}\pi\right)^{x_2} \left(\frac{1}{4}(1 - \pi)\right)^{y_2} \left(\frac{1}{4}(1 - \pi)\right)^{y_3} \left(\frac{1}{4}\pi\right)^{y_4}.$$

The integral in (6.34), connecting the density functions of the incomplete and the complete data, here translates to summing over all pairs $0 \le x_1, x_2 \le 125$ such that $x_1 + x_2 = 125$. That is,

$$f_Y(\mathbf{y}|\pi) = \sum_{x_1 + x_2 = 125} f(\mathbf{x}, \mathbf{y}|\pi). \qquad (6.50)$$

Since the last three cells in the complete data are given, the expectation in the E-step only involves the first two cells,

$$E[\log f(\mathbf{X}, \mathbf{Y}|\pi')|Y, \pi] = \sum_{x_1 + x_2 = 125} f(\mathbf{x}|\mathbf{y}, \pi) \log f(\mathbf{x}, \mathbf{y}|\pi'). \qquad (6.51)$$

Moreover, x_1 and x_2 split the cell sum 125 in proportions $\frac{1}{2} : \frac{\pi}{4}$, or, equivalently, into proportions $2 : \pi$. That is, if we let $\pi^{(k)}$ denote the estimate of π after k cycles, the E-step involves calculating the expectations of the corresponding random variables with $\pi^{(k)}$ inserted,

$$E[X_1|Y, \pi^{(k)}] = 125 \frac{2}{2 + \pi^{(k)}}, \qquad (6.52a)$$

$$E[X_2|Y, \pi^{(k)}] = 125 \frac{\pi^{(k)}}{2 + \pi^{(k)}}. \qquad (6.52b)$$

The M-step takes the current expectations $x_1^{(p)}$ and $x_2^{(p)}$ of x_1 and x_2 and produces a new estimate $\pi^{(p+1)}$ of π using maximum likelihood as if data was complete. Recall from Example 6.2 that the log-likelihood of the complete data is

$$l(\pi|X, Y) = c + x_1 \log \frac{1}{2} + (x_2 + y_4) \log \frac{\pi}{4} + (y_2 + y_3) \log \frac{1 - \pi}{4}, \quad (6.53)$$

where c is a constant not depending on π. Taking the derivative over π and setting it to zero yields a new parameter estimate

$$\pi^{(p+1)} = \frac{x_2^{(p)} + y_4}{x_2^{(p)} + y_2 + y_3 + y_4}. \quad (6.54)$$

By initiating in $\pi^{(0)} = 0.5$, the EM-algorithm reaches the estimate $\pi \approx 0.6268$ already after five iterations. □

If we are lucky enough to deal with a distribution belonging to the *exponential family*, the expectation in the E-step can be determined on a closed form. However, for more complicated distributions, the calculations often have to be solved numerically. There are various numerical procedures used for this, including *Monte Carlo* and *Newton–Cotes* methods (see for instance [1]).

A nice property of the EM-algorithm is that the likelihood is guaranteed not to decrease between steps, and a typical convergence criterion is to stop when the likelihood function no longer increases. There is no guarantee, however, that the global maximum ever is reached. The resulting maximum may very well be local. Another attractive feature of the EM-algorithm is that the M-step only involves maximum likelihood estimation of the complete data, which often results in rather straightforward computations. In more complicated situations, however, the algorithm becomes less useful. Extensions of the algorithm include the *Generalized* EM algorithm (GEM) [6], and the *expectation conditional maximization* (ECM) algorithm [26]. The GEM-algorithm does not attempt to maximize the auxiliary function, but is satisfied if the new parameter value increases Q in each iteration. In the ECM-algorithm, the parameter vector is split into two subvectors, which are then maximized separately and alternatively in a zig-zag manner.

Which stopping rule to use to terminate the EM-algorithm is not necessarily obvious. If the computations have converged, we terminate naturally; but if convergence is too slow, we might decide to quit anyway. The two most straightforward criteria are to stop either when the changes in the parameter estimates or in the likelihood function are small enough. Various convergence issues of the EM-algorithm are investigated in [33].

6.5 The Baum–Welch Algorithm

The problem of training hidden Markov models (HMMs) translates to finding the parameter settings that maximize the probability of the observed sequence for the given model. If the state sequence is known, parameters can be estimated using regular maximum likelihood estimates such as described in Sect. 6.3. In practice, this simply means using the observed relative frequencies of the various events as estimates of their corresponding probabilities in the HMM. If the state sequence is unknown, however, the problem becomes NP-complete, and the best we can do is to search for a local maximum.

The *Baum–Welch algorithm* [4] is a special case of the EM-algorithm [6] described above, applied to HMMs. Just as the EM-algorithm the Baum–Welch algorithm iterates over an expectation (E) and a maximization (M) step, utilizing something called the *forward–backward algorithm* described next.

The Forward–Backward Algorithm

The forward–backward algorithm [28] is a useful procedure that can be used both for reestimation of parameters in the Baum–Welch algorithm, and for calculating the probability of a particular feature predicted by the Viterbi algorithm, such as the exon probabilities computed in Sect. 2.2.4. The name of the algorithm comes from the fact that it utilizes both the forward and the backward variables described in Sect. 2.1.2.

We recall from Sect. 2.1.2 that an HMM consists of a hidden process X_1^T and an observed process Y_1^T. The hidden process is a Markov chain jumping between states in some state space $S = \{s_1, \ldots, s_N\}$, while the observed process, which is not necessarily Markov, emits values taken from some set $V = \{v_1, \ldots, v_M\}$ depending on the underlying state sequence. The hidden process is initialized according to the initial distribution $\pi = \{\pi_1, \ldots, \pi_N\}$, and jumps between states according to the transition probabilities $a_{ij}, i, j \in S$. In state j, the observed process emits a symbol Y_t according to the emission probability $b_j(Y_t | Y_1^{t-1})$. The parameters to be estimated are thus the initial, transition, and emission probabilities $\{\pi_i, a_{ij}, b_j : i, j \in S\}$. Recall further that a forward variables $\alpha_i(t)$ represent the joint probability of the current state i and the observed sequence up to time t,

$$\alpha_i(t) = \mathbf{P}(Y_1^t, X_t = i) = \sum_{j \in S} a_{ji} b_i(Y_t | Y_1^{t-1}) \alpha_j(t-1). \tag{6.55}$$

Similarly, the backward variables $\beta_i(t)$ correspond to the conditional probability of the observed sequence after t, given the current state i and the observed sequence up to t,

$$\beta_i(t) = \mathbf{P}(Y_{t+1}^T | Y_1^t, X_t = i) = \sum_{j \in S} a_{ij} b_j(Y_{t+1} | Y_1^t) \beta_j(t+1). \tag{6.56}$$

Using the definition of conditional probabilities, we can write the joint probability of the complete observed sequence and the state at some time point t as

$$\mathbf{P}(Y_1^T, X_t = i) = \mathbf{P}(Y_1^t, X_t = i)\mathbf{P}(Y_{t+1}^T|Y_1^t, X_t = i) = \alpha_i(t)\beta_i(t). \qquad (6.57)$$

The forward variables $\alpha_i(t)$ account for everything up to and including time t, while the backward variables $\beta_i(t)$ represent the remainder of the sequence. Moreover, the probability of the observed sequence, or the *likelihood* of the observed sequence, can be written as

$$\mathbf{P}(Y_1^T) = \sum_{i \in S} \mathbf{P}(Y_1^T, X_t = i) = \sum_{i \in S} \alpha_i(t)\beta_i(t). \qquad (6.58)$$

This leads to the very useful *forward–backward variables*, representing the probability of the state at time t, given the observed sequence,

$$\gamma_i(t) = \mathbf{P}(X_t = i|Y_1^T) = \frac{\mathbf{P}(Y_1^T, X_t = i)}{\mathbf{P}(Y_1^T)} = \frac{\alpha_i(t)\beta_i(t)}{\sum_{j \in S} \alpha_j(t)\beta_j(t)}. \qquad (6.59)$$

We see that γ is a probability measure since

$$\sum_{i \in S} \gamma_i(t) = 1. \qquad (6.60)$$

The forward–backward variables in (6.59) are the quantities used to update the parameter estimates in the Baum–Welch training algorithm, detailed in the next section.

The Baum–Welch Algorithm

As mentioned earlier, the Baum–Welch algorithm is a kind of EM-algorithm for estimating the parameters in HMMs from a set of unlabeled training sequences. For simplicity, we will assume that we are given a single training sequence Y_1^T, but the arguments are easily expanded to multiple training sequences.

If we knew the values of the HMM parameters $\theta = \{\pi_i, a_{ij}, b_j : i, j \in S\}$, we could calculate the forward and the backward variables directly as in Sects. 2.1.4 and 2.1.5. Instead, we start out with an initial "guess" of the parameter values. The Baum–Welch algorithm then alternates between the following expectation (E) and maximization (M) steps:

1. *E-step*: calculate the forward and the backward variables for Y_1^T using the current parameter settings.
2. *M-step*: compute new parameter estimates using the forward–backward algorithm.

The forward–backward algorithm generates the variables γ_i in (6.59). A related property is the probability of making a given transition, from state i to j say, at time t

$$\xi_{ij}(t) = \mathbf{P}(X_t = i, X_{t+1} = j | Y_1^T) = \frac{\mathbf{P}(Y_1^T, X_t = i, X_{t+1} = j)}{\mathbf{P}(Y_1^T)}. \tag{6.61}$$

The numerator in (6.61) can be rewritten as

$$\mathbf{P}(Y_1^T, X_t = i, X_{t+1} = j) =$$
$$= \mathbf{P}(Y_{t+2}^T | Y_1^{t+1}, X_{t+1} = j)\mathbf{P}(Y_{t+1} | Y_1^t, X_{t+1} = j)\mathbf{P}(X_{t+1} = j | X_t = i)\mathbf{P}(Y_1^t, X_t = i)$$
$$= a_{ij}b_j(Y_{t+1} | Y_1^t)\alpha_i(t)\beta_j(t+1) \tag{6.62}$$

such that we get

$$\xi_{ij}(t) = \frac{a_{ij}b_j(Y_{t+1} | Y_1^t)\alpha_i(t)\beta_j(t+1)}{\sum_{k \in S} \alpha_k(t)\beta_k(t)}. \tag{6.63}$$

The variables $\gamma_i(t)$ and $\xi_{ij}(t)$ are related through

$$\gamma_i(t) = \sum_{j \in S} \xi_{ij}(t). \tag{6.64}$$

If we take the sum of $\xi_{ij}(t)$ over t, we get a property that can be seen as the expected number of transitions from i to j,

$$\sum_{t=1}^{T-1} \xi_{ij}(t) = \text{expected number of transitions from } i \text{ to } j. \tag{6.65}$$

Similarly, by summing $\gamma_i(t)$ over t we get the expected number of transitions out of state i,

$$\sum_{t=1}^{T-1} \gamma_i(t) = \text{expected number of transitions from } i. \tag{6.66}$$

These properties can now be used for reestimation of the HMM parameters in order to maximize the likelihood of the training data. The reestimation formulas are given by

$$\hat{\pi}_i = \gamma_i(1) = \text{expected frequency in state } i \text{ at time } t = 1, \tag{6.67a}$$

$$\hat{a}_{ij} = \frac{\sum_{t=1}^{T-1} \xi_{ij}(t)}{\sum_{t=1}^{T-1} \gamma_i(t)} = \frac{\text{expected no. of transitions from } i \text{ to } j}{\text{expected no. of transitions from } i}, \tag{6.67b}$$

$$\hat{b}_j(c) = \frac{\sum_{\substack{t=1 \\ Y_t=c}}^{T-1} \gamma_j(t)}{\sum_{t=1}^{T-1} \gamma_j(t)} = \frac{\text{expected no. of times in } j \text{ observing } c}{\text{expected no. of times in } j}. \qquad (6.67c)$$

The Baum–Welch procedure can be summarized as follows:

1. Start out with some initial guess of the parameters, either based on previous knowledge of the model, or chosen at random.
2. Calculate the forward and backward algorithms for the given parameters (E-step).
3. Calculate the γ_i- and ξ_{ij}-variables.
4. Reestimate the parameters using the formulas for $\hat{\pi}_i$, $\hat{a}_{ij}$, and $\hat{b}_j$ above.

Steps 2–4 are iterated until the estimates converge according to some criterion, for instance when the difference in the likelihood $\mathbf{P}(Y_1^T)$ in (6.58) is sufficiently small.

An alternative approach to the one just described is to use the Viterbi algorithm (see Sect. 2.1.6) instead of the forward–backward algorithm. That is, starting with an initial guess of the parameters, we determine the optimal state path $X_1^*, \ldots, X_T^*$ for the observed training sequence and the current parameter settings, using the Viterbi algorithm. In the reestimation step, we then estimate the parameters, using maximum likelihood directly. That is, the parameters are estimated by the corresponding relative frequencies in the hidden and observed sequences. The procedure is iterated until the state path no longer changes. A major difference from the standard Baum–Welch approach is that when using the Viterbi algorithm instead of the forward–backward procedure, we no longer maximize the true likelihood $\mathbf{P}(Y_1^T|\theta)$, since we use the optimal path to reestimate the parameters rather than summing over all paths. However, it may be argued that if we use the Viterbi algorithm as definition of what is optimal, we should use it for training as well [7].

While the Baum–Welch algorithm or gradient descent methods described in the next section are the most commonly used approaches to HMM parameter estimation when using unlabeled training sequences, there is the problem of local maxima. We are never guaranteed to reach the global maximum, and if we choose the initial estimates unwisely we may very well end up in a local maximum instead. Alternative methods such as Gibbs sampling or simulated annealing, may be better at avoiding these [7, 9], as they all work by introducing randomness into the fit, so that even if the algorithm reaches a local optimum it still has the ability to leave again. These methods are overviewed in Sects. 6.10 and 6.9, respectively.

6.6 Gradient Ascent/Descent

Gradient descent is an optimization algorithm used to minimize differentiable continuous functions. It is an iterative procedure where in each step we move in the direction of the steepest negative slope, the *descent* of the function *gradient*. For instance, if we are standing on a hillside, the gradient is a vector pointing in the direction of the steepest slope, and the steepness of the slope is given by the magnitude of the vector. The word "descent" indicates that we always search for the minimum of a function, but naturally the "gradient ascent" method is completely analogous. In search for the function maximum, we move along the steepest *positive* slope instead of the negative.

Definition 6.1 *Gradient*
Let $\mathbf{x} = (x_1, \ldots, x_p)$ be a p-dimensional real-valued vector, $\mathbf{x} \in \mathbb{R}^p$, and $f : \mathbb{R}^p \to \mathbb{R}$ a real-valued differentiable function. Then the *gradient* of f is defined as the vector of partial derivatives of f

$$\nabla f(\mathbf{x}) = \left(\frac{\partial f}{\partial x_1}, \ldots, \frac{\partial f}{\partial x_p}\right). \tag{6.68}$$

The goal of the gradient descent method is to find the point $\mathbf{x}$ that minimizes (or maximizes) an objective function f. This is done iteratively by moving in small steps from the current point to a new point that lies in the gradient direction. The procedure of the gradient descent algorithm basically goes as follows:

1. Choose an initial point $\mathbf{x}^{(0)}$, either randomly or near the expected minimum.
2. Calculate the next point

$$\mathbf{x}^{(n+1)} = \mathbf{x}^{(n)} - \Delta\mathbf{x}^{(n)}, \tag{6.69}$$

where $\Delta\mathbf{x}^{(n)}$ is the update term of step $n + 1$ given by

$$\Delta\mathbf{x}^{(n)} = \eta_n \nabla f(\mathbf{x}^{(n)}), \tag{6.70}$$

and $\eta_n > 0$ a small positive step size, called the *learning rate* parameter.
3. If $f(\mathbf{x}^{(n+1)}) < f(\mathbf{x}^{(n)})$ return to step 2 and continue the iteration. Otherwise, terminate and return $\mathbf{x}^{(n)}$ as the computed minimum.

Example 6.7 A simple gradient descent example.
Assume that we want to find the minimum of the function

$$f(x, y) = 3x^2 + 2y^2 - 3x + y + 4 \tag{6.71}$$

over $x, y \in \mathbb{R}^2$. The gradient of this function is given by

$$\nabla f = \left(\frac{\partial f}{\partial x}, \frac{\partial f}{\partial y}\right) = (6x - 3, 4y + 1). \tag{6.72}$$

For a learning rate η, the updates in each iteration become

$$x^{(n+1)} = x^{(n)} - \eta(6x^{(n)} - 3),$$
$$y^{(n+1)} = y^{(n)} - \eta(4y^{(n)} + 1).$$

Say that we start in some arbitrarily chosen point $(x^{(0)}, y^{(0)}) = (3, 2)$ and use learning rate $\eta = 0.2$. Algorithm 5 illustrates an implementation of the gradient descent method. Here we have chosen not only to test that the function decreases,

---

Algorithm 5 Illustrating the gradient descent method

$x_{new} = 3$
$y_{new} = 2$
$\eta = 0.1$ /* learning rate */
$\varepsilon = 0.00001$; /* precision */
$f(x, y) = 3x^2 + 2y^2 - 3x + y + 4$
$d = -1$
while $(d < 0$ and $|d| > \varepsilon)$ **do**
$\quad x_{old} = x_{new}$
$\quad y_{old} = y_{new}$
$\quad x_{new} = x_{old} - \eta(6x_{old} - 3)$
$\quad y_{new} = y_{old} - \eta(4y_{old} + 1)$
$\quad d = f(x_{new}, y_{new}) - f(x_{old}, y_{old})$
end while

---

but that the decrease is sufficiently large. The results of the iterations are given in Table 6.2. We see that at the fifth step, the convergence is within the given precision $\varepsilon = 0.00001$. $\qquad\square$

The gradient descent method is a common ingredient in the backpropagation algorithm described below, which is used to train artificial neural networks. Training a neural network involves estimating the weights associated with each computational unit (neuron), by using some kind of error function that measures the difference between the observed outputs and desired *target* outputs. The gradient descent method can be used to iteratively adjust the weights with the aim to minimize this error function.

Table 6.2 The iteration results of the Algorithm 5

n	x	y	$f(x, y)$	Δx	Δy
0	3.0	2.0	32.0	3.0	1.8
1	0.0	0.2	4.28	−0.6	0.36
2	0.6	−0.16	3.1712	0.120	0.072
3	0.48	−0.23	3.1268	−0.024	0.0144
4	0.50	−0.25	3.12507	0.0048	0.0029
5	0.50	−0.25	3.12500	−0.00096	0.00058
6	0.50	−0.25	3.12500	0.000192	0.00012

Example 6.8 Single-layer neural networks
Consider a simple single-layer neural network with a linear activation function (see Sect. 2.4.2) taking real-valued inputs $\mathbf{x} = (x_1, \ldots, x_N)$ and producing an output on the form

$$y(\mathbf{x}) = \sum_{i=0}^{M} w_i \phi_i(\mathbf{x}). \tag{6.73}$$

The w_i are the input weights and ϕ_i are some functions transforming the inputs (sometimes called *basis functions*).

Assume that we want to train the weights $\mathbf{w} = (w_1, \ldots, w_M)^T$ using D training patterns $\{(\mathbf{x_1}, t_1), \ldots, (\mathbf{x_D}, t_D)\}$ of paired input vectors $\mathbf{x}_j$ and corresponding desired target values t_j of $y(\mathbf{x}_j)$. Assume further that we use the following error function to measure the distance between observed outputs and target values

$$E(\mathbf{w}) = \frac{1}{2} \sum_{j=1}^{D} (y(\mathbf{x}_j) - t_j)^2. \tag{6.74}$$

The error function in (6.74) is called a *sum-of-squares* function, which measures the sum of the squared deviations between the observed values and the desired target values. In this particular example, with a linear network and this error function, the value of the weights can actually be solved analytically. By solving the equation system

$$\frac{\partial E}{\partial w_i} = 0 \Big|_{i=1}^{M} \tag{6.75}$$

we get the *least squares estimates* (rendering the smallest, or least value, of the sum-of-squares deviations)

$$\hat{\mathbf{w}} = (\boldsymbol{\Phi}^T \boldsymbol{\Phi})^{-1} \boldsymbol{\Phi}^T \mathbf{T}. \tag{6.76}$$

T is a $(D \times 1)$-vector of target values for the D training patterns, $\mathbf{w}$ an $(M \times 1)$-vector of weights, and $\boldsymbol{\Phi}$ a $(D \times M)$-matrix with entry $\phi_i(\mathbf{x}_j)$ in row j and column i.

If we were to use the gradient descent method instead, we would calculate the gradient of the error function E

$$\nabla E = \left(\frac{\partial E}{\partial w_1}, \ldots, \frac{\partial E}{\partial w_M} \right), \tag{6.77}$$

where

$$\frac{\partial E}{\partial w_k} = \sum_{j=1}^{D} (y(\mathbf{x}_j) - t_j) \phi_k(\mathbf{x}_j). \tag{6.78}$$

The updated set of weights would then be given by

$$w_k^{(n+1)} = w_k^{(n)} - \eta \sum_{j=1}^{D} \left(\sum_{i=0}^{M} w_i^{(n)} \phi_i(\mathbf{x}_j) - t_j \right) \phi_k(\mathbf{x}_j) \tag{6.79}$$

for some learning rate parameter η. $\qquad\square$

One difficulty with the gradient descent algorithm is to choose the appropriate learning rate at each iteration. The gradient gives us the direction, but not the size of the update. One needs to take care here, since too large steps may lead to oscillations, while too small steps may lead to slow convergence. Another problem is that with an update function such as in (6.70) convergence may be very slow, in particular if the search space contains plateaus where the error function is almost constant. An approach that has proven successful in improving the convergence rate is to include a *momentum* term in the update [31],

$$\mathbf{x}^{(n+1)} = \mathbf{x}^{(n)} - \Delta\mathbf{x}^{(n)} + \mu\Delta\mathbf{x}^{(n-1)}, \tag{6.80}$$

where $0 < \mu < 1$ is a momentum parameter. That is, in addition to the current gradient, the momentum adds a fraction of the update made in the previous step. In the absence of local minima, the gradient keeps pointing in the same direction, in which case the momentum term simply increases the step size. However, if the error surface has a complex curvature, such as including long narrow valleys, the standard gradient descent might oscillate between the sides and only progress very slowly toward the minimum. The momentum term helps to smooth out these oscillations.

6.7 The Backpropagation Algorithm

In artificial neural networks, such as those described in Sect. 2.4, the parameters to be trained are typically the weights associated with the network units in each layer. A single-layer neural network can be trained by comparing the observed outputs to the desired target values of the training data, and adjust the weights accordingly to minimize the deviation. As long as the activation function is differentiable, we can use methods such as the gradient descent method directly. In a multilayer network this is not possible, however, since we do not know how to assign target values to the hidden units. If the observed output differs from the target value, we have no way of knowing which of the hidden units that are responsible for the deviation. A popular solution is then to use the *backpropagation algorithm*, which is an iterative method used in particular for training multilayered feed-forward neural networks. The name "backpropagation" is an abbreviation of "backward propagation of errors," and indicates how the error recorded, when comparing the observed outputs to the target values, is propagated backward through the network in order to modify the weights of the layers recursively.

Consider a two-layer network such as that in Fig. 6.3. The network consists of N input units $(x_1, \ldots, x_N)$, a hidden layer of M units $(z_1, \ldots, z_M)$, and K output units $(y_1, \ldots, y_K)$. The weight of the edge between input unit i and hidden unit j is denoted $w_{ij}^{(1)}$, while the weight of the edge between hidden unit j and output unit k is denoted $w_{jk}^{(2)}$. The threshold of the activation function, called the *bias*, is invoked in the input layer by adding an extra input node $x_0 = 1$ with weights $w_{0j}^{(1)}$ on the edges between x_0 and the units in the hidden layer. Similarly, we add an extra hidden node $z_0 = 1$ with weights $w_{0k}^{(2)}$ between this node and the output layer. The input values of each layer are integrated into a single value using some kind of integration function. The integrated values are then transformed by an activation function to produce an output value.

We assume here that the integration functions of the input layer in the hidden layer are weighted sums on the form

$$a_j = \sum_{i=0}^{N} w_{ij}^{(1)} x_i \quad \text{and} \quad b_k = \sum_{j=0}^{M} w_{jk}^{(2)} z_j. \tag{6.81}$$

Moreover, we assume the activation function to be the *sigmoid* function

$$\phi(a) = \frac{1}{1 - e^{-a}}. \tag{6.82}$$

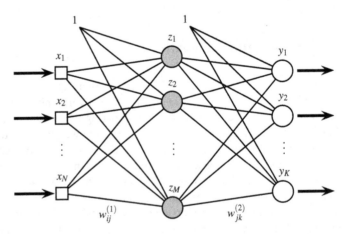

Fig. 6.3 A two-layer feed-forward network of N input nodes, M hidden nodes, and K output nodes. The biases in the hidden and the output layers are included as extra constant nodes in each layer

Presented with a set of D training patterns $\{(\mathbf{x}_1, \mathbf{t}_1), \ldots, (\mathbf{x}_D, \mathbf{t}_D)\}$, where each pattern consists of an input vector $\mathbf{x}_d = (x_1^d, \ldots, x_N^d)$ and a *target* vector $\mathbf{t}_d = (t_1^d, \ldots t_K^d)$ of desired outputs we want to estimate the weights in such a way that the error function

$$E(\mathbf{w}) = \frac{1}{2} \sum_{d=1}^{D} \sum_{k=1}^{K} (y_k^d - t_k^d)^2 \qquad (6.83)$$

is minimized with respect to the weights $\mathbf{w}$. If the network consisted of a single layer, we could determine the weights using the gradient descent method directly. The minimum would be achieved by calculating the gradient of the error function with respect to the weights, and solving the equation system

$$\begin{cases} \frac{\partial E}{\partial w_{11}} = 0, \\ \quad \vdots \\ \frac{\partial E}{\partial w_{NK}} = 0. \end{cases}$$

In multilayer networks, where we do not know the target values of the intermediate hidden layers, the backpropagation algorithm provides an efficient method for deriving the partial derivatives of all weights, including those connected with the hidden layer.

Assume for simplicity that we are given a single training pattern $\{\mathbf{x}, \mathbf{t}\}$. Given a set of initial values of the network weights, the backpropagation algorithm iteratively modifies the weight values in order to minimize the error E. The algorithm consists of three main steps:

1. *The feed–forward step*: run the input vector $\mathbf{x}$ of the training pattern through the network with the current weights and observe the output. During this step, we compute and store the outputs and the derivatives of the activation function of each individual unit.
2. *The backpropagation step*: run the network "backwards" by feeding a constant vector $\mathbf{1} = (1, \ldots, 1)$ into the output units and evaluate the partial derivatives of the error function with respect to the weights.
3. *The gradient descent step*: modify the weights using the gradient descent method.

The backpropagation algorithm iterates over these steps, by running the training pattern through the network repeatedly and modifying the weights in each cycle, until the error becomes sufficiently small. The following description, detailing the individual steps of the algorithm, is inspired by the very nice presentation given in [30].

The Feed–Forward Step

In order to simplify calculations, we add an extra layer of K units $\mathbf{y}' = (y_1', \ldots, y_K')$, call them the *extended output units*, to the network (see Fig. 6.4). We use these added units to calculate the error term for each pair of observed values and target values.

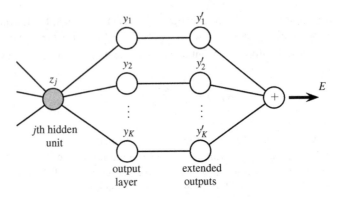

Fig. 6.4 The two-layer network is extended to include the calculation of the error function E

In the feed–forward step the input vector $\mathbf{x}$ is fed into the extended network, and the vectors $\mathbf{z} = (z_1, \ldots, z_M)$, $\mathbf{y} = (y_1, \ldots, y_K)$, and $\mathbf{y}' = (y_1', \ldots, y_K')$ are calculated according to

$$z_j = \phi\left(\sum_{i=0}^{N} w_{ij}^{(1)} x_i \right), \quad j = 1, \ldots, M, \tag{6.84a}$$

$$y_k = \phi\left(\sum_{j=0}^{M} w_{jk}^{(2)} z_j \right), \quad k = 1, \ldots, K, \tag{6.84b}$$

$$y_k' = \frac{1}{2}(y_k - t_k)^2, \quad k = 1, \ldots, K. \tag{6.84c}$$

We have included the bias of each layer by adding the nodes $z_0 = y_0 = 1$. Their corresponding weights are to be estimated in the same manner as the rest of the network weights.

Next, to facilitate the backpropagation step, we compute and store the derivatives of the activation function in each unit. The derivative of the sigmoid function is given by

$$\frac{d\phi(a)}{da} = \phi(a)(1 - \phi(a)). \tag{6.85}$$

Thus, in addition to the computed vectors above, we also compute and store the following values:

$$\text{In } z_j : \quad z_j(1 - z_j)$$
$$\text{In } y_k : \quad y_k(1 - y_k)$$
$$\text{In } y_k' : \quad (y_k - t_k)$$

The Backpropagation Step

Starting with the output units, we introduce a useful notation, corresponding to the *backpropagated error* of output unit k

$$\delta_k^{(2)} = y_k(1 - y_k)(y_k - t_k). \tag{6.86}$$

Using the chain rule, the partial derivatives for the second-layer weights between the hidden layer and the output units take the form

$$\frac{\partial E}{\partial w_{jk}^{(2)}} = \frac{\partial y_k'}{\partial y_k} \cdot \frac{\partial y_k}{\partial b_k} \cdot \frac{\partial b_k}{\partial w_{jk}^{(2)}} = (y_k - t_k)\,y_k(1 - y_k)z_j = \delta_k^{(2)}z_j, \tag{6.87}$$

where $b_k = \sum_{j=0}^{M} w_{jk}^{(2)} z_j$ is the integration function of the hidden layer.
The weight $w_{jk}^{(2)}$ of interest only appears in the y_k' term in E and therefore we have

$$\frac{\partial E}{\partial w_{jk}^{(2)}} = \frac{\partial y_k'}{\partial w_{jk}^{(2)}}. \tag{6.88}$$

To simplify calculations, during the feed–forward computation these terms are computed and stored in the corresponding units as illustrated in Fig. 6.5. Thus, to achieve the values of the partial derivatives, we just multiply the corresponding stored terms.

The partial derivatives of the first layer of weights between the input units and the hidden layer are computed similarly. However, now we have to include the sum over the k extended outputs, since the partial derivative with respect to $w_{ij}^{(1)}$ appears in each of them. First, we denote the backpropagated error of hidden unit j as

$$\delta_j^{(1)} = z_j(1 - z_j) \sum_{k=1}^{K} w_{jk}^{(2)} \delta_k^{(2)}. \tag{6.89}$$

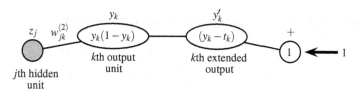

Fig. 6.5 Terms of the partial derivatives of the second layer of weights $w_{jk}^{(2)}$ are stored in the units during the feed–forward computation

Fig. 6.6 Computation of the partial derivatives of the first layer of weights $w_{ij}^{(1)}$

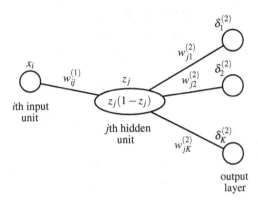

output layer

The partial derivatives in question become

$$\frac{\partial E}{\partial w_{ij}^{(1)}} = \delta_j^{(1)} x_i.$$

(6.90)

These are easily derived from the feed–forward computation as illustrated in Fig. 6.6, by taking the weighted sum over the backpropagated errors of the output layer $\delta_k^{(2)}$, and multiplying it by the stored terms in the hidden units and by the input value.

The Gradient Descent Step

The partial derivatives computed above constitute the gradient of the error function, which is a vector pointing toward the steepest slope of the error function. The gradient descent method, described in Sect. 6.6, is an optimization method the iteratively moves toward the nearest (possibly local) minimum by moving the weight values in the direction of the gradient. The moves are done in small steps, where the step length is given by a *learning rate* parameter $\eta > 0$.

If $w_{ij}(n)$ denotes the value of weight w_{ij} after n cycles, the weights are updated using the formula

$$w_{ij}(n+1) = w_{ij}(n) - \Delta w_{ij}(n).$$

(6.91)

The weight modifications in the two different layers are given by

$$\Delta w_{ij}^{(1)} = \eta \delta_j^{(1)} x_i, \quad i = 0, \ldots, N; \ j = 1, \ldots, M,$$

(6.92a)

$$\Delta w_{jk}^{(2)} = \delta_k^{(2)} z_j, \quad j = 0, \ldots, M; \ k = 1, \ldots, K.$$

(6.92b)

Note that the bias weights $w_{0j}^{(1)}$ and $w_{0k}^{(2)}$ are updated as well, and that the update computations include the constant bias nodes $x_0 = z_0 = 1$. It is very important that all partial derivatives are computed before the weights are updated, or else the update may not correspond to the gradient direction [30].

Several Training Patterns

The description above uses a single training pattern to illustrate the computations. In the presence of several training patterns $\{(\mathbf{x}_1, \mathbf{t}_1), \ldots, (\mathbf{x}_d, \mathbf{t}_d)\}$, we simply run the algorithm for each pattern separately and then combine the weight updates computed for each. That is,

$$\Delta w_{ij}^{(1)} = \Delta_1 w_{ij}^{(1)} + \Delta_2 w_{ij}^{(1)} + \cdots + \Delta_d w_{ij}^{(1)} \qquad (6.93)$$

where $\Delta_d w_{ij}$ is the update of weight w_{ij} given by the dth training pattern $(\mathbf{x}_d, \mathbf{t}_d)$. If the training set is large this computation may become very time consuming, however. In that case, instead of running all patterns through the network before updating the weights, a possible speed-up is to update the weights using the gradient of each separate pattern. The corrections may not correspond exactly to the gradient direction, but if the patterns are selected randomly, the direction will on average be the correct one. Another argument for this variant is that shallow local minima may be avoided [30].

6.8 Discriminative Training

In maximum likelihood (ML) estimation, each class of submodels is typically trained separately and then combined into the final model. While this results in a model that best fits the training data, it may not be optimal when it comes to actual prediction, in particular if the transitions between submodels are ambiguous [20]. The notion *discriminative training* simply implies that instead of using the parameter estimates that best "explain" the observed data, we try to optimize the discriminative ability of the model. That is, for a set of training sequences with known state labels, maximization is taken over some function that measures the classification accuracy of the model, rather than measuring the fit to data.

Assume that we are given a single (observed) training sequence $\mathbf{Y} = (Y_1, \ldots, Y_T)$ with known state labels $\mathbf{X} = (X_1, \ldots, X_T)$. The objective is to estimate the model parameters θ in such a way that the classification probability is maximized, or, likewise, such that the classification error is minimized. The most common objective functions for discriminative training are those of *conditional maximum likelihood* (CML) [16], *maximum mutual information* (MMI) [2], and *minimum classification error* (MCE) [15].

Conditional Maximum Likelihood

Assume that we have $p \geq 1$ pairs of training sequences with corresponding known state sequences $(\mathbf{X}_1, \mathbf{Y}_1), \ldots, (\mathbf{X}_p, \mathbf{Y}_p)$. Recall from Sect. 6.3 that the maximum likelihood estimate is given by

$$\hat{\theta}_{ML} = \underset{\theta}{\text{argmax}} \prod_{i=1}^{p} P(\mathbf{X}_i, \mathbf{Y}_i | \theta). \tag{6.94}$$

The solution to this equation maximizes the fit of the model to the training data. However, in terms of discrimination the resulting parameter estimates may not lead to the optimal model in terms of its discriminative ability. An alternative approach would be to use the *conditional maximum likelihood* (CML) estimator [16], that maximizes the probability of the true state sequence, given the observed sequence. The CML estimator is given by

$$\begin{aligned} \hat{\theta}_{CML} &= \underset{\theta}{\text{argmax}} \prod_{i=1}^{p} P(\mathbf{X}_i | \mathbf{Y}_i, \theta) \\ &= \underset{\theta}{\text{argmax}} \prod_{i=1}^{p} \frac{P(\mathbf{X}_i, \mathbf{Y}_i | \theta)}{P(\mathbf{Y}_i | \theta)} \\ &= \underset{\theta}{\text{argmax}} \prod_{i=1}^{p} \frac{P(\mathbf{X}_i, \mathbf{Y}_i | \theta)}{\sum_{\mathbf{X}} P(\mathbf{X}, \mathbf{Y}_i | \theta)}. \end{aligned} \tag{6.95}$$

The numerator in (6.95) is the regular ML estimator, and the denominator is simply the probability of the observed sequence.

This training criterion has for instance been used in the gene finding software HMMgene [20], which uses a *class* HMM (CHMM) for its predictions. The extension of a standard HMM to a class HMM is that in addition to emitting DNA bases in each step, each state emits class labels alongside the DNA sequence. In this setting, it is easier to see that the numerator of the CML estimator represents only valid paths through the model, while the denominator is the sum over all paths, valid and invalid. As a result, since $P(\mathbf{X}, \mathbf{Y} | \theta) \le P(\mathbf{Y} | \theta)$, the maximum is reached when the probability of all invalid paths is minimized.

In addition to optimizing the discriminative ability, another advantage is that while the ML method estimates one submodel at a time, CML estimates all parameters simultaneously. A serious disadvantage, however, is that the CML equation in (6.95) cannot be solved analytically [28]. The estimates have to be approximated using some kind optimization procedure, such as the gradient ascent method described in Sect. 6.6, which can be a very slow process.

Maximum Mutual Information

Another alternative to maximum likelihood is to use the *maximum mutual information* (MMI) between the observed sequences and the state sequences [2]. Mutual information is a concept used in both probability theory and information theory, and is a measure of the dependence between two variables. More specifically, the mutual information of two random variables X and Y measures how much information one variable contains about the other,

$$MI(X, Y) = E\left[\log \frac{\mathbf{P}(X, Y)}{\mathbf{P}(X)\mathbf{P}(Y)}\right]. \tag{6.96}$$

Note that when X and Y are independent they contain no information of one another, and, consequently, the mutual information is zero, $MI(X, Y) = 0$.

From this definition it is common to define the *instantaneous* or *pointwise* mutual information

$$MI(x, y) = \log \frac{p(x, y)}{p(x)p(y)}, \tag{6.97}$$

where $p(x, y)$ is the joint density function of X and Y, and $p(x)$ and $p(y)$ are the corresponding marginal densities.

For the purpose of parameter training on a set of labeled observation sequences $(\mathbf{X}_1, \mathbf{Y}_1), \ldots, (\mathbf{X}_p, \mathbf{Y}_p)$, the maximum mutual information estimator can be defined as

$$
\begin{aligned}
\hat{\theta}_{MMI} &= \operatorname*{argmax}_{\theta} \sum_{i=1}^{p} \log \frac{\mathbf{P}(\mathbf{X}_i, \mathbf{Y}_i | \theta)}{\mathbf{P}(\mathbf{X}_i | \theta)\mathbf{P}(\mathbf{Y}_i | \theta)} \\
&= \operatorname*{argmax}_{\theta} \sum_{i=1}^{p} \log \frac{\mathbf{P}(\mathbf{Y}_i | \mathbf{X}_i, \theta)}{\mathbf{P}(\mathbf{Y}_i | \theta)} \\
&= \operatorname*{argmax}_{\theta} \sum_{i=1}^{p} \log \frac{\mathbf{P}(\mathbf{Y}_i | \mathbf{X}_i, \theta)}{\sum_{\mathbf{X}} \mathbf{P}(\mathbf{X}, \mathbf{Y}_i | \theta)}.
\end{aligned} \tag{6.98}
$$

When the prior distribution $\mathbf{P}(\mathbf{X}_i | \theta)$ of the state space is independent of the observed sequence and of the model parameters, the maximum mutual information and the conditional maximum likelihood estimators coincide [16].

The MMI estimator has an advantage over ML and CML estimators when the prior information about the state distribution is significant. In theory, if the underlying distribution assumptions are correct, the ML and the MMI estimators should converge. However, since MMI not only tries to increase the likelihood of the correct state labels, but also decreases the likelihood of the incorrect labels at the same time, it generally produces a lower likelihood than the ML estimator.

Minimum Classification Error

One problem with the ML estimator is that when the assumed model differs from the true model, the resulting optimization has little to do with the performance of the model in terms of prediction. Using conditional maximum likelihood or maximum mutual information improves the accuracy of the model significantly in such situations [14]. However, an even more direct approach is to attempt to minimize the misclassification rate directly.

Suppose for a moment that we would like to classify an observation Y into one of two possible classes C_1 and C_2. In terms of Bayes decision theory, we utilize some sort of prior knowledge of the distribution of classes, and then use the posterior probability as a decision rule

$$\text{class}(Y) = \begin{cases} C_1 & \text{if } \mathbf{P}(C_1|Y) \geq \mathbf{P}(C_2|Y), \\ C_2 & \text{otherwise.} \end{cases} \tag{6.99}$$

Generalized to $N > 2$ classes, the decision rule becomes

$$\text{class}(Y) = \operatorname*{argmax}_{1 \leq i \leq N} \mathbf{P}(C_i|Y). \tag{6.100}$$

This is the so-called *maximum a posteriori* (MAP) decoder, and theoretically it is the decision rule that yields the minimum error rate [14].

Now we would like to incorporate the decision rule into some sort of misclassification measure. Again, for the two-class case, the simplest measure is the *Bayes discriminant*

$$d(Y) = \mathbf{P}(C_2|Y) - \mathbf{P}(C_1|Y), \tag{6.101}$$

and the optimal decision boundary is achieved by solving the equation $d(Y) = 0$ [15].

The generalization of the Bayes discriminant to $N > 2$ classes is not straightforward, and can be done in several ways. The *minimum classification error* (MCE) criterion attempts to approximate the misclassification rate by measuring the distance between the correct classification score and an average of the incorrect classification scores [29],

$$d_i(Y) = -\log \mathbf{P}(Y|C_i) + \log \left(\frac{1}{N-1} \sum_{j \neq i} \mathbf{P}(Y|C_j)^\eta \right)^{1/\eta} \tag{6.102}$$

$$= -\frac{1}{\eta} \log \frac{\mathbf{P}(Y|C_i)^\eta}{\sum_{j \neq i} \mathbf{P}(Y|C_j)^\eta} - \frac{1}{\eta} \log(N-1) \tag{6.103}$$

where η is a positive number. The misclassification error rate is then usually approximated by embedding the misclassification measure into a "smoothed zero-one function," such as the sigmoid

$$l(d) = \frac{1}{1 + e^{-\gamma d}} \tag{6.104}$$

where $\gamma \geq 1$.

In terms of sequence analysis and discriminative training on p training sequences $(\mathbf{X}_1, \mathbf{Y}_1), \dots, (\mathbf{X}_p, \mathbf{Y}_p)$, we swap the problem around into a maximization problem, and define the minimum classification error as

$$\hat{\theta}_{MCE} = \underset{\theta}{\text{argmax}} \sum_{i=1}^{p} \log \frac{\mathbf{P}(\mathbf{Y}_i | \mathbf{X}_i, \theta)}{\sum_{\mathbf{X}} \mathbf{P}(\mathbf{X}, \mathbf{Y}_i | \theta)}. \tag{6.105}$$

We see that this is essentially the minimum mutual information criterion, except that instead of summing over all possible paths in the denominator, we sum over all incorrect paths [23].

One downside with discriminative training is that its implementation is very complex. Unfortunately, no type of EM-algorithm exists for this problem. Some applications use gradient descent-based approaches instead (see for instance [21]), or extended Baum–Welch [29]. Moreover, compared to maximum likelihood training the meaning of the parameters is less intuitive, and the link to the underlying biological problem tends to get lost [24].

6.9 Gibbs Sampling

Gibbs sampling is an iterative technique that belongs to the large class of sampling algorithms known as *Markov chain Monte Carlo* (MCMC) methods. It was originally presented in [10], but carry large similarities with the much older, and very popular *Metropolis-Hastings algorithm* [11, 27]. In fact, most MCMC algorithms can be seen either as a special case or an extension of the Metropolis–Hastings algorithm. An example of the Metropolis algorithm is given in the next section.

We assume that we want to draw a sample from a distribution $P(\mathbf{Y})$ of a sequences $\mathbf{Y} = (Y_1, \ldots, Y_T)$ of length T. Suppose that drawing the sample directly from the distribution is infeasible for some reason (e.g., the distribution is unknown, or the sample space is too large), but that drawing from the *conditional* distributions $P(Y_t | Y_1, \ldots, Y_{t-1}, Y_{t+1}, \ldots, Y_T)$, $t = 1, \ldots, T$ is quite doable. The Gibbs sampling approach makes use of this, and generates a sample by cycling through all conditional distributions and keeping all but the current variable fixed. This cycling, which can be done randomly or in order, is repeated many times to achieve a sample that comes from approximately the desired distribution.

A Gibbs sampling algorithm for generating a random sample from the distribution P is illustrated in Algorithm 6. The algorithm is initialized by an arbitrary sequence $\mathbf{Y}^{(0)} = (Y_1^{(0)}, \ldots, Y_T^{(0)})$, and then each new sequence (e.g., each new "state" of a Markov chain) is generated by cycling through the conditional distributions of each sequence position. Note how the conditional distribution of Y_t utilizes the residues generated so far in the current cycle. That this procedure in fact generates a sample from the correct distribution P has been proved for instance in [13].

Gibbs Sampling for HMM Training

We mentioned at the end of Sect. 6.5 that while the Baum–Welch algorithm is the most commonly used algorithm for training hidden Markov models (HMMs) on unlabeled sequences, it suffers from the problem of avoiding local maxima. Simulated annealing

---

Algorithm 6 A Gibbs sampler

/* Initialize: */
$$\mathbf{Y}^{(0)} = (Y_1^{(0)}, \ldots, Y_T^{(0)})$$

/* Create K samples of $\mathbf{Y}$ */
for $j = 1$ to K **do**
  for $t = 1$ to T **do**
    Sample $Y_t^{(j)} \sim P(Y_t | Y_1^{(j)}, \ldots, Y_{t-1}^{(j)}, Y_{t+1}^{(j-1)}, \ldots, Y_T^{(j-1)})$.
  end for
  /* Return the current sample */
  return $\mathbf{Y}^{(j)}$
end for

---

described in the next section and Gibbs sampling are both generally better in this respect [9]. Here we give a brief overview, inspired by [5] and [7], of a Gibbs sampler for training of HMMs on a set of observed, unlabeled sequences.

Recall that an HMM consists of a hidden Markov chain $\mathbf{X} = (X_1, \ldots, X_T)$ which emits a sequence of observations $\mathbf{Y} = (Y_1, \ldots, Y_T)$. The model parameters to be estimated are the initial, transition, and emission probabilities $\theta = \{\pi_i, a_{ij}, b_j : i, j \in S\}$, where S is the state space that the Markov chain operates on. Suppose we are given a set of training sequences $\mathbf{Y}_1, \ldots, \mathbf{Y}_p$. If we knew the corresponding state paths $\mathbf{X}_1, \ldots, \mathbf{X}_p$, we could estimate the parameter set θ using maximum likelihood directly (see Sect. 6.3), which in this case simply reduce to frequency counts of the respective events. Thus, there is a direct translation between state paths and parameter estimates, and we can therefore treat the state paths as the parameters to be estimated in order to maximize the likelihood [7].

We initialize the Gibbs sampler by an arbitrary set of state sequences $\mathbf{X}_1^{(0)}, \ldots, \mathbf{X}_p^{(0)}$, and then the algorithm proceeds by iterating between two main steps. In cycle $j = 1, \ldots, K$, these two steps are

1. Remove the sequence pair $(\mathbf{X}_i^{(j-1)}, \mathbf{Y}_i)$ from the dataset, and reestimate the parameters on the remaining sequences,

$$(\mathbf{X}_1^{(j)}, \mathbf{Y}_1), \ldots, (\mathbf{X}_{i-1}^{(j)}, \mathbf{Y}_{i-1}), (\mathbf{X}_{i+1}^{(j-1)}, \mathbf{Y}_{i+1}), \ldots, (\mathbf{X}_p^{(j-1)}, \mathbf{Y}_p).$$

2. Resample the state path $\mathbf{X}_i^{(j)}$ for training sequence $\mathbf{Y}_i$.

These steps are iterated until the state paths no longer change. The parameters are reestimated after each new state path has been sampled, using regular maximum likelihood. The tricky part is how to sample the state paths. One approach is to use the forward-traceback procedure described in [7], which is illustrated in Algorithm 7, and which utilizes the HMM forward variables $\alpha_i(t)$ described in Sect. 2.1.4.

We use the forward variables for the traceback instead of the Viterbi variables, since we do not want to determine the *optimal* path, but rather sample a path from the distribution over all paths. The expression in (6.106) in Algorithm 7 may be better understood by looking at Fig. 6.7. Given the current state X_t at time t we want

Fig. 6.7 Stochastic traceback: for a given state X_t at time t the previous state is weighted according to its forward variable times the transition into the current state

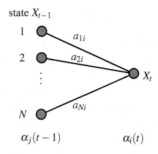

state X_{t-1}

to sample the previous state X_{t-1} according to its forward variable value times the transition probability into the current state. With this formulation, the more probable a path, the higher its probability of being sampled. However, as pointed out in [7], the path distribution sampled from is not *exactly* right, since we sample from the *previous* model rather than from the optimal model, but it is a reasonable approximation.

Algorithm 7 A stochastic traceback algorithm

/* Calculate the forward variables */

Initialize: $\alpha_i(0) = \pi_i$
for $(t = 1 \text{ to } T)$ and $(i = 1 \text{ to } N)$ **do**
  $\alpha_i(t) = P(Y_1^t, X_t = i) = \sum_{j \in S} a_{ji} b_i(Y_t|Y_1^{t-1}) \alpha_j(t-1)$
end for
Terminate: $\alpha_i(T+1) = P(Y_1^T, X_{T+1} = i) = \sum_{j \in S} a_{ji} \alpha_j(T)$

/* Sample a state path */
for $t = T+1 \text{ to } 1$ **do**
  Sample X_{t-1} from

$$P(X_{t-1}|X_t) = \frac{a_{X_{t-1},X_t} \alpha_{X_{t-1}}(t-1)}{\sum_{k \in S} a_{k,X_t} \alpha_k(t-1)} \qquad (6.106)$$

end for

6.10 Simulated Annealing

The general idea of *Monte Carlo simulations* is to approximate a target distribution by the empirical distribution obtained by drawing a large sample from the distribution. *Markov chain Monte Carlo* (MCMC) algorithms are an extension of Monte Carlo simulations, and are used when the target distribution cannot be sampled from directly, or when the sampling is computationally intractable. The idea is to construct a Markov chain instead that converges to the target distribution. *Simulated annealing* is an iterative approach, first introduced in [19]. The method combines the idea of MCMC methods with an *annealing schedule*, where an artificial *temperature* variable is iteratively decreased to zero. Simulated annealing applied to multiple sequence alignment is described in Sect. 3.2.8. Here we give a slightly more formal description of the general method, before we show how it can be applied to the training of HMMs. A more general, but still very hands-on, description can, for instance, be found in [13].

Suppose we have a large system that can switch between N possible "states" in a state space $S = \{s_1, \ldots, s_N\}$. For instance, N can be the number of configurations of a large graph, or, more relevant to us, the combinatorial possibilities of a multivariate variable. Suppose further that we want to minimize some function $f(s), s \in S$ of the state space. In statistical mechanics, this function typically represents some sort of *cost* or *energy* of the system. If N was small enough, we could conduct an exhaustive search and calculate $f(s)$ for all states $s \in S$, but this is often not the case. The flavor of MCMC methods would be to construct an aperiodic, irreducible Markov chain that converges to a distribution that places most of its mass in the state(s) that minimize f. A sample from that distribution would then have a high probability of being in the vicinity of such a state. A distribution that has the desired properties is the *Boltzmann distribution*

$$P_f(s) = \frac{1}{Z} e^{-f(s)/k\tau} \tag{6.107}$$

where $Z = \sum_{s \in S} e^{-f(s)/k\tau}$ is a normalizing factor called the *partition function*, τ is the temperature of the system, and k the *Boltzmann constant*. For large temperatures τ, the Boltzmann distribution places nearly equal probabilities to all states, while for $\tau = 0$ the probability mass is focused in the states with the smallest values of f (the lowest energy). Thus, if we could create a Markov chain that converges to P_f for a small value of τ, we would get an approximate solution to our optimization problem. The problem, however, is that the smaller the τ the longer the time to convergence.

This is where simulated annealing comes in. Instead of fixing τ and run our MCMC, we start out at a relatively high temperature, run the Markov chain for a while, and then iteratively lower the temperature until the process converges according to some convergence criterion. In effect, by decreasing the temperature (or *cooling* the system), the probability mass of the Boltzmann distribution gets more and more concentrated around the minimizing states. The temperatures $\tau_1 > \tau_2 > \tau_3 > \cdots$ and the corresponding times $t_1, t_2, t_3, \ldots$ that we run the Markov chain at each temperature constitute our *annealing schedule*. The most commonly used MCMC method is the *Metropolis algorithm* [27], illustrated in Example 6.9, but basically any MCMC algorithm could be used.

Example 6.9 The Metropolis algorithm
This example is inspired by the very nice description in [3]. Each cycle of the Metropolis algorithm involves a *proposal* distribution q_{ij} and an *acceptance* distribution r_{ij}, where q_{ij} denotes the probability of drawing state s_j when being in state s_i, and r_{ij} denotes the corresponding probability of accepting the proposed state change. We assume for now that the proposal distribution is symmetric, $q_{ij} = q_{ji}$, but this property can be relaxed. The Metropolis algorithm proceeds as follows:

1. Choose an initial state s_i at random.
2. Given the current state s_i, propose a new state s_j according to proposal probability q_{ij}.
3. Accept the new state s_j with probability r_{ij}, or reject it with probability $1 - r_{ij}$ in which case we stay in state s_i for another round.

Steps 2–3 are iterated until convergence. The most common acceptance distribution is given by

$$r_{ij} = \min\left\{1, \frac{P_f(s_j)}{P_f(s_i)}\right\}. \tag{6.108}$$

Using the Boltzmann distribution in (6.107) with energy function $f(s)$ results in acceptance probabilities

$$r_{ij} = \min\left\{1, e^{-\Delta_{ij}f/k\tau}\right\} \tag{6.109}$$

where $\Delta_{ij}f = f(s_j) - f(s_i)$. Note how the partition function Z conveniently cancels out in this expression. Step 3 above then translates to:

3. If $f(s_j) \leq f(s_i)$ accept s_j, and if $f(s_j) > f(s_i)$ accept s_j with probability $e^{-\Delta_{ij}f/k\tau}$.

That is, we always move if the new state has lower energy, but we can also move to a higher energy state occasionally. This last property makes it possible for the algorithm to move away from a local optimum, even after falling into one. The symmetry assumption of q_{ij} can be relaxed by using acceptance function

$$r_{ij} = \min\left\{1, \frac{P_f(s_j)q_{ij}}{P_f(s_i)q_{ji}}\right\}. \tag{6.110}$$

The Metropolis algorithm was constructed specifically for the Boltzmann distribution [27], but was later generalized into the much used *Metropolis–Hastings* algorithm [11]. The generalized algorithm works for any distributions $P_f(s)$, as long as there exists a function f that dominates the density of the distribution. □

How to choose the annealing schedule is nontrivial, since small enough changes in temperature guarantees that the process will converge eventually, but the smaller the temperature the longer the convergence time. On the other hand, if the cooling is too rapid, the risk of ending up in a local minimum, rather than a global, increases. One example of a cooling strategy is to use a logarithmic annealing schedule,

$$\tau_i = \frac{k}{\log \tau_{i-1}} \tag{6.111}$$

for some constant k. It can be proven that the algorithm converges almost surely (with probability one) in such a situation [10]. However, logarithmic cooling is almost as slow as an exhaustive search and often becomes impractical. A more common approach is therefore to use a geometric cooling scale

$$\tau_i = \gamma \tau_{i-1} \qquad (6.112)$$

for some constant $0 < \gamma < 1$. This is for instance used in the multiple alignment program MSASA [18] described in Sect. 3.2.8.

Simulated Annealing for Training of HMMs

Assume that we are given a set of observed sequences $\mathbf{Y}_1, \ldots, \mathbf{Y}_p$ where he corresponding state sequences $\mathbf{X}_1, \ldots, \mathbf{X}_p$ are unknown, and we want to estimate the HMM parameters $\theta = \{\pi_i, a_{ij}, b_j : i, j \in S\}$. We present here a simulated annealing variant to this training problem, described in [7] and [8]. As energy function, we use the negative log-likelihood function of our data,

$$-l(\theta|\mathbf{X}, \mathbf{Y}) = -\log \mathbf{P}(\mathbf{X}, \mathbf{Y}|\theta), \qquad (6.113)$$

such that the Boltzmann distribution in (6.107) becomes

$$P(\mathbf{Y}) = \frac{1}{Z} e^{-\frac{1}{\tau}(-l(\theta|\mathbf{X},\mathbf{Y}))} = \frac{1}{Z} \mathbf{P}(\mathbf{X}, \mathbf{Y}|\theta)^{1/\tau}, \qquad (6.114)$$

where $Z = \int \mathbf{P}(\mathbf{X}, \mathbf{Y}|\theta')^{1/\tau} d\theta'$. It is not clear how to sample from this distribution directly, but a useful approximate approach is described in [8]. In this approach, a state path is sampled randomly using the same forward-traceback procedure as in the Gibbs sampling section above. In order to apply the dependence on a temperature variable, we want to sample a path $\mathbf{X}_i$ for observed sequence $\mathbf{Y}_i$ based on the likelihood of the data, but with a slight modification

$$P(\mathbf{X}) = \frac{\mathbf{P}(\mathbf{X}, \mathbf{Y}|\theta)^{1/\tau}}{\sum_{\mathbf{X}'} \mathbf{P}(\mathbf{X}', \mathbf{Y}|\theta)^{1/\tau}}. \qquad (6.115)$$

The denominator, which is the normalizing factor Z, is simply the sum over all paths and can be obtained by a modified forward algorithm using *exponentiated* parameters: $\hat{\pi}_i = \pi_i^{1/\tau}$, $\hat{a}_{ij} = a_{ij}^{1/\tau}$, and $\hat{b}_j(y) = b_j(y)^{1/\tau}$. The algorithm then iterates over the two steps described for HMM training by Gibbs sampling above, and uses the same stochastic traceback as illustrated in Algorithm 7, only with exponentiated parameters.

References

1. Antia, H.M.: Numerical Methods for Scientists and Engineers. Birkhauser, Basel (2002)
2. Bahl, L., Brown, P., de Souza, P., Mercer, R.: Maximum mutual information estimation of hidden Markov model parameters for speech recognition. Proc. ICASSP-86 **1**, 49–52 (1986)
3. Baldi, P., Brunak, S.: Bioinformatics: The Machine Learning Approach. MIT Press, Cambridge (2001)

4. Baum, L.E.: An equality and associated maximization technique in statistical estimation for probabilistic functions of Markov processes. Inequalities **3**, 1–8 (1972)
5. Chatterji, S., Pachter, L.: Large multiple organism gene finding by collapsed Gibbs sampling. J. Comput. Biol. **12**, 599–608 (2005)
6. Dempster, A.P., Laird, N.M., Rubin, D.B.: Maximum likelihood from incomplete data via the EM algorithm. J. R. Stat. Soc. B. **39**, 1–38 (1977)
7. Durbin, R., Eddy, S., Krogh, A., Mitchison, G.: Biological Sequence Analysis. Probabilistic Models of Proteins and Nucleic Acids. Cambridge University Press, Cambridge (1998)
8. Eddy, S.R.: Multiple alignment using hidden Markov models. Proc. Int. Conf. Intell. Syst. Mol. Biol. **3**, 114–120 (1995)
9. Eddy, S.R.: Profile hidden Markov models. Bioinformatics **14**, 755–763 (1998)
10. Geman, S., Geman, D.: Stochastic relaxation, Gibbs distributions, and the Bayesian restoration of images. IEEE PAMI **6**, 721–741 (1984)
11. Hastings, W.K.: Monte Carlo sampling methods using Markov chains and their applications. Biometrika **57**, 97–109 (1970)
12. Hughey, R., Krogh, A.: Hidden Markov models for sequence analysis: extension and analysis of the basic method. Comput. Appl. Biosci. **12**, 95–108 (1996)
13. Häggström, O.: Finite Markov Chains and Algorithmic Applications. Cambridge University Press, Cambridge (2002)
14. Juang, B.-H., Chou, W., Lee, C.-H.: Minimum classification error rate methods for speech recognition. IEEE Trans. Speech Audio Proc. **5**, 257–265 (1997)
15. Juang, B.-H., Katagiri, S.: Discriminative learning for minimum error classification. IEEE Trans. Sig. Proc. **40**, 3043–3054 (1992)
16. Juang, B.-H., Rabiner, L.R.: Hidden Markov models for speech recognition. Technometrics **33**, 251–272 (1991)
17. Karlin, S., Altschul, S.F.: Methods for assessing the significance of molecular sequence features by using general scoring schemes. Proc. Natl. Acad. Sci. USA **87**, 2264–2268 (1990)
18. Kim, J., Pramanik, S., Chung, M.J.: Multiple sequence alignment using simulated annealing. Comput. Appl. Biosci. **10**, 419–426 (1994)
19. Kirkpatrick, S., Gelatt, C.D., Vecchi, M.P.: Optimization by simulated annealing. Science **220**, 671–680 (1983)
20. Krogh, A.: Two methods for improving the performance of an HMM and their application for gene finding. Proc. Int. Conf. Intell. Syst. Mol. Biol. **5**, 179–186 (1997)
21. Krogh, A., Riis, S.K.: Hidden neural networks. Neural Comput. **11**, 541–563 (1999)
22. Lawrence, C.E., Altschul, S.F., Boguski, M.S., Liu, J.S., Neuwald, A.F., Wootton, J.C.: Detecting subtle sequence signals: a Gibbs sampling strategy for multiple alignment. Science **262**, 208–214 (1993)
23. Majoros, W.H.: Methods for Computational Gene Prediction. Cambridge University Press, Cambridge (2007)
24. Majoros, W.H., Salzberg, S.L.: An empirical analysis of training protocols for probabilistic gene finders. BMC Bioinform. **5**, 206 (2004)
25. McLachlan, G., Krishnan, T.: The EM Algorithm and Extensions. Wiley, New York (1996)
26. Meng, X.L., Rubin, D.B.: Maximum likelihood estimation via the ECM algorithm: a general framework. Biometrika **80**, 267–278 (1993)
27. Metropolis, N., Rosenbluth, A.W., Rosenbluth, M.N., Teller, A.H., Teller, E.: Equations of state calculations by fast computing machines. J. Chem. Phys. **21**, 1087–1092 (1953)
28. Rabiner, L.R.: A tutorial on hidden Markov models and selected applications in speech recognition. Proc. IEEE **77**, 257–286 (1989)
29. Reichl, W., Ruske, G.: Discriminative training for continuous speech recognition. Eurospeech-95 **1**, 537–540 (1995)
30. Rojas, R.: Neural Networks: A Systematic Introduction. Springer, New York (1996)
31. Rumelhart, D.E., Hinton, G.E., Williams, R.J.: Learning internal representations by error propagation. In: Rumelhart, D.E., McClelland, R.J. (eds.) Parallell Distributed Processing, vol. 1, pp. 318–362. MIT Press, Cambridge (1986)

32. Tatusov, R.L., Altschul, S.F., Koonin, E.V.: Detection of conserved segments in proteins: iterative scanning of sequence databases with alignment blocks. Proc. Natl. Acad. Sci. USA **91**, 12091–12095 (1994)
33. Wu, C.F.J.: On the convergence properties of the EM algorithm. Ann. Stat. **11**, 95–103 (1983)

Chapter 7
Implementation of a Comparative Gene Finder

In this chapter we exemplify the implementation of a gene finding software by describing SLAM [1] in a little more detail. SLAM is a cross-species gene finder particularly adapted to eukaryotes, and works by simultaneously aligning and annotating two homologous sequences. The basic framework of SLAM is a generalized pair HMM (GPHMM), which is a seamless merging of pair HMMs (PHMMs) typically used for pairwise alignments, and generalized HMMs (GHMMs) that have been successfully implemented in single species gene finders such as Genscan [3]. SLAM was used by the Mouse Genome Sequencing Consortium to compare the initial sequence of mouse to the human genome [5], and by the Rat Genome Sequencing Consortium to perform a three-way analysis of human, mouse, and rat [4, 6].

7.1 Program Structure

The GPHMM in SLAM is implemented in the C-language, but the whole program is coordinated by a Perl script `slam.pl` that handles the shuffling of files between three main modules (see Fig. 7.1): repeat-masking, creating an approximate alignment, and the SLAM gene prediction module. Repeat-masking greatly assists in gene finding because, for the most part, interspersed repeats do not occur in coding exons. The repeat masking is performed using the program RepeatMasker [7]. RepeatMasker screens the DNA sequences for known repeats an low complexity regions. This is a very important step, as the low complexity of many repeats will confuse the gene finder, and on average almost 50 % of the human DNA will be masked by this program. The output of RepeatMasker is a file with detailed information about the detected repeats, and modified sequence files where the repeats have been masked by the letter 'N.' SLAM has the ability to work with sequences that have been masked for repeats, but more importantly, sequences for which repeats have been annotated. In the latter case the program takes advantage of repeat types that are known to rarely occur within coding exons.

© Springer-Verlag London 2015   311
M. Axelson-Fisk, *Comparative Gene Finding*, Computational Biology 20,
DOI 10.1007/978-1-4471-6693-1_7

Fig. 7.1 The SLAM
modules

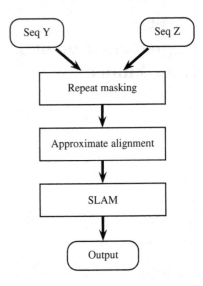

The approximate alignment created is used to reduce the search space for the dynamic
programming algorithm. In effect the two input sequences are globally aligned using
the alignment program AVID [2], and then the alignment is "relaxed." How this is
done is described in a little more detail below. The SLAM module is the main part,
in which the gene finding and alignment is performed using a GPHMM. The input
to the SLAM module consists of two (repeat-masked) sequences, an approximate
alignment and parameter files for the pair of organisms to be analyzed. The sequences
are provided in FASTA format and the format of the approximate alignment file is
a column of integers, where the first column is the base position in the first input
sequence, and the resulting columns are left and right coordinates of the matching
window in the second sequence. Note that a base position in the first sequence can
have several matching regions in the second sequence. The SLAM module in Fig. 7.1
consists of the following main steps:

1. Process command line arguments.
2. Read in parameter files.
3. Read in sequence files.
4. Pair up candidate exon boundaries in the input sequences.
5. Run the GPHMM.
6. Print output.

7.1.1 Command Line Arguments

The synopsis of SLAM is:
```
slam [opts] seqY seqZ -p pars -a aat -org1 o1 -org2 o2
```

Description:

> seqY seqZ
> > The FASTA files of the two orthologous sequences to be analyzed.
>
> -p pars
> > Specifies the directory path where the parameter files are located.
>
> -a aat
> > Specifies the name of the approximate alignment file.
>
> -org1 o1 -org2 o2
> > Specifies which organisms to be analyzed.

SLAM comes with a number of options [opts]:

-verbose	Run in verbose mode (quite a bit of output).
-debug	Run in debug mode (lots of output). For debugging purposes.
-oneseq	Single species gene finding.
-acceptorAG	Require acceptors to have the minimal AG consensus.
-donorGT	Require donors to have the minimal GT consensus.
-nocns	Turn off prediction of CNSs.
-indep	Independent exon scoring is used. For debugging purposes.
-okstops	In-frame stop codons are allowed but at very low probabilities.
-withMatrix	Allocates, initiates and uses the backtrack matrix. Quite heavy memory-wise.
-geneFile gf	A kind of gene mapping (see Sect. 4.5).

The verbose and debug options are for developers use mainly and result in a long number of checks that each substage of the program is running correctly. The oneseq option "turns off" the second sequence and runs an ordinary GHMM similar to that of Genscan in Sect. 2.2.4.

While most donors and acceptors contain the consensus dipeptides 'GT' and 'AG' respectively, variations exist, especially in lower organisms. Therefore the splice site model by default allows for other consensus sequences, even if their probabilities will be very small. The options acceptorAG and donorGT forces the splice site model to assign probability zero to splice sites that deviate from the consensus dipeptides.

For long sequences the dynamic programming matrices require a lot of memory. While the matrices for the forward, backward and Viterbi algorithms are necessary, the backtracking matrix can be reconstructed afterwards. This saves space but adds to the running time, and therefore, if the memory is sufficiently large, or the input sequences are short, storing the backtrack matrix may be more efficient. The default in SLAM is to reconstruct the backtrack matrix after the GPHMM algorithms are done, but with the withMatrix option the backtrack matrix construction is turned on.

The geneFile option functions similarly to the gene mapping programs Projector and GeneMapper described in Sect. 4.5. The file gf contains the known annotation for seqY and forces the GPHMM to only predict genes in seqZ that matches this annotation.

7.1.2 Parameter Files

SLAM expects a whole bunch of parameter files, achieved from training data, and the wrapping Perl script `slam.pl` expects them to lie in a subdirectory named `Pars` in the program file directory. The structure of `Pars` is typically

```
        Pars/
    organism1_organism2/
  bin1/  bin2/  bin3/  bin4/
        parameterfiles
```

The parameter files are stratified into a number of bin directories according to the % GC-content of the input sequences. For human and mouse the binning is:

bin1: $[0, 43]$
bin2: $[43, 51]$
bin3: $[51, 57]$
bin4: $[57, 100]$

Which bin that is read is determined by the Perl script `slam.pl`. The parameter files in each bin directory are

```
exon.len.init.dat        IG.freq.dat
exon.len.intl.dat        introns.freq.dat
exon.len.sing.dat        initprobs.dat
exon.len.term.dat        stopprobs.dat
IG.len.dat               transitions.dat
IG.lenBetwCNS.dat        aminoMargins.dat
intron.len.dat           mddDon.dat
intron.lenBetwCNS.dat    mddAcc.dat
exon.freq.dat            pairStats.dat
PAM1
```

The (`.len.`) files contain the length distributions of the corresponding features. For the exons, these are empirical distributions where the files are structured as:

```
1222
0.002790071
0.002826134
0.002862142
0.002898041
0.002933941
0.002969688
0.003005384
0.003041009
...
```

The first number m indicates the number of entries in the file, and the subsequent rows indicate the empirical probabilities of exon length $n = 1, 2, \ldots m$. As intron and intergene (IG) lengths are modeled with the geometric distribution, the corresponding length files only contain the average length of the state. For instance,

```
3276
```

The frequency files .freq. contain the base compositions of the different features. For instance, introns and intergenes are modeled by a second-order Markov model (tripeptides) and consist of 16 rows of counts

```
Order 2, Period 1
42578   15937   22163   29396
20576   12094   2504    18230
24172   14420   18518   19802
24180   14737   21160   32083
. . .
```

The first row indicates the order and period of the model, in this case the Markov model is non-periodic. The following 16 rows gives the tripeptide counts in the order {A, C, G, T}. That is, tripeptide AAA occurs 42578 times in the training set, AAC 15937 times, and so on. These counts are read in and transformed into probabilities in SLAM.

The initprobs.dat and transitions.dat files contain the initial and transition probabilities for all states in the state space, and stopprobs.dat the empirical distribution over the three stop codons. The pairStats.dat contains the gap probabilities of the PHMMs for the different I-states. PAM1 contains the PAM1 matrix, which is converted to the desired PAM order by SLAM according to which organism pair that is considered. For instance, human and mouse uses a PAM20 matrix. The aminoMargins.dat contains the empirical distribution of the 20 amino acids. The mddDon.dat and mddAcc.dat files contain sets of splice site sequences used to construct the MDD (maximal dependence decomposition) splice site detector described in Sect. 5.4.3 for donors and acceptors, respectively. For instance, mddDon.dat is on the form

```
order 5
upstream 3
downstream 6
size 3763
THRES 700
conserv 4 5
ignore 10
GAGGTGAGT   GAGGTGAGT
CAGGTGAGA   CAGGTGAGT
AACGTGAGC   AATGTGAGT
CAGGTACCT   CAGGTACTG
AAGGTGGGC   AAGGTGGGC
ATGGTGAGC   ATGGTGAGC
   . . .
```

Here order gives the order of the model, upstream and downstream gives the location of the exon–intron boundary (here it is between position 3 and 4), size gives the number of lines in the file, THRES the threshold for when a set is large enough to be split, conserv any eventual conserved positions (positions 4 and 5 consist of the consensus dipeptide 'GT'), and ignore any eventual positions to be ignored by the MDD (none here since the given position is beyond the motif). Each subsequent lines contain aligned pairs of splice site sequences, one from each organism. These are read in by SLAM and the MDD is created according to Sect. 5.4.3. The alignment of signal sequences allows the future implementation of a comparative MDD, but currently SLAM builds two separate MDDs, one for each organism.

7.1.3 Candidate Exon Boundaries

Before starting the HMM algorithms SLAM runs through the input sequences to detect potential exon boundaries. There are four types of boundaries considered: translation start and stop sites, and donor and acceptor splice sites. Keeping the forward and the reverse strands separate leaves us with eight boundary types: fStart, fStop, fDon, fAcc, bStart, bStop, bDon, bAcc where 'f' stands for the forward and 'b' for the backward strand. The start and stop boundaries are detected by their exact patterns: ATG for start codons, and TAA, TAG, or TGA for stop codons. The stop codons are scored according to their respective usage in the training set. The splice sites are scored using the Maximal Dependence Decomposition (MDD) model described in Sect. 5.4.3.

The potential boundaries are then paired up according to the corresponding exon types

Left bdy	Right bdy	Exon type
Start	Stop	Single
Start	Donor	Initial
Acceptor	Donor	Internal
Acceptor	Stop	Terminal

For instance, each potential start site is paired up with a number of potential stop and donor sites and vice versa. However, to avoid having to run through all possible combinations, unallowed pairs are removed based on a numbers of restrictions on the potential exon. These restrictions include length limits on the exon, in-frame stop codons, or if the candidate exon sequence would fall outside the approximate alignment. Each potential boundary is stored with information about its coordinate position, its score, and its potential boundary pairings. Then, when the HMM enters a potential exon state only the exons corresponding to the correct pairings are considered.

7.1.4 Output Files

SLAM produces a number of output files and prints a summary of the predicted gene structures to the screen. The screen output takes the form:

```
Seq Gene Exon Dir Type Start Stop Len Fr Ph Pr

Y                 cns  1486  1541 56  (82 % identity)
Z                 cns  2028  2083 56

Y    1    1   +   init 1634  1705 72  1  0
Z    1    1   +   init 2169  2240 72  2  0  0.83

Y    1    2   +   intl 2672  2774 103 1  0
Z    1    2   +   intl 2829  2931 103 2  0  0.92

Y                 cns  3167  3269 103 (69 % identity)
Z                 cns  3945  4047 103

Y    1    3   +   intl 3344  3459 116 0  1
Z    1    3   +   intl 4112  4227 116 0  1  0.84

Y                 cns  5530  5656 127 (90 % identity)
Z                 cns  7067  7193 127
```

There are two lines for each predicted exon or CNS, one for each organism. The columns are defined as follows:

Seq: First (Y) or second (Z) sequence.
Gene: Gene index.
Exon: Exon index within the gene.
Dir: Strand direction, forward +, reverse −.
Type: Feature type, one of sing/init/intl/term or cns.
Start: The left coordinate of the feature seen from the forward direction.
Stop: The right coordinate of the feature.
Len: Feature length = Stop − Start + 1.
Fr: Frame: if the "rightmost" base of one of the exon's codons ends at position k, the frame is (k mod 3).
Phase: The number of extra bases at the start of the exon sequence, before reaching a new codon.
Pr: Exon probability.

Note that the Start and Stop coordinates only indicates the left and right coordinates in the sequence seen in the forward direction. If the exon is on the reverse strand (Dir −), the start coordinate signifies the end of the exon and vice versa. Since the CNSs have no phase or frame, the percent identity of the alignment is given instead.

Besides the screen output SLAM generates a number of output files. The base name of the output files will be the same as the base name of the corresponding input sequences, and files with the following extensions are generated:

```
.gff    GFF (general feature format) file of the predictions.
.rna    A multiple FASTA file of predicted, spliced mRNAs (without UTRs).
.pep    A multiple FASTA file of predicted protein sequences.
.aln    Alignments of the predicted proteins in the two sequences.
.cns    Alignments of the predicted CNSs in the two sequences.
```

7.2 The GPHMM Model

Recall from Sect. 4.4.2 that the SLAM state space can be divided into I-states and E-states, where the I-states are the intergene and the intron states, and the E-states are the exons (see Fig. 7.2). Moreover, the I-states are themselves composed of two substates, namely a pair of independent I-state modeling unrelated intronic or intergenic regions, and a CNS state modeling conserved noncoding stretches occurring within the I-states.

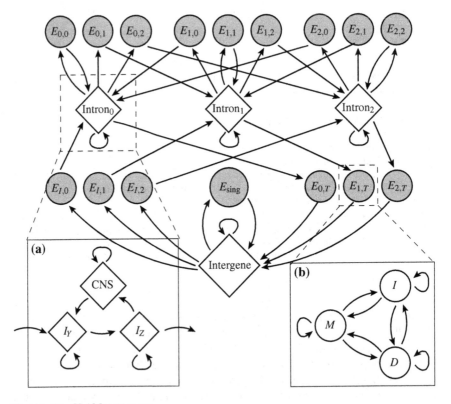

Fig. 7.2 The SLAM state space

7.2.1 Modeling Intron and Intergenic Pairs

Introns and intergene sequences are modeled using second-order Markov models, where the tripeptide frequencies are read in from a parameter file. The sequence lengths are modeled using the geometric distribution, and the average state lengths are supplied in the parameter files. Introns and intergenes would typically be modeled by standard PHMMs, but there are several issues with this approach, however. In the single species case, if the state length l is geometric, the self-transition probability of the state can be estimated by $a_{ii} = 1 - \frac{1}{l}$, and the probability of transitioning out of the state is given by $1 - a_{ii}$. In the two species case the situation is more complicated as the model does not generate single sequences, but an aligned pair of sequences, possibly containing gaps. Given the average state lengths for each of the sequences, the computation of emission and transition probabilities is not as straightforward as in the single species case.

Moreover, an intrinsic property of PHMMs is that they generate pairs of sequences of approximately the same length. This can be shown using a simple law-of-large-numbers argument. Assume, for instance, that we have a paired sample $(Y_1, Z_1), \ldots, (Y_n, Z_n)$ where (Y_i, Z_i) are independent random variables taking values in $\{(0, 1), (1, 0), (1, 1)\}$, such that the difference $D_i = Y_i - Z_i$ takes values in $\{0, 1, -1\}$. The central limit theorem gives that for the sum $Y - Z = \sum_i (Y_i - Z_i)$ the ratio

$$\frac{Y - Z}{\sigma \sqrt{n}} \tag{7.1}$$

is approximately normally distributed $N(0, 1)$ with mean 0 and variance 1. That is, on average the differences in lengths will be approximately 0 and with small variances. This is not a good model, however, in comparative sequence alignment, as this constraint is often violated. For instance, the intron lengths in human and mouse have a ratio of longer/shorter $= 1.5$ [5].

SLAM moves around both issues above by introducing a three-state model in each I-state, with substates denoted by I_Y, I_Z and CNS, and where the I-sequences consist of long, independent stretches of intronic or intergenic sequence (I_Y and I_Z) intervened by short, but highly conserved regions (CNS). The transition probabilities of the independent states are estimated using the shifted geometric distribution presented in Sect. 5.2.1

$$a_{ii}^{(Y)} = \frac{l_Y}{1 + l_Y}, \tag{7.2a}$$

$$a_{ii}^{(Z)} = \frac{l_Z}{1 + l_Z}, \tag{7.2b}$$

$$1 - a_{ii} = 1 - a_{ii}^{(Y)} a_{ii}^{(Z)}. \tag{7.2c}$$

The CNS-states are allowed to be of fairly similar lengths in the two sequences and are therefore modeled by regular PHMMs, with the self-transition set by a single average length parameter l_{CNS}

$$a_{ii}^{(CNS)} = \frac{l_{CNS}}{1 + l_{CNS}}. \tag{7.3}$$

Assuming that the two I-sequences are independent works well for organisms as far apart as human and mouse. For more closely related organisms this assumption may cause problems, however. On the other hand, in that case the law-of-large-number constraint in PHMMs will usually not be violated and a standard PHMM for the entire I-state may suffice.

7.2.2 Modeling Exon Pairs

The main body of an exon pair is scored using a PHMM at the amino acid level. Consider the amino acid sequences Y_1^P and Z_1^Q of a potential matching exon. Let c_Y denote the codon coding for amino acid a_Y at position Y_{p+1}, and, similarly, let c_Z be the codon coding for amino acid a_Z at Z_{q+1}.
For such a pair we need to evaluate the probability

$$\mathbf{P}(c_Y, a_Y, c_Z, a_Z | Y_1^p, Z_1^q) =$$
$$= \mathbf{P}(c_Y | Y_1^p, Z_1^q) \mathbf{P}(a_Y | c_Y, Y_1^p, Z_1^q) \mathbf{P}(a_Z | a_Y, Y_1^p, Z_1^q) \mathbf{P}(c_Z | a_Z, Y_1^p, Z_1^q). \tag{7.4}$$

Codon usage tables give us the $\mathbf{P}(c_Y)$ and $\mathbf{P}(c_Z | a_Z)$ probabilities. Furthermore, we have that $\mathbf{P}(a_Y | c_Y) = 1$ (or 0) and $\mathbf{P}(a_Z | a_Y)$ is given by the PAM matrix. The dependency on the previous sequence is best modeled using a 5th-order Markov model (2nd order at the amino acid level), but in the case of the PAM entry, this dependency is ignored. Gap probabilities are obtained from the PAM matrix as well. The probability of the exon pair is given by summing (7.4) over all amino acid pairs in the alignment. The optimal alignment (within the approximate alignment) can obtained using the Viterbi algorithm in the standard manner.

Once the HMM algorithms have processed the input sequences and we have an optimal prediction from the Viterbi algorithm, we can use the forward and the backward variables to compute the *posterior* exon probabilities. Say for instance that the Viterbi algorithm has predicted an exon of type s_e with coordinates $[l_y, r_y]$ and $[l_z, r_z]$ in the Y and Z sequences, respectively. If we denote the exon durations as $d = r_y - l_y + 1$ and $e = r_z - l_z + 1$, the joint probability of the observed data and the predicted exons can be written

$$\mathbf{P}\left(Y_1^T, Z_1^U, \cup_{l=1}^L (X_l = s_e, d_l = d, e_l = e, p_{l-1} = l_y - 1, q_{l-1} = l_z - 1)\right). \tag{7.5}$$

If we let s_{i-} and s_{i+} denote the uniquely determined I-states appearing directly before and after the predicted exon s_e, the probability in (7.5) can be written as

$$\alpha_{i-}(l_y - 1, l_z - 1)\, a_{i-,e}\, f_{s_e}(d, e)\, b_{s_e}\!\left(Y_{l_y}^{r_y}, Z_{l_z}^{r_z} \,|\, Y_1^{l_y-1}, Z_1^{l_z-1}\right)\beta_{i+}(r_y, r_z). \qquad (7.6)$$

If we normalize this expression by $\mathbf{P}(Y_1^T, Z_1^U)$, which is achieved by the forward algorithm, we get the probability of the predicted exon pair, given the data. This probability has the same interpretation as in the one organism case.

7.2.3 Approximate Alignment

In order to reduce the search space for the dynamic programming algorithms, SLAM makes an approximate alignment of the two input sequences. The assumption is that there exists a "true" alignment, which is unknown to us, and in the approximate alignment we state bounds on where we believe this alignment to lie. Approximate alignments are necessary in the GPHMM framework in order to handle input sequences on the order of hundreds of kilobases. In effect, the SLAM approximate alignment of the two input sequences Y_1^T and Z_1^U is a lookup table in which each base $Y_t, 1 \le t \le T$ is mapped to a window of bases $Z_{u-h}, \dots, Z_{u+h}$. The HMM variables are then set to 0 whenever a coordinate pair (t, u) falls outside this window. For a window size h per Y-base, the memory requirements gets reduced from $O(TUN_I)$ to $O(hTN_I)$ and the number of computations from $O(TUN_I^2 D^4)$ to $O(hTN_I^2 D^4)$, where N_I is the number of I-states in the model, T and U the lengths of the input sequences, respectively, and D the maximum length of an exon. However, while small windows improve the computational complexity, it increases the dependency on the accuracy of the approximate alignment. On the other hand, relaxations of the approximate alignment allow for more robustness at the expense of computational complexity. The approximate alignment is constructed in three steps.

1. Create a global alignment of the input sequences.
2. Relax each base-to-base match to a window of fixed size.
3. Extend the approximate alignment further around the candidate exon boundaries.

The global alignment is produced using the alignment program AVID [2]. The alignment is then relaxed by extending each match (Y_t, Z_u) to a window of fixed size $2h$, $(Y_t, [Z_{u-h}, Z_{u+h}])$. However, while the alignment of exons usually tend to be rather unambiguous, the part of the splice site sequences that extend into the surrounding exons might have rather weak alignments. Therefore, to ensure that exons are not excluded because of this, the approximate alignment is relaxed even further around potential splice sites.

7.3 Accuracy Assessment

A vital step in the software development is to assess the accuracy of the program, mainly in order to detect problems, but also in order to be able to benchmark the method against other methods. The two most common accuracy measures in gene prediction are the *sensitivity* (*SN*) and the *specificity* (*SP*) measures, usually defined as

$$SN = \frac{TP}{TP + FN}, \tag{7.7a}$$

$$SP = \frac{TP}{TP + FP}, \tag{7.7b}$$

where the values *TP* (true positives), *TN* (true negatives), *FP* (false positives), *FN* (false negatives) are defined as follows

TP = the number of coding bases predicted as coding,
TN = the number of noncoding bases predicted as noncoding,
FP = the number of noncoding bases predicted as coding,
FN = the number of coding bases predicted as noncoding.

The sensitivity (*SN*) is thus the proportion of bases predicted as coding among all truly coding bases, while the specificity (*SP*) measures the proportion of bases correctly predicted as coding among all bases predicted as coding. When prediction is perfect both *SN* and *SP* are equal to 1, but usually prediction accuracy is a trade-off between the two. For instance, a program that tends to overpredict will get a high sensitivity, but a low specificity, while in a program that is more conservative we will see the opposite relationship. Thus, in order to illustrate the accuracy, both measures need to be presented.

Measures that attempt to capture both the sensitivity and the specificity in one single measure include the *correlation coefficient* (CC) defined as

$$CC = \frac{(TP \cdot TN) - (FN \cdot FP)}{\sqrt{(TP + FN) \cdot (TN + FP) \cdot (TP + FP) \cdot (TN + FN)}}. \tag{7.8}$$

While combining both sensitivity and specificity this measure has the undesirable property that if no coding sequence has been predicted, the measure is undefined. A similar measure that avoids this problem is the *approximate correlation* (AC) defined as

$$AC = \frac{1}{2} \left(\frac{TP}{TP + FN} + \frac{TP}{TP + FP} + \frac{TN}{TN + FP} + \frac{TN}{TN + FN} \right) - 1. \tag{7.9}$$

In addition to measuring accuracy on the nucleotide level, it may be of interest to measure accuracy on the exon level as well. Sensitivity and specificity of entire exons are simply given by

$$SN_E = \frac{TE}{AE}, \tag{7.10}$$

$$SP_E = \frac{TE}{PE}, \tag{7.11}$$

where *TE* (true exons) is the number of exactly correctly predicted exons, and *AE* and *PE* are all annotated exons and all predicted exons, respectively. However, considering only exactly correct exons may be a bit crude, and therefore it may be more informative to divide exon predictions into 'correct' for correctly predicted exons, 'partial' for exon predictions with one correct boundary, 'overlap' for predicted exons with no correct boundaries but overlapping a true exon, and 'wrong' for predicted exons that have no overlaps with true ones. Useful measures may then be various proportions of these categories, such as missed exons (ME) and wrong exons (WE) defined as

$$ME = \frac{\text{number of missed exons}}{\text{number of annotated exons}}, \tag{7.12}$$

$$WE = \frac{\text{number of wrong exons}}{\text{number of predicted exons}}. \tag{7.13}$$

7.4 Possible Model Extensions

The model presented above has a number of undesirable assumptions built into it. The structure of the state space forces every gene to have the exact number of exons in each organism, an assumption that is sometimes violated, especially in the beginning or the end of genes. For instance, it is not uncommon that an exon in one organism has been split into two in the other, leaving the resulting protein more or less unchanged. Another model restriction is that it does not allow for frameshifts. That is, paired exons have to have the same phase. When comparing human and mouse, this is not a serious restriction, but may become a problem when comparing more distantly related organisms.

A first extension to the model could therefore be to allow for frameshifts. Relaxing the no-frameshift assumption means that an intron state is no longer characterized by the preceding exon in either organism, but rather by the pair of preceding exons in both organisms. Thus, allowing for frameshifts multiplies the number of intron states by 3, resulting in a ninefold increase in computational complexity. Furthermore, there is a ninefold increase in the number of exon states, with each current exon being replaced by 9 exon–exon states.

It is becoming more and more common to include regulatory regions in the gene predictions, both as they may improve the prediction accuracy of the exons, and vice versa that identifying the coding region may aid in locating its regulatory elements. SLAM currently predicts conserved noncoding sequences (CNSs), and these tend to appear in regulatory regions such as the 5′ and 3′ UTRs. The intergene CNSs can fairly easy be converted into modeling UTRs and other regulatory elements such as promoters or polyA-signals. None of these models take into account gene duplications, gene rearrangements, gene overlaps or alternative splicing, however, something that would be very valuable to be able to predict in the future.

References

1. Alexandersson, M., Cawley, S., Pachter, L.: SLAM: cross-species gene finding and alignment with a generalized pair hidden markov model. Genome Res. **13**, 496–502 (2003)
2. Bray, N., Dubchak, I., Pachter, L.: AVID: a global alignment program. Genome Res. **13**, 97–102 (2003)
3. Burge, C., Karlin, S.: Prediction of complete gene structures in human genomic DNA. J. Mol. Biol. **268**, 78–94 (1997)
4. Dewey, C., Wu, J.Q., Cawley, S., Alexandersson, M., Gibbs, R., Pachter, L.: Accurate identification of novel human genes through simultaneous gene prediction in human, mouse, and rat. Genome Res. **14**, 661–664 (2004)
5. Mouse Genome Sequencing Consortium: Initial sequencing and comparative analysis of the mouse genome. Nature **420**, 520–562 (2002)
6. Rat Genome sequencing consortium: Genome sequence of the BrownNorway rat yields insights into mammalian evolution. Nature **428**, 493–521 (2004)
7. Smit, A.F.A., Hubley, R., Green, P.: RepeatMasker. http://www.repeatmasker.org

Chapter 8
Annotation Pipelines for Next-Generation Sequencing Projects

Next-generation sequencing technologies has caused an explosion in the availability of genomic sequence data. This creates both opportunities and challenges, not the least within the bioinformatics field. The opportunities include the possibility to sequence and analyze a wide variety of organisms, spanning distant parts of the tree of life. The challenges include dealing with the shorter sequence lengths, the reduced data quality, and training and quality control issues when dealing with completely novel sequences. In this chapter we present the various issues and aspects involved in building a genome annotation pipeline, particularly aiming at next-generation sequencing data.

8.1 Introduction

The Genomes OnLine Database (GOLD) [113] aims at being the most comprehensive resource containing information about sequencing and metagenomics projects. As of January 2015, there were 56,774 sequencing projects in the database, out of which 6,649 are completed, 23,552 permanent drafts, and another 26,573 ongoing sequencing projects. This can be compared to September 2011, when GOLD contained information about 11,472 sequencing projects, out of which 2,907 were completed and 8,565 ongoing. The amount of sequence data is thus increasing fast in marks a new era in numerous related fields.

First-generation sequencing projects (Sanger sequencing) typically focused on established model organisms, benefiting much from the large bulk of pre-existing data knowledge. In terms of genome annotation, comprehensive datasets of species specific gene models could be used both to train the gene prediction tools and to quantify their level of accuracy. Second-generation sequencing projects, however, usually do not have access to such information. They commonly involve evolutionary isolated organisms without any known close relatives and without pre-existing data, which limits the ability to train gene prediction models and to measure the accuracy of the resulting annotations. The challenges for the bioinformatics field are

© Springer-Verlag London 2015

M. Axelson-Fisk, *Comparative Gene Finding*, Computational Biology 20,
DOI 10.1007/978-1-4471-6693-1_8

manifold, including the development of computational tools to store and manage the large quantities of data, adapted means to analyze and visualize the annotations, and metrics to measure the accuracy of the results. Also, as sequencing costs continue to drop, numerous small research groups, often with apt biology knowledge but limited bioinformatics skills, can now sequence their favorite organisms. Therefore there is an urgent need for user friendly, portable, and easily adapted annotation pipelines, that can proceed through the many steps of the annotation process without requiring an expert level of mathematical modeling and computer programming skills of the user.

In this chapter we discuss different aspects of genome annotation in next-generation sequencing projects. We begin by giving a historical view of DNA sequencing at large, and the corresponding bioinformatics development, followed by going through the different steps in an NGS annotation pipeline. We round off with an illustration of annotation pipelines from the viewpoint of the MAKER annotation pipeline suite [20, 21, 65].

8.2 History of DNA Sequencing

DNA sequencing refers to any method or technology used to determine the order of the chemical building blocks, the nucleotide bases, in a given stretch of DNA. Everything alive on this planet is defined by its genomic sequence, which is why DNA is often referred to as the "molecule of life". The order of the chemical components of the DNA is extremely important, as it holds the recipe for everything we are, and if we could decode this recipe we could learn what underlies the diversity of life. Differences and similarities in the DNA sequence, both within and between species, can teach us many invaluable things. Some history of DNA sequencing is as follows:

- 1869: The DNA molecule is isolated.
- 1944: DNA is the carrier of inheritance.
- 1953: The DNA molecule is a double-helix.
- 1965: The first RNA molecule is sequenced.
- 1968: The first DNA sequence is published.
- 1970: The discovery of type II restriction enzymes.
- 1975: Sanger's 'plus and minus' method.
- 1977: The first complete genome, bacteriophage ϕX174.
- 1977: Sanger's chain-terminating *dideoxy method*.
- 1977: Maxam and Gilbert's chemical degradation sequencing method.
- 1983: Polymerase Chain Reaction (PCR)
- 1986: Automated Sanger sequencing machines are manufactured.
- 1990: The Human Genome Project is launched.
- 1995: The first complete genome of a free-living organism, the bacterium *H. influenzae*.
- 1996: Pyrosequencing method published.

- 2000: Whole-genome assembly of the fruit fly *Drosophila melanogaster* [106].
- 2000: The first next-generation sequencing method, MPSS.
- 2001: Human genome draft sequence.
- 2004: Human genome project completed.
- 2004: Massively parallel sequencing technologies publicly available
- 2006: Illumina NGS method on the market.
- 2010: RNA-Seq.

In many ways, the DNA sequencing history begins in 1869 when Friedrich Miescher was able to isolate a phosphate-rich microscopic substance in the pus of surgical bandages. Since this substance resided within the cell nuclei he chose to call it 'nuclein', which was later changed to *nucleic acid*, and eventually to *deoxyribonucleic acid* or *DNA*. The existence of nuclein was unknown at the time, and the focus of Miescher's study were the protein components of the white blood cells (leukocytes) extracted from the pus. However, the material he came across exhibited no chemical properties of like those of proteins, and Miescher realized that he had discovered an unknown substance [28].

Meanwhile, in the 1850s Gregor Mendel set out to investigate how visible traits were transferred between generations. His genetic model system was the pea plant, as its fertilization could be easily controlled by transferring pollen between plants. After several years of tedious experiments, in 1865 Mendel proposed his three famous laws of inheritance, and without knowing anything about DNA or genes, he hypothesized that each parent contributes some particular matter, which he called 'elementen', to the offspring. Although controversial at first, Mendel's principles grew in acceptance during the decades that followed, but the chemical nature of the hypothesized matter remained unknown for quite some time.

Scientists knew that the chromosomes somehow were the carriers of inheritable traits, and that the chromosomes consisted of both DNA and proteins. But since the proteins appeared more varied than the DNA both in chemical composition and physical properties, proteins seemed a better choice as genetic material. Therefore it came as a great surprise, when Avery et al. in 1944 suggested that it in fact was the DNA, and not the proteins, that was the carrier of inheritance [5]. The mystery, however, was how this inheritance could be copied and passed on between generations.

The DNA double helix structure was published by Watson and Crick in 1953 [173]. The realization that the DNA was structured in two intertwined chains, rather than a triple-strand that had previously been suggested, where the chains mirrored each other in so-called *base pairs* resolved the puzzle of how the genetic blueprint of an organism could be stored, copied, and passed on between generations. This is now considered one of the most important scientific discoveries of the 20th century, and awarded Crick and Watson, together with Maurice Wilkins, the Nobel prize in 1962. Still, knowing the larger structure was one thing, however, and determining the precise order of the building blocks quite another. Due to various technical difficulties, it would take several decades until the first DNA fragments could be "read".

Due to various technical difficulties, the first experimental sequencing of DNA was at a pace of only a few bases per year, and it would take several decades until larger DNA fragments could reliably determined. The first nucleic acid structure to be published was that of a tRNA molecule isolated from yeast in 1965 [64]. The method used specific enzymes, *ribeonucleases* (or *RNase* for short), to cleave the RNA molecule into shorter fragments, but no such enzyme corresponding to DNA was yet known to exist. In 1968, the first DNA sequence was published, representing the cohesive ends of a bacterial virus, the *bacteriophage lambda* [177]. The sequence was only partially determined, however, and the complete sequence of 12 base pairs was published in 1971 [178]. The use of oligonucleotide primers and chain termination was introduced in the process, enabling a generalization of the sequencing approach. Furthermore, in 1973 Gilbert and Maxam reported a 24 base pairs partial sequence of the *lac* operator, a protein binding site in bacteria, using a 'wandering-spot analysis' technique [46]. Also, paving the way for gel-based sequencing through electrophoresis, in 1970 type II restriction enzymes that could cleave DNA at specific short (4–6 bp) sequences were discovered [73, 146].

The era of automated sequencing of longer DNA fragments started in 1975 when Fredrick Sanger introduced the "plus and minus" method using DNA polymerase with radiolabeled nucleotides [134]. The method could sequence as many as 80 bases in a single run, but struggled with resolving the length of repeated stretches of a single base (homopolymers). The plus-and-minus method was used to sequence the first complete genome, that of the bacteriophage ϕX174 [133], consisting of 5386 bp of which 95 % are coding for a total of 11 proteins. Having access to an entire genomic sequence, it became apparent that large portions of the genome were translated in more than one reading frame, containing sets of overlapping genes, something that was previously unheard of.

In 1977 two new sequencing methods methods were published almost in parallel. Maxam and Gilbert presented a sequencing method using chemical nucleotide-specific cleavage, which was similar to the plus-and-minus method, but without suffering from the homopolymer issue [95]. However, it was best used to sequence shorter oligonucleotides, typically smaller than 50 bp in length. The real breakthrough came with Sanger's *dideoxy method* or *chain-termination method*, presented in 1977 [135], and awarding him the Nobel prize in Chemistry in 1980. The new Sanger method had adopted the primer-extension strategy used on the lambda phage mentioned above [178], and had also solved the homopolymer issues of the plus and minus method. Requiring fewer toxic chemicals and lower amounts of radioactivity than the Maxam-Gilbert approach, it soon became the method of choice. Since then, with several technical advances and refinements that sped things up and automatized the sequencing process, the Sanger method has come to dominate the sequencing world for several decades, and is what we now refer to as the *first generation* sequencing technology. Ultimately, in mass production form, Sanger sequencing produced the first draft of the human genome in 2001 [79, 168]. One indispensable technical advance was the development of the *polymerase chain reaction* (PCR), that is used to amplify DNA fragments in the sequencing process. Although disputed, the discovery is generally attributed to Kary Mullis in 1983 [7], something that awarded him the Nobel prize in Chemistry in 1993. Although a great advance with the Sanger method, DNA sequencing remained rather cost and labor intensive, until 1986 when a company named Applied Biosystems started to manufacture automated DNA sequencing machines using Sanger sequencing with fluorescent dyes to tag each nucleotide. Following this, in 1995 the first complete genome of a free-living organism, the bacterium *Haemophilus influenzae* was sequenced [43], which marked the first published use of *whole-genome shotgun sequencing*. With these and numerous other advances in sequencing technology, vastly increasing the speed and reducing the cost, the dream of sequencing the entire human genome started to appear within reach.

The *Human Genome Project* (HGP), launched in 1990, was a publicly funded international project with the main objective to determine the DNA sequence and all the genes of the euchromatic human genome (containing most of the genetically active material). The HGP project plan was initially set to 15 years, but a private biotech company (Celera Genomics) rallied the HGP consortium into higher gear, and both groups simultaneously published a human genome sequence draft in 2001 [79, 168]. The finished draft was then announced in 2003, two years ahead of schedule, and the completion of the project was published in 2004 [70]. In reality, however, the human genome sequence is still not complete. Technical difficulties has left several million bases of gaps, both large and small, typically in repeat-rich heterochromatic regions. To help achieve the goals of the HGP, a number of model organisms were sequenced. These organisms include the common human gut bacterium *Escherichia coli* [13], the fruit fly *Drosophila melanogaster* [1], the laboratory mouse *Mus musculus* [102], the baker's yeast *Saccharomyces cerevisiae* [51], and the nematode *Canaeorhabditis elegans* [161]. These organisms and a few more are now included in the Gene Ontology Reference Genome Project [157], which aims at completely annotate twelve reference genomes in a unified framework to enable cross-species analyzes and phylogenetic studies.

Being a huge achievement with several revolutionary developments, the Human Genome Project only marked the beginning of the genomic era. The immense resources required to complete the project clearly indicated the need for even faster and cheaper technologies in the near future. In 2001 the National Human Genome Research Institute (NHGRI) of the US National Institute of Health (NIH) gathered 600 researchers to plan the next phase within genomics research [139]. Among other things, their discussions led to a list of "technological leaps" that were rather mind provocative and verging on science fiction at the time, and mainly intended to "provoke creative dreaming". One such list item was to reduce sequencing costs by four to five orders of magnitude, allowing the sequencing of an individual human genome for less than $1,000 [26]. This stimulated the development of *Next-Generation Sequencing* (NGS).

NGS technologies all share three major developments: they do not require bacterial cloning, the sequencing reactions are produced massively in parallel and the output is detected directly without the need for electrophoresis. As a result, enormous amounts of reads can be produced, from thousands to millions of sequences simultaneously, allowing the sequencing of entire genomes at an unprecedented speed.

The major drawback, however, is that the produced reads are much shorter than for Sanger sequencing, making both the assembly and the annotation process much more difficult and requiring the development of new and better suited computational algorithms.

The first NGS technology, MPSS (Massively Parallel Signature Sequencing), was developed in the 1990s by Lynx Therapeutics [18]. However, because of its complexity no machines were sold to independent laboratories. The first method to be commercialized was published in 2005 by 454 Life Sciences (now Roche) [94]. Their 454 Genome Sequencer was a parallelized version of *pyrosequencing*, which reduced sequencing cost sixfold compared to the Sanger method, and produced around 20 Mb of 110 bp sequence. One year later Solexa (later merged with Lynx Therapeutics and acquired by Illumina) released a method based on reversible dye-termination and engineered polymerases [10]. Around the same time Applied Biosystems (now Life Technologies) released their SOLiD (Sequencing by Oligo Ligation Detection) method [166], which applies *sequencing-by-ligation*. Both Solexa/Illumina and SOLiD sequencers generated much larger quantities of sequence compared to the 454 method ($\sim$1–3 Gb), but the reads were only about 35 bp in length. In 2010, the founder of 454, Jonathan Rothberg, released the Personal Genome Machine (PGM) under the flag of Ion Torrent (now Life Technologies) [129]. PGM resembled the 454 system, but used semiconductor technology and did not require optical detection using fluorescence and camera scanning, resulting in a higher speed, lower cost, and smaller machines. The first PGM generated up to 270 Mb of 100 bp reads. Numerous other NGS methods have been developed, including the Qiagen-intelligent bio-systems *sequencing-by-synthesis* [71], Polony sequencing [140], and Heliscope single molecule sequencing [123]. The latter is verging on the *third-generation* sequencing technologies. That is, next-generation sequencing is already regarded as second generation technology, and a new generation of sequencers are already emerging. Third-generation sequencing can be defined as methods that are capable of sequencing long sequences without amplification of the DNA template, and where sequence detection occurs in real time, reducing the sequencing time to minutes or hours rather than days [138]. The leader of the field is currently Pacific Biosciences that released the PacBio RS in 2010, which can generate several thousands of several kilobases long reads [36], something that makes it very suitable for completing *de novo* assemblies produced by second generation techniques.

Whether to prefer higher throughput or longer reads usually depends on the application. The higher throughput of Illumina and SOLiD have made them more suitable than 454 for protein-DNA interactions studies such as ChIP-Seq [114], while 454 has been the preferred technology for *de novo* genome assemblies and metagenomics studies. With various advances in both hardware and software, Illumina systems can now produce several hundreds bp long reads, which makes it applicable to genome assembly as well. Illumina also claims to have broken the $1,000 genome barrier of NHGRI with their HiSeq X Ten release, but this is when sequencing machines run at factory-scale performing population-scale genome sequencing [167]. Illumina is currently the leading NGS platform, offering the highest throughput, lowest cost, and reasonable long reads. And while several other technologies seem very promising, such as nanopore sequencing [25], which is also considered a third-generation technology because it sequences single molecules in real time, it remains to be seen if they can compete with Illumina. Regardless, a lot of sequence is produced in various forms, and more will come. The challenges on the computer system and analysis methods are immense, and the need for skilled bioinformaticians, computational biologists and system biologists is greater than ever.

8.2.1 The Origin of Bioinformatics

Since the early days of DNA sequencing improvements and advances in sequencing technology has led to a deluge of biological data, and computers have become indispensable to handle all these data. Fortunately computer technology has managed matched these developments, particularly regarding CPU and disk storage. Also, the development of Internet and the World Wide Web (WWW) has revolutionized the accessibility and exhangeability of data. However, the needs and demands for high-throughput bioinformatics tools are greater than ever, for data processing, storage, management, and interpretation, which in turn puts great demands on the education and training of suitable scientists to work in these areas.

The term *bioinformatics* was coined in 1970 by Ben Hesper and Paulien Hogeweg as "the study of informatic processes in biotic systems" [60], placing it on equal footing with fields such as biophysics (physical processes in biological systems) and biochemistry (chemical processes in biological systems). Since then the definition has been altered and updated many times, and now comes in a variety of shapes and forms depending on the foundation of the scientist giving it.

Initially, when the first short stretches of DNA sequences were identified, everything was handled manually, from generating the data to keeping records. But in 1977, with the ϕX174 genome of nearly 5400 bp [133] available, the manual management and analysis started to become intractable. To facilitate the needs, McCallum and Smith [96] reported a computer program that very well may mark the beginning of modern bioinformatics, as we would define it today. The software, programmed in COBOL, compiled the ϕX174 sequence from 60 manually entered punch cards, and allowed editing of the sequence, sequence searches and automatic translation in all reading frames. Subsequently, Roger Staden, who was involved in the computer analysis of ϕX174 constructed a suite of interactive programs in 1977 "designed specifically for use by people with little or no computer experience" [150]. The program suite, named the Staden Package, included tools for DNA sequence assembly, sequence editing, and sequence analysis, and is still in use today [151].

Another pioneer of bioinformatics was the physical chemist Margaret Oakley Dayhoff, most noted for her work on nucleic acid and protein sequence databases, and for the widely used amino acid substitution matrix PAM (Point Accepted Mutations) described in Sect. 3.1.3 [30]. In 1969, Dayhoff had collected all known protein sequences at the time, and published them in the *Atlas of Protein Sequence and Structure* [29]. Her work became the basis for the first public comprehensive,

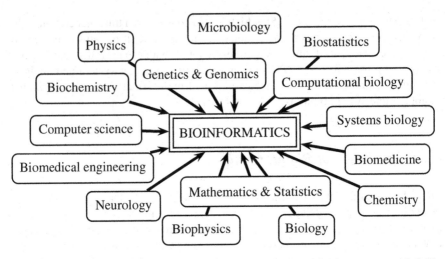

Fig. 8.1 A bioinformatics network. The bioinformatics definition varies between research fields and overlaps with numerous other fields

computerized, and publicly available protein database, and served as model for Gen-Bank and many other molecular databases. She is also attributable for the one-letter code for amino acids, which aimed at reducing file sizes in the punch-card era, but which is still employed.

Bioinformatics nowadays is a broad interdisciplinary field, that aims at managing and interpreting biological data, combining computer science, mathematics, statistics, and engineering to do so. It consists of several important subfields as it is both a collective term for biological studies that uses computer programming as part of their method, and a term for the development of algorithms and tools needed for the analysis of biological data. There is, thus, no clear consensus of the field definition, and there are significant overlaps with numerous other disciplines as illustrated in Fig. 8.1.

The Oxford English Dictionary defines bioinformatics as "the branch of science concerned with information and information flow in biological systems, especially the use of computational methods in genetics and genomics." The National Institute of Health (NIH) attempted in 2000 to define and distinguish bioinformatics from computational biology, recognizing that "no definition could completely eliminate overlap with other activities or preclude variations in interpretation." Their definitions were:

- Bioinformatics: "research, development, or application of computational tools and approaches for expanding the use of biological, medical, behavioral or health data, including those to acquire, store, organize, archive, analyze, or visualize such data."
- Computational biology: "the development and application of data-analytical and theoretical methods, mathematical modeling and computational simulation techniques to the study of biological, behavioral, and social systems."

The application areas of bioinformatics are manifold, and continues to grow, not least with all the new 'omics' fields (e.g. proteomics, transcriptomics, and metagenomics), that emerge from the growing availability and varied types of sequence data, and that embed a significant portion of often very specialized bioinformatics into their definition. Examples of bioinformatics research areas, particularly connected to NGS data, include:

- Sequence assembly
- Gene and functional element prediction
- Functional analysis of genes
- Comparative genomics
- Protein structure prediction
- Gene and protein expression analysis
- Analysis of gene regulation
- Systems biology
- Network analysis and protein-protein interactions
- Databases construction and management
- Clinical diagnostics
- Evolutionary biology
- Storage and management of big data

The future for bioinformaticians thus looks very bright, with many new and challenging areas in great need of competent and creative scientists.

8.3 Next-Generation Sequencing (NGS)

Sanger sequencing, also called *chain-termination* sequencing or the *dideoxy method*, was first introduced in 1977 [135]. It utilizes *DNA polymerase*, which is the enzyme the cell uses for DNA replication, to synthesize DNA templates, and is based on the findings that inclusion of *dideoxynucleotides* (ddNTPs) into the DNA synthesis chain inhibits the DNA polymerase activity and halts further strand extension. By bringing both regular deoxynucletides (dNTPs) and chain-terminating dideoxynucleotides (ddNTPs) to the mix, the result after many sequencing reactions is a number of DNA fragments of varying length. The fragments can then be heat denatured (separating the double DNA strand into two single strands) and length-separated by *gel electrophoresis*, in which an electric field pulls the molecules across a gel such that shorter molecules move faster than the longer. Labeling the DNA fragments with radioactive phosphorus or a fluorescent dye, and then exposing an X-ray film to the gel, gives rise to an image of dark bands that indicate the sizes of the fragments. Initially, the Sanger method involved running four separate DNA synthesis reactions on each sample, one for each nucleotide. That is, in each reaction all four regular dNTPs were added, but only one of the four chain-terminating ddNTPs. Each reaction then resulted in a number of DNA fragments of varying length, all terminating in the same nucleotide. Then by running the four samples in different lanes in parallel on the gel, the fragments lined up according to size, and the sequence could be "read" by knowing which lane ended in which nucleotide. The procedure is sensitive enough to separate DNA fragments differing by only one nucleotide in length. This process was first automated in 1986 when Applied Biosystems presented a dye-terminator variant [147]. By labeling each of the four ddNTPs with a *different* fluorescent dye, each terminated fragment contained a nucleotide specific dye at the 3' end, and all fragments fluorescing the same color all had the same terminating nucleotide. As a result, all four nucleotides could be processed in the same experiment, and the synthesized DNA fragments could be length separated in the same gel lane by placing a detector at the end of the gel, recording the color of the attached fluorescent dye as the fragments passed by. Another advantage was that the sequencing data could be sent directly to a computer. Another important breakthrough to the sequencing automation came in 1996, when the gel electrophoresis was replaced by *capillary* electrophoresis. In brief, by applying a high voltage charge to the sequencing reaction, the DNA fragments can be length-separated based on their charge, avoiding the tedious process of loading gels. In 1998, the 96 capillary system was announced, which allowed the sequencing of about 900 kb per day, compared to about 1 kb per day with the original chain-termination method. Also, with fragment lengths of up to 1,000 bp Sanger sequencing dominated the market up until the introduction of next-generation sequencing. However, while being a huge advancement for biological research, and being the prevailing method for several decades, the output was still limited. The main obstacles of Sanger sequencing are the use of gels or polymers as separation media, the limited number of samples that can be handled in parallel, and the difficulties to automatize the sample preparation. DNA sequencing platforms utilizing Sanger sequencing are considered the 'first generation' of sequencers. Their limitations triggered the develop the next-generation of sequencing technology.

8.3.1 NGS Technologies

Next-generation sequencing (NGS), or *second-generation sequencing*, has the universal characteristics of being massively parallel, meaning that the amount of sequence data generated from a single sample is immensely larger than that of Sanger sequencing. Sequencing projects that take years with Sanger methods can now be completed in weeks. One problem is that the produced reads are much shorter (around 100 bp compared to 1000 bp with Sanger sequencing) and less accurate than Sanger sequences, but these shortcomings are somewhat made up for by the much higher degree of sequencing *coverage*. Coverage, or *read depth*, is the average number of reads covering each nucleotide in the assembled sequence. Note, however, that this is still only an *average*, some nucleotides may still be completely uncovered.

The workflow of the most common NGS techniques is similar, even if the details differ. The first step is the sample preparation, in which the DNA sample is conversed into a sequencing library. This is done by using a wide variety of protocols, but the common feature is that the DNA (or RNA) is fragmented into smaller segments (50–500 bp) and fused (ligated) to small DNA oligonucleotide 'linkers' (or *adaptors*) that can be identified by specific primers. Most imaging systems cannot detect single fluorescent events, which is why the templates are amplified using *polymerase-chain reaction* (PCR), and attached/immobilized to a solid surface or support. This immobilization results in a spatial separation of the templates that allows the running of millions of sequencing reactions in parallel. One problem with PCR amplification, however, is that it creates mutations in the clones that may appear as sequence variants. An alternative approach, developed by Pacific Biosciences, is a single molecule technique in which the DNA template and a single active DNA polymerase are immobilized directly without amplification to the bottom of an optical waveguide, which is sensitive enough to detect a single fluorescently labeled nucleotide as it is incorporated by the DNA polymerase.

The second step in the NGS sequencing is the actual sequencing of the immobilized templates. This is done in different ways with different pros and cons regarding accuracy, speed, and read length. The Illumina (formerly Solexa) systems are currently dominating the market, with SOLiD (Life Technologies) in second place, and 454 (Roche), SMRT (Pacific Biosciences) and Complete Genomics sharing the remaining market. The Illumina/Solexa platform [10] uses the *sequencing-by-synthesis* approach, which similarly to Sanger sequencing uses a kind of chain-termination nucleotides. The main difference is that the termination is reversible. The attached nucleotide is blocking the elongation of the chain, its fluorescent dye is recorded and removed, and the block is removed chemically to allow the synthesis to continue. This cycle is iterated until the entire DNA template is sequenced. The SOLiD platform by Applied Biosystems (now part of Life Technologies) [166] performs a *sequencing-by-ligation* technique, utilizing *DNA ligase*, rather than DNA polymerase. DNA ligase is an enzyme that catalyzes the joining of separate DNA fragments. Instead of using modified nucleotides, a pool of all possible oligonucleotides of a fixed length (typically 8–9 bp) are added to the mix, each flourescently labeled according to which position that will be sequenced. Sequence-by-ligation has the advantage of being easy to implement and only using off-the-shelf reagents. Disadvantages include very short read lengths, and a problem with palindromic sequences. The 454 Genome Sequencer by Roche Diagnostics [94], is based on *pyrosequencing*. Pyrosequencing utilizes the fact that when a new base is incorporated by DNA polymerase a pyrophosphate group is released, which can be detected as emitted light using fiber-optic technology. Single molecule real time (SMRT) sequencing by Pacific Biosciences is considered a *third* generation technology [36]. It is also based on the sequencing-by-synthesis approach. A unique feature is that no clonal amplification is required due to a highly sensitive fluorescence detection system. During synthesis a nucleotide is incorporated by the DNA polymerase, its fluorescent signal is detectable for a short time and then the dye tag is cleaved off and diffuses out of the observation area. Another advantage is that the sequencing occurs in real time, which reduces the sequencing time to hours rather than days or weeks. The resulting read lengths are comparable to or even longer than those of Sanger sequencing, but the technology still struggles with high error rates. However, on the plus side, the artifacts of PCR amplification are avoided.

The process of sequencing a novel genome for the first time is called *de novo* sequencing. A common ingredient in such sequencing projects is the generation of *mate-pair* libraries. Mate-pairs are sequence reads generated in pairs from each end of a DNA fragment, keeping careful records of the fragment lengths, which can be several kilobases. Thus, mate-pair information can be utilized to span long repeat regions and gaps in the original assembly, and is used by most of the current *de novo* assemblers. This will be discussed more in the next section.

8.3.2 Genome Sequence Assembly

With current sequencing technologies it is impossible to sequence long DNA fragments directly. For capillary methods (Sanger sequencing) the upper limit is about 1,000 bp, and for next-generation technology much lower. Thus, longer sequences must be divided into smaller fragments, and then reassembled into the sequence of the DNA template. Two main principles are used for this: *chromosome walking*, and *shotgun sequencing*. In chromosome walking the entire DNA fragment is sequenced in consecutive overlapping segments, piece by piece. The procedure starts from a short piece of known sequence, called a *primer*, and then the first 1,000 bases are identified. Then a new primer is generated, complementary to the final 20 bases of the last sequenced segment, and another 1,000 bases are identified, and so on until the entire DNA template is sequenced. However, since for large-scale sequencing this method becomes intractable, it is typically mostly used to close gaps, or to sequence a disease gene located near a specific marker. For large-scale sequencing typically shotgun sequencing is used, a method named by its analogy with the random firing pattern of a shotgun, and which was developed in the 1970s by Frederick Sanger. In shotgun sequencing, the DNA template is broken up into smaller random fragments, which are then sequenced individually to obtain sequence *reads*. By repeating the shotgun fragmentation several times, multiple overlapping reads are obtained for the same DNA template. The problem of sequence assembly is then to reconstruct the original DNA template by joining the reads together as in a huge, redundant jigsaw puzzle. The reads are joined together into larger contiguous sequences, or *contigs*, by means of their overlaps, the contigs into larger *scaffolds*, and ultimately the scaffolds are merged into the complete DNA template.

8.3.2.1 Sanger Sequencing Assembly

For capillary sequence assembly, or Sanger sequencing, the two main types of algorithms used are the overlap-layout-consensus (OLC) approach, and the *de Brujin graphs* approach. There are several reviews and comparisons of these methods

[44, 87, 99], and there are numerous successful implementations including Arachne [8], Atlas [59], Celera Assembler [106], CAP3 [66], Euler [121], PCAP [67], Phrap [33], RePS [170] and Phusion [104]. The OLC method comprises three steps: the overlap step, the layout step, and the consensus step. In the overlap step all read pairs are compared, in both strand orientations, to detect overlaps between the reads and to create an overlap graph, also called an assembly graph. The nodes in the graph represent the reads, and an edge connecting two nodes signifies an overlap of two reads. The overlap step is heavily computer intensive as it scales quadratically with the number of reads. However, Arachne has a sort-and-extend strategy that vastly improves the situation [8]. In the layout step the overlap graph is compressed using principles of graph theory. An assembled sequence corresponds to identifying a path through the graph that visits every node exactly once, also known as a *Hamiltonian circuit*. This problem is *NP-hard*, meaning that except for very small graphs, it is practically impossible to search for an optimal solution. As a result, OLC algorithms use various heuristics to simplify the overlap graph, typically by merging "clearly overlapping" reads into contigs. Such clearly overlapping reads can be identified as they tend to form highly connected clusters, or subgraphs, in the overlap graph. The merging ends when reaching a *fork*, which is a node that is connected to two or more nodes that do not share an overlap. Forks typically signify the boundary between repeated and unrepeated segments. Greedy strategies, such as used by the OLC-based assemblers Phrap [33] or CAP3 [66], simply merge strings in the order of the highest scoring overlap, until a single string remains. There is no guarantee that the optimal solution is reached, however, as the result depends heavily on the scoring scheme and in particular on how equally scoring overlaps are ordered.

An alternative approach to the layout step in OLC is to utilize de Brujin graphs [31]. De Brujin graph assemblers take the reads and cut them into even smaller equally-sized k-tuples, i.e., sequence segments of length k. The collection of $(k - 1)$-tuples occurring in these k-tuples (two overlapping in each) constitute the nodes of the de Brujin graph, and an edge between two nodes correspond to an overlap in an existing k-tuple. For instance, for a 3-tuple CGT, the corresponding 2-tuple nodes connected by an edge would be CG and GT. The assembly problem now entails finding the shortest path (or circuit) that visits every edge, also known as the *Eulerian path* [120]. Finding a Eulerian path is much less demanding than finding the Hamiltonian path, and also, if a Eulerian path exists it is the optimal one. Moreover, the overlap step, that of pairing all reads in the search for overlap, can be eliminated altogether when using de Brujin graphs.

Repetitive sequences cause major problems in sequence assembly, often creating erroneous circuits and tangled graphs. After the initial merging of reads, the resulting contigs typically belong to one of two categories: true contigs and repeat contigs. True contigs consist of unambiguously assembled reads. Repeat contigs are contigs that correspond to several different regions in the genome, and can often be identified through an unusually high coverage rate. The repeat contigs are typically put aside for later use, while the true contigs are joined into larger *scaffolds*, typically using *mate-pair* or *paired-end* information. These terms are often used interchangeably, but represent different protocols. In both cases the reads come in pairs, representing opposite ends of the same DNA fragment. The main difference is the fragment size, where paired-end fragments are typically shorter (<1 kb) than mate-pair fragments (2–5 kb). However, both types can be used to aid the assembly. By keeping track of the distance between the pair of reads, the assemblers can make further links between reads or contigs. This again is done by using a graph-approach, in which now the contigs are the nodes, and the edges correspond to the paired information. The problems here include finding all connected components in the overlap graph or de Brujin-graph, resolving the strand orientation of each contig to make the assembly consistent, and laying out the graph on a coherent line (or circle). Again, these problems are NP-complete, but good heuristics exist. In addition to using mate-pair information to cover longer regions, another approach is to combine different sequencing technologies. For instance, the major bulk of the sequencing can be done using NGS technology, but then be complemented by capillary sequencing. Several assemblers can combine mixed inputs, or can exploit physical or genetic map information.

In the final step of the assembly process, the consensus step, the final genome sequence is determined by using the graph generated in the previous steps. Ideally a single scaffold remains, and the assembly algorithm can resolve the consensus sequence by means of the ingoing reads. However, often the assembly still has gaps due to unresolvable repeats or to insufficient or conflicting mate-pair information. The resulting assembly is then a fragmented one, composed of a number of scaffolds. This is the case with many of the recent assemblies, especially when considering larger eukaryotic genomes [176]. Essentially, only the human and the mouse have reached the status of *finished genomes* [79, 102]. Subsequently published genomes, such as for the rat [124], the dog [88], the rhesus macaque ([125], and the cow [155] are all only at the level of drafts with about a 6–8 fold coverage using Sanger sequencing, and with an N50 measure (see below) of about 20–200 kbp [176].

8.3.2.2 Next-Generation Sequencing Assembly

When the first NGS technologies were released almost a decade ago, it appeared doubtful that their short read lengths could be assembled properly to be suitable for larger scale genome projects. However, early investigations showed that re-sequencing and *de novo* sequencing with read lengths as short as 20–30 bp still could produce useful, however highly fragmented, assemblies of both prokaryotic and eukaryotic genomes [174]. The tools developed for Sanger sequencing data, however, turns out to be less suited for the shorter NGS read datasets. The increased number of reads makes the problem computationally expensive, and with the shorter read lengths, sequencing errors have a much larger impact and unique overlaps are more difficult to find. Moreover, sequence repeats are harder to resolve as they often extend beyond the read length. Therefore, over the past decade numerous novel assembly strategies, specialized to NGS reads, have been presented. Already several years ago, Zhang and colleagues [182] managed to list 24 distinct, academic *de novo* assemblers, and more is coming.

The assembly procedure for NGS data can be classified into two main areas: reference-based assembly, in which the sequence reads are mapped to an existing reference genome, and *de novo* assembly, where there is no reference and the reads have to be pieced together by means of sequence similarity and library information alone. During reference-based assembly, the sequenced DNA fragments are aligned to a genome sequence from the same organism or a close relative, into a growing assembly. This approach is used in re-sequencing applications where the objective is to detect genetic variation between individuals, such as identifying single-nucleotide polymorphisms (SNPs), or between healthy and disease cells, such as in various cancers [122]. *De novo* assembly is used when there is no reference, for instance, when sequencing novel species, metagenomics samples, or transcriptome samples. While mapping reads to a reference is a relatively simple task, the absence of a reference sequence makes matters much more challenging. The following in this section is dedicated to *de novo* assembly.

Several problems arise when turning from traditional sequencing strategies to NGS technologies. The shorter read length in NGS data compared to Sanger sequencing results in numerous locations in which there are not enough overlaps between reads to cover the sequence confidently. One remedy is to increase the level of coverage. Mathematical modeling shows that for Sanger sequencing of a mammalian-sized genome, a 3× coverage, i.e. an average of three overlapping reads per nucleotide, is sufficient [78]. However, with NGS read lengths only about a tenth of Sanger reads, the requirement increases tenfold or more. Also, the higher rates of sequencing errors also puts higher demands on the coverage. In practice, while Sanger sequencing project may have used 7–10× coverage, NGS projects tend to use 50× or higher, due to necessity but also due to the falling costs. No amount of coverage can however solve the repeats issue.

Repetitive regions are the major bottlenecks when assembling NGS short reads, in particular in complex eukaryotic genomes. The repeat elements often extend longer than each read, making it difficult to link it to the nonrepetitive adjacent sequences and resolve the multitude of positions the read sequence can originate from. Some assemblers use paired-end or mate-pair information to resolve repeats and close gaps in the assembly, such as SSAKE [172], SOAPdenovo [84], ABySS [141], and Velvet [181]. Some assemblers simply mask out repeats, while others attempt to utilize them in the later stages of the scaffolding process.

With the inflated number of sequenced reads comes the challenge of a significantly increased computational complexity. Some of the major differences between NGS assembly algorithms lie in how they attempt to reduce this complexity problem and how they handle repeats. To simplify the assembly task, the assembly algorithms format the input reads into specific graph data structures. Similarly to Sanger sequencing assemblers, most NGS assemblers are using the overlap-layout-consensus (OLC) or de Brujin graphs for their initial data formatting. The first NGS assembler to employ de Brujin graphs was the Euler assembler [120], followed by significant improvements in speed and accuracy in assemblers such as Velvet [181] and ALLPATHS [50], and by introducing message passing interface parallelization in ABySS [141]. Examples of overlap-based assemblers include CABOG [98] and the

MSR-CA pipeline, however MSR-CA utilizes a de Brujin graph to combine reads that map to the same nodes and edges, significantly reducing the number of reads that need to be considered.

One drawback with de Brujin graphs is when cutting up the reads into k-mers, one looses the information of the longer contiguous sequence, and repeats longer than k can simply not be resolved. Some assemblers attempt to solve this by adding read path information, at the cost of computational complexity. Sequencing errors pose another problem. In a de Brujin graph a single base change in a read changes k of its k-mers into ones that may be rare in other reads. However, many assemblers make use of that feature and the topology of the graph to detect and correct such errors [142]. Another alternative is to utilize the concept of a *string graph* [105], where, similarly to the OLC method, an overlap graph is generated by considering all pairs of reads. A difference, however, is that the edges in the graph represent the sequence information and the nodes correspond to the beginning or end of overlaps. This way, reads contained in other reads can be discarded, as they contain no additional information. Also, *overhangs*, representing the part of an overlapping read that is not covered in the overlap, that contain several smaller ones can be discarded, saving memory in comparison to the overlay graph approach. Non-branching paths in the string graph are merged into one edge, and, similarly to de Brujin graphs, the genome assembly solution corresponds to finding the shortest non-branching path that passes through all edges (or nodes in the de Brujin graph). The subsequent steps are then similar to the OLC approach. One key point with the string graph is that it shows that cutting up the reads into k-mers, as done in the de Brujin approach, is unnecessary. While the string graph does not loose the read information, the disadvantage include the pairing of all reads in the overlap step. So far, the string graph approach has appeared useful mainly for smaller genome assemblies. The String Graph Assembler (SGA) [142] is the first assembler that has made assembly of larger mammalian genomes practical for the string graph approach. However, with the promise of improved read lengths, this approach might become more attractive in the future. Each assembly strategy has its own pros and cons, which can make it the method of preference for certain applications, while less suitable for others. There are several good reviews with more details on the different assembly methods. For instance, for *de novo* genome assembly, a detailed review of can be found in [117], and a benchmarking comparison between the main strategies in [182].

Proving that NGS technology can be used to sequence large genomes, a major milestone was reached in 2010 when the *de novo* assembly of the giant panda genome was published [84]. It was the first genome of such complexity to be published using next-generation sequencing methods. Besides generating interesting data, the project provided a proof of concept that NGS technology in fact can be used to decipher a genome sequence of such a complexity. However, although being a monumental accomplishment, it still contains significantly more gaps than previous mammalian draft genomes using Sanger sequencing. This proves the need for proper metrics, adapted to the new type of sequencing data, in order to to characterizes sequencing and assembly quality in large genome projects.

8.3.2.3 Measures of Assembly Quality

When measuring the quality of an assembly, there are two aspects to consider: *contiguity* and *accuracy*. Generally there is a trade-off between the two. When it comes to the contiguity and completeness of an assembly, one of the most important measures is the *N50 summary statistic*. The N50 statistic is defined as the largest contig length such that 50% of all the assembled nucleotides reside in contigs of that length or longer. It is computed by simply length-ordering all the assembled contigs and scaffolds, and, starting from the longest, summing the contig lengths until the sum equals 50% of the total assembly length. The N50 number then corresponds to the shortest length in this list. Naturally, such an N-statistic can be computed for any percentage level, but the N50 is the common choice. It basically corresponds to the mean contig length, but with greater emphasis placed on the longer contigs. Generally, an N50 measure of around the median gene length in that organism is considered a decent target for annotation, as about 50% of the genes will then be completely contained in the assembled contigs. Typically two different N50 statistics are computed, for contigs and for scaffolds, respectively. Note that the N50 relates to the *assembled* length, and not to the actual genome size. Therefore, comparisons of N50 measures between assembled organisms are usually not informative. When the genome size is known (or estimated) the NG50 statistic can be used instead of the N50, relating to the actual genome size instead of the assembled size. Moreover, while a higher N50 generally means a better assembly, a poor assembly with erroneously joined reads may also result in a high N50. Note also that the procedure for which contigs to include or exclude in the computation is not strictly defined and may vary between projects. Commonly *singletons*, i.e., contigs consisting of a single read or read pair, are discarded, but often contig lengths below a certain threshold are also excluded from the assembly. As an example of why not to blindly rely on the N50 measure, the great panda genome reported a contig N50 of 40 kbp and a scaffold N50 of 1.3 Mbp [84]. However, these numbers were computed on an assembly including only two-thirds of the highest-quality data, and the resulting sequence was still fragmented in 3,805 scaffolds, which can be compared to the dog assembly that had less than 100 scaffolds [88].

In addition to the N50 statistics it is important to report the amount of gaps. In the final assembly the scaffolds consist of linked contigs with the gaps filled with 'N's. Thus, two assemblies may have the same scaffold N50 but may differ heavily in the amount of gaps. Another important measure is the percent genome coverage, referring to the percentage of the genome that is contained in the assembly. A genome coverage of about 90–95% is generally considered good, depending on the level of repeats in the genome, as these typically are difficult both to sequence and to resolve in the assembly. Gene coverage can also be measured, representing the amount of genes included in the assembly. Typically the gene coverage is substantially higher than the genome coverage, since the repetitive regions usually are gene poor.

Regarding the accuracy, or correctness of an assembly, there is no standard metric. Several attempts to devising such metrics have been made, but they are typically computed in relation to a reference sequence. However, with the growing wealth of *de novo* genome sequences, there is a need for accuracy metrics that are not based on the alignment to a reference genome. The *correct contiguity* measure (CC50) gives a measure of the long-range connectivity of the assembly [35]. For two positions x_i and x_j, where $i < j$, in the reference sequence, a scaffold pair y_k and y_l, where $k < l$ are said to be *correctly contiguous* if y_k align to x_i and y_k to x_j in the assembly. The CC50 is then the longest distance between any correctly linked pair y_k and y_l such that the proportion of correctly contiguous pairs is at least 50 %. In other words, the CC50 measures the distance at which 50 % of the contigs are situated correctly in reference to one another. Note that a correctly contiguous pair need not be covered by the same contig or scaffold path, and that there may be numerous assembly errors in between them.

One can always expect that there is a trade-off between a high N50 measure and the sequencing accuracy. It has also been pointed out that the assembly quality is sensitive to the number of sequence errors only when the coverage is low [182]. The Assemblathon is a contest that aims to improve methods and metrics for genome assembly by letting scientific teams compete with their softwares on the genomes made available by the organizers. The first Assemblathon took place in 2011 [35], in which a simulated read set was used, created by subjecting a human genome sequence to simulated evolution. The three most successful softwares were ALLPATHS-LG [50], SOAPdenovo [84] and SGA [142], although there was no assembly program that was far ahead of the other. One issue with the Assemblathon dataset was that the repeat regions were about a half of the original human DNA, meaning that the repeat issues were not fully tested, and all methods were expected to do worse on more realistic data. In contrast, the Genome Assembly Gold-Standard Evaluations (GAGE) [132] evaluated genome assemblies and assembly algorithms on real data from high-throughput sequencing machines, providing a snapshot of the current status of the field. In contrast to the Assemblathon all protocols and parameter settings used in the project were complete transparent.

The second Assemblathon, which took place in 2013, provided sequence data from three vertebrate species: a bird, a fish and a snake [15]. From over 100 different metrics, ten measures were chosen to assess the overall assembly quality. Among others the amount of gene-sized scaffolds assembled, which is of interest for gene finding purposes, was selected as a metric. Also, the CEGMA set of 458 core genes [115, 116] was mapped to the assemblies to estimate how many genes that might be present in the assembly. The summary of Assemblathon 2 was that many of the algorithms produced useful assemblies, but there is still a lot of variation between the results indicating much room for improvement.

8.3.3 NGS Applications

Since the beginning a decade ago there has been many technical improvements of the NGS technology, which have led to its widespread use, breaking barriers and revolutionizing many application fields such as genomics, transcriptomics, metagenomics, proteogenomics, gene expression analysis, noncoding RNA discovery, SNP detection, and protein binding sites detection [39]. The NGS technology has had a major impact on basic research, inspiring scientists to address an increasingly diverse range of biological problems, such as variant discovery by re-sequencing genomes, transcriptome signature studies (RNA-Seq) [171], genome-wide profiling of epigenetic marks (ChIP-Seq) [175], and species classification and novel gene discovery by metagenomics studies [119].

Since this book focuses on computational gene prediction, the NGS application areas most relevant to us are gene prediction in *de novo* genomes, RNA-Seq, and metagenomics. Gene prediction in *de novo* genomes have to take into account the new types of data, such as the difficulties with short contigs, the various types of sequencing errors, and the parameter training and accuracy measure issues of novel genomes. RNA-Seq is relevant because it can be utilized to guide the gene prediction in *de novo* assemblies, and metagenomics with its additional issues to perform gene prediction in multi-species sequence datasets. In what follows we give a brief overview of metagenomics and RNA-Seq, followed by a little more thorough account for gene prediction in *de novo* genomes.

8.3.3.1 Metagenomics

Metagenomics, or *environmental genomics* as it is also called, is a fairly new field surfacing on the backwaters of NGS sequencing, and can be defined as "the application of modern genomics techniques to the study of communities of microbial organisms in their natural environments" [24]. It emerges from the ability to sequence any given environment sample at large scale, without an intermediate laboratory culture, and its ultimate goal is to get a more comprehensive understanding of the ecosystem. The term *metagenomics* literally means "beyond the genome." It stems from the idea that the gene set obtained in an environmental sample can be regarded as a *metagenome*, which in many ways can be treated as a single genome [57, 127].

An early attempt to *shotgun* metagenomics was reported in 2002 [17], in which uncultured marine viral communities revealed high levels of diversity through genome sequencing. However, real progress was made in the field in 2004, in terms of two different large-scale environmental sequencing projects. One project large-scale sequenced microorganisms in seawater samples from the Sargasso Sea [169], studying gene content, diversity, and relative abundance in the sample, and the other, by sampling acidophilic biofilm, sequenced a number of bacterial and archaeal genomes that previously had resisted culturing attempts [165]. Since then the metagenomics field has grown into its very own discipline, with applications as diverse as ecology and environmental sciences to chemistry and human health [93]. Notable examples include the sequencing of the human gut microbiome [47] and the metagenomic analysis of biomass deconstruction of the cow rumen [61]. With these studies and other similar projects it has become evident that in the understanding of the biology of higher organisms, it is not enough to understand its genetics with all its genome products and signaling networks, we also need to understand its microbiome. A very illustrative example is the genome sequencing of the great panda [84]. Although being classified as a carnivore, the diet is primarily made up of bamboo. However, sequence analysis has shown that while having all the necessary genetic components of a carnivorous digestive system, the great panda lacks the necessary enzymes for complete digestion of cellulose. Thus, the unusual dietary restriction of the great panda does not seem to be dictated by genetics, but rather must depend on its gut microbiome composition. The human body consists of the order of 10^{13} cells [12], while it contains more than 10^{14} microorganisms with a collective *microbiome*, which in turn constitutes more than 100 times as many genes as in the human genome [47]. Thus, humans can be seen as "super-organisms" with a fusion of human and microbial metabolism, and to understand all processes in the human body we need to map both.

With the emergence of next-generation sequencing, sequence-based metagenomics has dramatically accelerated. Single genome studies using capillary sequencing technology have many advantages, particularly regarding the assembly and the downstream bioinformatics analyzes. However, the organism under study needs to be cultured before the sequencing can take place, which is a major limitation in microbiology, as only a few percent of existing microbes can undergo culturing. Using NGS technology, environmental samples of microbial communities can be cloned and sequence directly, without the intermediate step of a laboratory culture. However, current analyzes still rely heavily on computational tools originally designed for capillary sequenced microbial genome projects. One of the biggest challenges of metagenomics, besides the general NGS problems of large datasets and short read lengths, is the high species complexity in the samples.

Some of the main bioinformatical steps involved in a metagenomics study are: sequencing, assembly, taxonomic binning and classification, and functional annotation. In the sequencing step it is important to extract DNA that is representative to the sample, and large enough amounts of high-quality nucleic acids for library construction and sequencing. In the assembly step, if the aim is to recover the genomes or obtain full-length coding sequences (CDSs) of the uncultured organisms, then a read assembly will be performed to obtain longer contigs. As with NGS in general, metagenomic assemblies can be reference-based or *de novo*. Reference-based assembly works well if closely related reference genomes are available, while more distant relations will create a more fragmented assembly. *De novo* assemblies are typically based on de Brujin graphs, such as in Velvet [181] or SOAPdenovo [84], and typically require large computational resources. Current assembly programs, however, are designed to deal with single genome sequences and should be applied with caution, since microbial communities typically include high diversity also on the strain or species level, which might lead to suppression of contig formation in heterogeneous regions. Examples of softwares trying to deal with this are MetaVelvet [108] and Meta-IDBA [118], which both attempt to identify subgraphs within the assembly graph that correspond to related genomes.

Taxonomic binning involves the partitioning of sequences into "species bins", or *operational taxonomic units* (OTUs). This step is very important in metagenomics, as information of the taxonomic origin gives access to the evolutionary history and the ecological roles of the microbes in the given community. The binning step is very challenging, however. When analyzing metagenomics samples from soil, water, or intestinal tracts, the sequence coverage rarely reaches levels that will make an assembly practically useful. The majority of reads remain unassembled, and will thus be classified solely based on its rather short sequence. The size of the data poses the same problem here as in any other step involving NGS data. Moreover, the novelty of the microbes involved also hampers the binning process, as no reference matches will be found in the sequence databases. In addition to all this, the diversity of available binning tools presents a challenge in itself. Binning algorithms can be divided into similarity-based and composition-based methods. The similarity-based methods vary in choice of reference database, type of search algorithm, or how the database matches are processed into a taxonomic assignment. Popular composition-based tools include MEGAN [69], SOrt-ITEMS [100], MG-RAST [49], CARMA [112], and MetaPhyler [89]. Composition-based methods also use reference databases for their sequence partitioning, but can in addition be divided into supervised and unsupervised methods, based on whether they use a reference database or not in their initial parameter training procedures. Popular tools include PhyloPhytia [97], NBC [128] and Phymm [16]. There are also a number of binning tools that combine compositional and similarity information, such as PhymmBL [16] and MetaCluster [82].

Before the functional annotation can take place the genes need to be identified. As we have seen in previous chapters in this book, many methods and tools have successfully been developed for gene prediction in completed genomes or long enough contig sequences. However, again due to the fragmented nature of the data, as well as the high level of sequencing errors compared to finished genomes, these tools are not well suited for metagenomics projects. A number of tailor-made metagenomic prediction tools have been developed, including FragGeneScan [126], MetaGene-Mark [97], MetaGeneAnnotator (MGA) [110] and Orphelia [63]. FragGeneScan is described in a bit more detail below.

After the gene prediction, the predicted coding sequences, which can include partial gene stretches, are typically matched against databases of annotated data. Due to the size of a typical metagenomic dataset, manual curation is typically not possible. However, automated annotation, such as running a BLASTX search for each open reading frame (ORF), is also very expensive computationally, while at the same time less demanding approaches have little success on the short read lengths involved. Thus, faster and more robust algorithms are badly needed. Coding sequences that do not receive a functional classification are referred to as *ORFans*. Some of these might simply be false predictions, while some might be true genes with yet unknown biochemical function, or genes that although lacking known sequence homologies might structurally match known protein families or folds. After the functional annotation, a natural next step is the reconstruction of biochemical and regulatory pathways. However, this lies outside the scope of this section, instead confer [32].

The field of metagenomics will only continue to grow, and the need for suitable and efficient analysis and data management tools grows with it. Metagenomics holds great promises in revealing the massive microbial diversity present in our environment, and perhaps provide new insights and interesting molecules for future therapeutic and biotechnological applications. For a good review on the current status of metagenomics analytical tools and databases, see [75].

FragGeneScan: Gene Prediction in Metagenomics Data

Metagenomic gene prediction is of utmost importance as the mapping of the genetic components in a microbial community can help elucidate the activities and interactions of these components, and from there the metabolic and signaling pathways can be reconstructed and identified. However, the gene prediction problem in metagenomics data is as challenging as ever. Traditional gene prediction tools developed for application to finished sequences and complete genomes show a significant decrease in performance when applied to metagenomics data. The main reasons are the fragmented nature of the data, the high diversity of the sequences in the sample, and the higher error-rate in the sequence reads. Due to the high diversity and the short read lengths, the resulting assembly usually consists of a significant portion of unassembled reads. This is then the data set we are left with to perform gene prediction on. Moreover, not only are we facing novel data with very limited information on how to train our algorithms. The datasets can consist of a large number of different species, ranging over different phyla, and with high diversity in all genome characteristics including sequence composition and genome structure. On top of it we can expect a significant rate of sequencing errors that will distort our view. The most common gene prediction approach in metagenomics is the similarity-based, performing homology searches of potentially coding sequences against known protein databases. However, novel genes are not detected this way, although this is one of the main objectives of metagenomics, and in the highly diverse and largely unexplored microbiomic world we might expect quite a few of those.

FragGeneScan [126] is designed to take on these challenges, and combines codon usage with sequence error models in a hidden Markov model (HMM). NGS reads can have error rates up to 3%, some of which can cause frameshifts and thus alter and disrupt the gene prediction [62]. FragGeneScan is an open reading frame (ORF) detector that can predict partial gene fragments and correct for frameshifts caused by insertion and deletion errors in the reads. The underlying HMM combines measures for codon usage bias, sequencing errors and start and stop codon patterns. In order for a gene to be predicted its putative coding region must be longer than 60 bp, and the region must be bounded either by a start or a stop codon or by the read boundary. Whether the putative regions matching these criteria are actually predicted is then up to the underlying HMM. The FragGeneScan state space is illustrated in Fig. 8.2.

The figure only shows the forward strand with four main states. The full model consists of seven states, three for each strand and a joint intergene state. The reverse strand can be included by simply adding a mirror image of the forward strand, joined at the intergenic state. The states in Fig. 8.2 correspond to intergene (large diamond), start and stop codons (circles), and a gene state model (shaded area). The gene state model in itself consists of six match states (diamonds), six insertion states

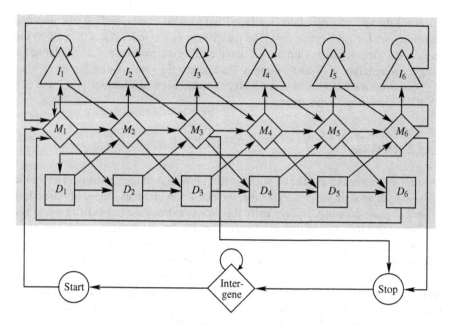

Fig. 8.2 A simplified version of the FragGeneScan state space, with only the forward strand showing. The main model consists of four superstates, the intergene (*large diamond*), the inner gene state (*shaded*), and the start and the stop codons (*circles*). The gene state model in itself consists six match states (*diamonds*), six insertion states (*triangles*), and six deletion states (*squares*)

(triangles), and six deletion states (squares). The insertion and deletion (indel) states, which resemble the structure in multiple alignment HMMs (see Sect. 3.2.6), are there to account for indel sequencing errors that may cause frame shifts. The resulting gene state correspond to a six-periodic inhomogeneous HMM. When dealing with finished sequences, where the sequencing error rate is expected to be low, one can simply set the transition probabilities to the indel states to 0. The match states use a second-order (trinucleotide) Markov model to account for the codon structure, while the intergene state uses a first-order (dinucleotide) Markov model. The stop state simply consists of frequencies for the three possible stop codons TAA, TAG, and TGA, based on the training data. The start state is more complex. In bacteria the true start codon is often surrounded by numerous putative ones. FragGeneScan handles this by scoring the 63 bp surroundings of each putative start codon using a position-specific scoring matrix (see Sect. 5.4.1), which considers the AT-content, the Shine-Dalgarno box (see Sect. 5.3.2), an a triple-A downstream box [136]. The probability of each potential start codon is computed using a naive Bayesian classifier (see Sect. 5.4.5) by fitting two Gaussian distributions to the training set of real and false start codons, respectively.

A set of 139 microbial complete genomes were used to train the parameters of the FragGeneScan HMM in [126], and a linear regression model was applied to train the parameters for varying GC-content. When applied to a new sequence, FragGeneScan computes the GC-content of the input read and uses the corresponding precomputed parameter set for gene prediction. The parameters of the indel states depend on the sequencing method used. Currently, there are different parameter sets available for Sanger sequencing, 454, and Illumina. Parameters for four sequencing error rates has been estimated: 0.5 and 1 % for Sanger sequencing and Illumina, and 1 and 3 % for 454 sequencing, respectively.

8.3.3.2 RNA-Seq

The *transcriptome* is the set of all possible RNA products in an organism, in a specific cell tissue, or in a single cell. Typically, the definition includes the set of transcripts as well as their quantity, and typically the data is extracted from cells in a specific developmental stage or physiological condition. The key issues of *transcriptomics*, or *expression profiling* as it also is called, are to determine all the transcripts of a species, determine their positions and the underlying gene structures in the genome, and to quantify their expression levels under different conditions. This has traditionally been performed using high-throughput DNA microarray technology. One disadvantage of microarrays, however, is that they require an existing reference genome, which limits the detection of transcripts to already known sequences.

Next-generation sequencing has provided a powerful alternative to microarray analysis. This technology, termed *RNA-Seq* (RNA sequencing), or *whole transcriptome shotgun sequencing*, has the advantage that transcripts can be characterized without any prior knowledge of the origin of the genomic sequence, which is particularly convenient when considering novel genomes. RNA-Seq has higher resolution than microarrays and can identify novel transcripts and isoforms, alternative splice sites, allele-specific expression, and detect rare transcripts, all in the same experiment. Also, it does not require sequence probes or primers (specific short oligonucleotides) to hybridize with, which is useful as they tend to bias the sample [171].

RNA-Seq uses NGS technology for single-end or paired-end sequencing, and the produced reads are either mapped onto a reference genome or a transcriptome, or assembled *de novo*. The result is both a map of the transcriptome structure and of the gene expression levels. The resolution is down to single-base precision, meaning that the exact transcription boundaries can be detected. The background signal relative to microarrays is very low, because RNA-Seq reads can be unambiguously mapped to the corresponding genomic region. Moreover, while microarrays have a limited detection range and lack sensitivity both for very low and for very high expression levels, RNA-Seq does not have an upper limit for transcript quantification, resulting in a large dynamic detection range.

RNA-Seq is, however, faced with similar challenges as for other applications of NGS technology. Due to the short read length, larger RNA molecules become fragmented and must be reassembled. Moreover, different fragmentation techniques generate different biases in the dataset. For instance, while *RNA fragmentation* experiences depletion of the transcript ends, *cDNA fragmentation* is biased toward the 3' end of the transcripts [101, 107]. The bioinformatic challenges are also similar to other NGS applications, regarding the storing and processing of large quantities of data, and the handling of low-quality reads and sequencing errors. Moreover, in addition to the problems of resolving alternative splicing, that of *trans*-splicing becomes an issue. *Trans*-splicing refers to splicing events that takes place between sequences originating from distant positions in the genome, or between exons from two different genes. Therefore, due to the short read lengths, *trans*-splicing will cause many reads to match multiple locations. Moreover, while RNA-Seq is capable of detecting rare transcripts, considerable sequencing coverage is required for this. However, coverage is harder to compute for transcriptomic data than for genome sequencing, because the true quantities of the transcripts are usually not known, and because the transcription levels vary across the genome and over sequencing conditions. Despite these challenges RNA-Seq provides an unprecedented opportunity regarding the detection of novel and rare transcripts, quantification of splicing diversity, and the capturing of transcriptome dynamics across different tissues and sequencing conditions.

Naturally, RNA-Seq data can also aid in gene prediction. Sequencing of RNA products and the reconstruction of full-length cDNAs have been considered as the gold standard for the discovery and annotation of complete gene structures in eukaryotic genomes [54]. However, before the introduction of NGS technology, this task was very labor- and cost-intensive. With the advents of RNA-Seq, genome annotations can be substantially improved both by correcting already existing predictions, and by discovering novel genes and transcript variants. The use of RNA-Seq in gene prediction can be done either by its inclusion of the gene prediction algorithm in a homology-based manner, or directly by producing spliced alignments of the RNA-Seq reads to a reference genome and a reconstruction of the transcripts. An example of the latter is the genome annotation update of the cucumber genome *Cucumis sativus* var. *sativus* L. The draft genome was originally published in 2009 [68], in which the genome sequence was assembled using a combination of Sanger and Illumina sequencing. The gene prediction was performed by integrating multiple *de novo* predictions with spliced alignments of protein and transcript sequences to the genome, resulting in a consensus gene set of 26,682 genes. In 2011 the cucumber annotation was updated using RNA-Seq [86] by mapping the RNA-Seq reads onto the cucumber genome sequence using Bowtie [80] and TopHat [162], and then reconstructing the transcripts using Cufflinks [163]. In this update, the RNA-Seq reads came from 10 cucumber tissues, and the reannotation resulted in 23,248 identified protein coding genes of which 8,700 were modified gene structures and 5,265 were novel genes.

8.4 NGS Genome Sequencing Annotation Pipelines

Sequencing has become easy and cheap, while at the same time the annotation process has become harder. There are several reasons for this. The assembly difficulties for the shorter read lengths discussed above results in shorter contigs, which complicates the gene annotation as the gene models tend to be more cut up. A novel genome that lacks known evolutionary close relatives makes the training, optimizing and configuration of prediction tools difficult. The use of ESTs and RNA-Seq data holds promise, but merging different sources into a consensus training or prediction set is nontrivial, as there is no reference sequence to compare to. A myriad of assembly, analysis and visualization tools is emerging, and in fact not much bioinformatics and computational biology skill is needed to produce a genome annotation. However, interpreting the results is not nearly as easy. Here we briefly discuss the procedure and issues with NGS annotation pipelines. A more thorough review of the process is given in [180].

8.4.1 Assembly Quality

First, one needs to decide if the assembly is suitable for annotation. Standard draft assemblies that meet the minimum standards for submission to public databases, typically contain large portions of poor quality sequence and may even include contaminating sequences [23]. A much better target for annotation is a high-quality draft assembly, which has an overall coverage of at least 90 % and efforts have been made to filter out contaminations. A good guide of assembly quality is the N50 summary statistic discussed earlier. Recall that the N50 measure defines a contig (or scaffold) length where 50 % of the assembly resides in contigs (scaffolds) of at least that length. In other words, the higher the N50 the better the assembly. A rule of thumb is that the N50 scaffold length needs to be at least around the average gene size of the organism in question, because in such an assembly about 50 % of the genes will be covered by a single scaffold. Average gap size and average gap number per scaffold are also useful measurements, as too many gaps will cut up the genes, interrupt exons, and distort the annotation.

8.4.2 Repeat Masking

Some of the biggest challenges with NGS assembly and annotation are posed by repetitive sequences, i.e sequences that appear in identical or highly similar copies throughout the genome. In particular eukaryotes tend to be very repeat rich. Of the human genome, for instance, nearly half is covered by repeats, and the maize has over 80 % of its genome residing in transposable elements [164]. Moreover, the study of repetitive elements is in itself an interesting research area, both in terms of the biological meaning and evolutionary history of repeats, and in terms of developing methods detection methods that evolve with the sequencing technologies.

Repetitive sequences can roughly be divided into two categories: low-complexity repeats and interspersed repeats. Low-complexity repeats consist of stretches where a handful of bases are repeated in numerous subsequent copies. Such sequences contain very little information but can be very long, with hundreds or thousands of repetitions, especially around the centromere or at the telomeres of a chromosome. Examples of low-complexity repeats are mono-nucleotide runs like AAAAAA, tandem repeats like AACTGAACTGAATCTG, and different types of satellite repeats. Interspersed repeats are more complex in both their structure and function, as they can contain real genes and have the ability to change location (transpose), often duplicating the sequence surrounding genes in the process. Interspersed repeats need to be removed in some manner before annotation, as they tend to confuse the prediction algorithms. Prohibiting gene prediction in interspersed repeats is sometimes called *hard-masking* of repeats, and the practice is to replace interspersed repeat nucleotides with the letter 'N' in the genomic sequence file. Low-complexity repeats can also be confusing, as they often have a GC-content that differ from noncoding sequence and therefore may resemble the statistical pattern of real genes. A real gene cannot consist of low-complexity sequence alone, however, but portions of it can. Therefore, low-complexity repeats are *soft-masked*, whereby the corresponding genomic sequences is transferred into capital letters in the sequence file. The gene prediction tool can then allow the prediction of genes that are partially covered by the soft-masked region. Repeat masking is typically done by using a software called *RepeatMasker* [145], that makes use of species specific repeat libraries and performs a homology search between the relevant libraries and the input sequence. Repeats tend to be poorly conserved across species, however, thus novel genomes typically contain novel repeats not present in the libraries. While low-complexity repeats are relatively easy to detect also in a novel genome, more complex repeats, such as segmental duplications, transposable elements, and processed pseudogenes are more difficult to handle. In particular transposable elements are difficult, as they are structured similarly to true genes, and may contribute extra exons to the prediction and thereby corrupting the final gene structure.

Methods to detect transposable elements, or mobile DNA, typically fall into one of four different categories: homology-based methods, *de novo* methods, structure-based methods, and comparative methods [11]. The homology-based approach is the most common as it capitalizes on the knowledge of previously detected transposons. Another advantage is that transposons present in a single copy alone cannot be detected by any other method. The homology-based methods are, however, naturally biased toward previously detected transposon families and elements of recent activity. *De novo* methods attempt to identify mobile DNA without prior information about structure or similarity to already known transposable elements. The main advantage is that novel elements can be detected. However, these methods identify repeated elements in general, which can include highly conserved, duplicated genes in addition to the targeted transposons. The output must therefore be carefully post-processed to remove real protein-coding genes from the repeats library. Structure-based methods uses knowledge about the architecture of different transposable element families, and focus on common structural features necessary for the process of transposition. This category of methods share the advantage with homology-based methods of being able to detect elements of low copy numbers. A limiting factor here, however, is that each specific type of transposons need to be modeled and implemented separately. An innovative fourth category involves a comparative approach proposed in [22], where transposition events are detected as large insertions in whole-genome multiple alignments of related species. One advantage over the previous three method categories is that this method is not constrained neither to known homologies, nor to structures or repetition of the element in question. The disadvantage is that the method relies on the quality of the multiple alignment, which typically is poor in transposon-rich regions. Also, a transposable element that is more ancient than any of the aligned genomes, and thus appearing in all of them, will not be detected by this method.

Processed pseudogenes are even harder to handle. Processed, or *retrotransposed*, pseudogenes are pieces of mRNA that are reverse transcribed back into DNA and inserted into the genome sequence. Thus, it originates from a gene in the genome in question, and will highly resemble the structure of a real gene. While originating from a mature mRNA, a retrotransposon lacks the upstream promoter, and is not viable as a gene in itself. However, they sometimes contribute exons to existing genes, via alternative splicing, and should then be part of the gene annotation [6]. For a discussion on methods for constructing novel repeat libraries for novel genomes, see [81].

8.4.3 Gene Annotation

Once the assembly is deemed acceptable and repeats are masked, the step annotation follows. The term *genome annotation* is typically used for two different purposes, *structural annotation* and *functional annotation*. The structural annotation involves locating the functional elements in the genome sequence, which includes resolving the coding region boundaries, the regulatory elements, and the the resulting expressed product. The functional annotation, on the other hand, seeks to attach biological meaning to the structural annotation, such as the biological and biochemical function, the involvement of the gene product in signaling pathways and interactions, and gene expression information. In what follows we solely focus on the structural annotation.

Gene *annotation* and gene *prediction* are often used interchangeably, which is somewhat misleading. Gene prediction softwares usually only predict the protein-coding portion of the genes, and typically only report the highest scoring exon-intron structure for each potential gene. Moreover, a gene prediction typically contains both complete and partial genes. A proper gene annotation, however, should contain complete gene models only, in combination with detailed information about the supporting evidence trails and quality control metrics. The gene models should contain transcript boundary information with untranslated regions (UTRs), regulatory features such as the promoter and the polyA-tail, and information about alternative splicing. In addition to all this, in a *functional annotation* a function prediction is attached to each gene model. Gene annotation is thus much more complex than mere gene prediction, and an annotation pipeline must be able to handle several different types of information, often quite heterogeneous in their form, and combine this information into a coherent consensus set of gene models. Moreover, the output must be detailed enough to work as input to genome browsers and annotation databases.

Gene annotation pipelines typically go through three main step: inclusion of homology information and external data, gene prediction, and the final annotation. In the first step most pipelines include previously known information such as proteins, ESTs and RNA-Seq data. The information comes both from the organism in question, as well as from related organisms. Since the computational cost for aligning EST and RNA-Seq data is rather high, and since protein sequences are much more conserved than their underlying nucleotide sequences, other organisms typically only contribute with protein data. A good resource for protein sequences is the UniProt Knowledgebase (UniProtKB) [159]. It consists of two sections, Swiss-Prot that contains manually curated protein records, and TrEMBL that contains automatically annotated records awaiting review [14]. The NCBI taxonomy browser [9, 137] can be used to assemble additional ESTs and proteins from related organisms. The sequence data is aligned to the assembly, typically using rapid alignment programs such as BLAST [4, 76] or BLAT [74]. The matching regions are filtered, based on sequence similarity and, as EST data for highly expressed genes can be highly redundant, the filtered dataset is clustered to identify gene-specific sequence groups. After the filtering and clustering, the remaining sequences may be realigned to the assembly to aid the exon boundary accuracy of the final annotation. As BLAST does not have a splice site model, it is typically better to use spliced alignment tools such as Splign [72] or Exonerate [144]. The running time is considerably higher for these tools than for BLAST and BLAT, but the improvement is significant.

As mentioned above in Sect. 8.3.3.2, RNA-Seq data can be used to improve the exon boundaries further and to identify alternatively spliced isoforms. However, the computational complexity may be difficult to conquer due to the sheer size of such transcriptome datasets. Therefore, RNA-Seq reads are typically handled in one of two ways. The reads can be assembled *de novo*, using tools like SOAPdenovo [85] or Trinity [52], and then aligned to the genome much in the same way as ESTs are. The alternative is to align the RNA-Seq reads to the genome directly using tools like TopHat [162], GSNAP [179] or Scripture [53], and then assemble the alignments, rather than the reads, using tools like Cufflinks [163].

The next step in the annotation process is typically to run one or several ab initio gene prediction softwares on the sequence, such as those described in Chap. 2. As we have emphasized earlier, the main advantage with ab initio gene finding is that no external evidence is needed to run on a new genome. A major obstacle, however, is that typically such software tools need to be parameter trained on known gene models that exemplify the structure and composition we are looking for. Given a proper training set and a high-quality sequence assembly, the sensitivity of ab initio gene predictors can climb well above 90 %. However, the accuracy in novel genomes is typically much lower. Many programs provide pre-compiled parameter sets for a number of well-annotated genomes (such as human, mouse, yeast, fruit fly, *C. elegans*, *A. thaliana*, etc.), but for a novel genome there might not be a suitable close relative among those. Or it may be unknown what a suitable close relative would be for the genome in question. Also, depending on what kind of organism that is under study, even a close relative may differ significantly in terms of sequence composition and gene model structure. Parameter training can be performed using known proteins,

ESTs and RNA-Seq data if available, but requires a significant portion of work. EST clusters and RNA-Seq assemblies need to be post-processed to identify gene model structures and splice sites, possibly requiring a lot of manual labor and specialized software tools. Another approach is to use the CEGMA program [115], which is an HMM-based program that utilizes a subset of highly conserved, universal eukaryotic genes to train on. These issues, among others, are what automated pipelines, such as MAKER [21] described below, attempt to address.

The last step of the annotation process is the actual annotation, where a final set of predicted gene models is produced. Traditionally this has been done through manual curation, which results in high-quality predictions, but which is very time consuming and only works for small datasets with a limited number of gene models consider. Automated pipelines typically combine the alignments of external evidence, such as ESTs, proteins and RNA-Seq data, with the results of several different ab initio gene finders. The combination of different gene tracks is done by some kind of *combiner* or *chooser algorithm* that produces a consensus gene set. Such evidence-combining approaches range from simply using majority-voting of the ingoing tracks, to more sophisticated modeling schemes. Examples of combiners include JIGSAW [2], the EVidenceModeler (EVM) [56], GLEAN [40] and its successor Evigan [90]. JIGSAW pretty much accepts any raw exon predictions from any source, evidence-based or ab initio, and combines them using a dynamic programming algorithm similar to that in generalized HMMs (GHMMs) described in Sect. 2.2. The program utilizes the confidence scores provided by each prediction method, when available, weights each exon contribution using a decision tree, and chooses the highest scoring path as the final prediction set. Like ab initio gene finders, JIGSAW requires a set of known gene models for each new sequence to be analyzed to train the algorithm parameters. JIGSAW has, among others, been used to annotate the rice genome [158] and *Cryptococcus neoformans* [91]. Combining multiple evidence improves the accuracy of gene predictions significantly also in well-annotated genomes. When applied to the human genome, JIGSAW was exactly correct for about 75 % and partially correct for about 97 % of the human genes [3]. The EVidenceModeler (EVM) [56] uses a nonstochastic weighted approach to combine ab initio predictions and protein and transcript alignments. Besides the genome sequence and the different gene prediction tracks, EVM takes as input a list of weight values, accounting for both the abundance and the source of evidence, to be applied to each type of prediction. The reported set of consensus gene structures is a resulting high scoring path through an acyclic directed graph. EVM can either be trained on a training set or be provided with a set of weights directly, and was used for the genome analysis of the mosquito *Aedes aegypti* [109] among others. GLEAN [40] was developed to produce a reference gene set for the honey bee *Apis mellifera* and uses latent class analysis to automatically combine disparate gene prediction evidence in the absence of known genes. GLEAN evaluates the gene predictions from different sources by estimating the error frequencies in each source, and takes a weighted average for a final consensus prediction. GLEAN was also used for the original annotation of the cucumber *Cucumis sativus* genome [68]. Evigan [90], a successor of GLEAN, uses a Bayesian network to weigh and integrate evidence from various sources. Both GLEAN and

Evigan uses unsupervised learning. Even after producing a consensus gene set, the predictions may need postprocessing. Also, if integrated into a pipeline the gene predictions and evidence alignments can be combined during run time. Moreover, the postprocessed gene models can be refined by choosing the ones most consistent with external evidence such as ESTs, RNA-Seq and proteins. This is the approach taken by various gene annotation pipelines such as PASA [55] and MAKER [21]. PASA constructs maximal alignment assemblies by clustering overlapping alignments of ESTs and full-length cDNAs. It has been used to refine and update the Arabidopsis gene annotation [55] among other things. The PASA pipeline can both update existing gene model annotations, by comparing them to the generated alignment clusters, as well as predict novel gene models based on the full-length cDNAs. MAKER is an annotation pipeline described in detail in Sect. 8.4.5 below.

When the pipeline has produced a gene annotation, it is often useful to visualize the results in some manner. For this purpose there are numerous handy genome browsers. However, in order to use these one needs to produce the gene annotation output in the specific format that the browser requires. The Generic Model Organism Database (GMOD) [156] is an organization that attempts to standardize the gene annotation process and creates tools for creating, managing, analyzing and visualizing gene annotations. To be able to utilize GMOD tools the annotation output needs to be in GFF3 (Generic Feature Format version 3) format [45]. This can be a complex task, however, as each included feature (e.g. ESTs, repeats, protein alignments, gene predictions etc.) must include various detailed information in order to be accepted by the browser. However, once in the correct format the annotation files can be visualized directly, using tools like GBROWSE [34] or JBROWSE [143], to produce local data views just like those in for example the UCSC Genome Browser [160]. Moreover, the JBROWSE browser can be embedded into Wikis for web-based community use, which simplifies the process of updating and refining annotations.

8.4.4 De Novo Annotation Assessment

Assessing the accuracy of the genome annotation is a vital part of any genome project as incorrect annotations will propagate throughout subsequent experiments and projects. However, in novel genome projects where no reference genome is available, assessment is less than straightforward. A first approach is to use tools like InterProScan [103] or Pfam [42] to quantify the proportion of the annotated gene models that include known protein domains. While the relative number of domains vary between organisms, the estimated domain content may still provide a reasonable estimate of accuracy. For instance, the domain content for well-annotated proteomes such as human, the fruit fly *D. melanogaster*, the roundworm *C. elegans*, the plant *A. thaliana*, and the yeast *S. cerevisiae* ranges between 57 and 75 %, while a poorly trained gene finder typically produces frequencies as low as 5–25 % [65].

A low domain percentage can thus indicate a poor overall annotation quality. It does not, however, say anything about the accuracy of the actual annotation at hand, but only gives an estimate of the gene coverage. In the same manner as EST and RNA-Seq data can be useful for training of ab initio programs, it can also be used to refine the annotation. For instance, if the external evidence contradicts the predicted exon–intron boundaries of a gene, the gene deserves an additional examination. Manual assessment of specific genes is usually superior and fairly straightforward. However, with large amounts of data this process too needs to be automatized, which is a considerably more complex task than gene prediction. For one thing, one needs a reliable assessment measure for comparison and reference. For this purpose, the Sequence Ontology Project [37] has developed several quality control metrics for gene annotation projects. One such measure is the Annotation Edit Distance (AED) [38], which measures how compatible the annotation and the corresponding supporting evidence are. When a reference annotation is available, the most common accuracy measures are *sensitivity* (SN) and *specificity* (SP) described in Sect. 7.3. The sensitivity is the proportion of the reference annotation that is correctly predicted, while the specificity is the proportion of the predicted annotation that is correct. Both these measures are needed to give a comprehensive measure of accuracy, but as they strive in opposite directions, attempts to combine the two into a single measures are done in measures like the *correlation coefficient* (CC) and *approximate correlation* (AC). See Sect. 7.3 for details. Naturally, instead of comparing a given genome prediction to a reference annotation, one can compare two different predictions of the same genome. The AED is such a measure that combines the sensitivity and specificity in order to measure the level of agreement between two annotations, or between an annotation and supporting evidence. The AED is computed as

$$AED = 1 - \frac{SN + SP}{2}$$

where the sensitivity (SN) and the specificity (SP) is computed as in Sect. 7.3 with the supporting evidence used in place of the reference annotation. AED $= 0$ indicates that the two annotations are in complete agreement, while AED $= 1$ means that there is no congruence. Computed this way, the AED can both be used to identify questionable annotations as well as measure the level of changes between two subsequent annotations.

Once the annotation errors are identified, they need to be corrected, which is yet another task that is far from obvious. The most direct approach is to edit the exon-intron boundaries manually by use of some kind of genome browser. For instance, browsers such as Apollo [83], Argo [41] or Artemis [130] allow direct drag-and-drop actions where the edits are written back to the underlying annotation files. Another popular approach for more efficient annotation auditing is to make use of community-driven annotation, in so-called *annotation jamborees*. The term was coined in 2000 when over 40 scientists met for two weeks to jointly refine the gene models and functionally annotate the *D. melanogaster* genome [58]. Moreover, by providing internet means to search, browse and manually edit the annotation, such jamborees can nowadays meet virtually. A successful recent example is the swift curation and analysis of three different ant genomes, all distributed over the web [111, 149, 154].

8.4.5 *MAKER: An Annotation Pipeline for Next-Generation Sequencing Projects*

Database resources such as Ensembl [27] has for long provided a golden standard in terms of genome annotation. However, the amount novel genome data produced today is exceeding their capacity, both in terms of data size and organism range. When each lab can sequence their own favorite organism, there is an urgent need for efficient, portable and easy-to-use annotation pipeline software to handle the data. However, in order to be applicable, there are numerous criteria to fulfill. A pipeline needs to contain a diverse set of softwares for data management, filtering, repeat masking, sequence alignment, gene prediction, and consensus annotation. It has to be easy configurable and trainable on new training data, efficient in handling large data sets from a wide array of sources, and has to produce an output that is both comprehensive and database ready. Preferably, a pipeline should also provide means to view and edit the annotations manually. The ultimate goal of an annotation pipeline is to provide an automatic mean that can match, or even exceed, the level of accuracy of a human annotator, so that the annotation process can keep up with the rate at which genomic sequence is produced.

A necessary ingredient in any annotation project is a combiner software that can make use of a wide variety of evidence sources. MAKER2 [65] is an annotation pipeline specialized for NGS data. Since it builds heavily on MAKER [21], we begin our description there. MAKER is a combiner annotation package that combines various sources of evidence for genome annotation. It is not a gene predictor by itself, but makes use of gene predictions and other sources to produce a final consensus annotation. MAKER is designed to work for researchers with limited bioinformatics knowledge working on small annotation projects, but is scalable to virtually any project size. The combiner can be used for *de novo* annotation of novel genomes, for updating existing annotations, or simply for combining a variety of evidence sources. The output is compatible with other GMOD [156] programs such as GBROWSE [34] or JBROWSE [143] by providing the output in feature-rich GFF3-format [45]. MAKER also supports distributed parallelization on computer clusters, which means that it is scalable to virtually any data size.

The MAKER procedure is divided into five main steps: the compute phase, the filter/cluster step, polishing, synthesis, and annotation. In the compute phase the input sequence is masked for repeats, ab initio gene prediction is performed, and external homology evidence is aligned to the input sequence. The repeat masking is performed in two steps: first RepeatMasker [145] is run to identify all types of repeats matching entries in the RepBase library. The users can create their own species specific repeat libraries and add them to the search. Moreover, MAKER comes with an internal library of transposable elements and viral proteins. This library is matched against the genomic sequence using an internal repeat masking software called RepeatRunner, which utilizes BLASTX [48] to identify mobile elements. Such mobile elements tend to be missed by RepeatMasker, even when the repeat libraries are genome specific [148]. After repeat masking, MAKER runs a number of ab initio gene predictors. In its default setting MAKER is configured to use SNAP [77] as gene predictor, which is an hidden Markov model (HMM)-based gene prediction software similar to Genscan [19] (described in Sect. 2.2.4), but slightly more flexible in terms of allowing for user-defined feature models and state spaces. In addition to SNAP, MAKER supports the use of the gene prediction softwares Augustus [152, 153], FGENESH [131] and GeneMark-ES [92]. After the gene prediction step, BLAST [4] is run to align proteins, ESTs, and mRNAs to the genome sequence. Specifically, BLASTX and BLASTN are used to align species-specific proteins, ESTs and mRNAS, respectively, and TBLASTX is used to translate and align ESTs and mRNAs from related organisms.

In the filter/cluster step, low-scoring predictions and low-identity alignments of the BLAST hits from the compute phase are filtered out, and the remaining hits are clustered into overlapping sets expected to correspond to common gene transcripts. Both the filtering and the clustering criteria are set by default but can be modified by the user. Since BLAST is not splice-site aware, the tool Exonerate [76] is used in a polishing step on the remaining data to refine the alignment clusters into spliced alignments. The BLAST hits are realigned around splice sites, which forces the alignments to occur in order.

After polishing, the next step is the synthesis, in which the gene predictions and the polished EST and protein alignment clusters are combined to generate *hints* to the location and boundaries of the protein coding regions. In this step MAKER attempts to mimic a human annotator by recognizing internal exons with differing boundaries, and matching protein alignments to consistent EST splice forms, in order to detect potential alternative splicing. Regions outside gene clusters are labeled intergenic and regions that fall between putative exons are labeled introns. MAKER then computes a score for each nucleotide based on the supporting evidence of the alignments and the gene predictions. The scores and the nucleotide labels are passed back into the default gene predictor SNAP, which modifies its HMM state space accordingly and is rerun on the genomic sequence. In regions lacking external evidence MAKER uses the SNAP predictions directly. The final step is the annotation, whereby the synthesized predictions are checked against all existing ESTs and mRNAs, UTR-regions are included when available, and alternatively spliced forms are recorded.

The input to MAKER is the genomic sequence, and three configuration files containing information about external executables to be used, database locations, and various computational parameters. The internal database of transposable elements and viral proteins is provided with the installation package. An organism specific repeat library can be provided by the user, but is optional. If nothing is known about the organism in question, MAKER uses the internal database only to mask mobile elements. Similarly, the user can provide additional protein and EST/mRNA files. The training of MAKER is a two-step process. First, SNAP is trained using the CEGMA [115] with its subset of highly conserved, universal eukaryotic genes. These genes are aligned using pairwise and profile-HMM alignment. The resulting gene models then serve as an initial training set to SNAP. Second, MAKER is run on a randomly selected subset (a few megabases) of genomic sequence, and the resulting annotations are passed back into SNAP to further refine the HMM state space.

MAKER2 [65] is an extension of MAKER made to facilitate the annotation of second generation sequencing projects. The major additions to the original MAKER programming include the integration of the AED metric [38] described in the previous section, support for the inclusion of RNA-Seq data, and a gene model pass-through capability. Besides annotating novel sequencing projects, by including several ab initio predictions and/or additional evidence from new sequence sources such as RNA-Seq data and others, MAKER2 is useful for the re-annotation of existing projects.

MAKER-P [20] is yet another extension of MAKER, adapted to better suit the annotation of plant genomes, which typically are large and repeat-rich, and where noncoding RNA and pseudogene detection is needed to a greater extent than in animal genomes. The major extensions include pseudogene and ncRNA prediction, and a tailor-made optimization of the computer cluster parallelization to suit large and repeat-rich genomes.

References

1. Adams, M.D., Celniker, S.E., Holt, R.A., Evans, C.A., Gocayne, J.D., Amantides, P.G., Scherer, S.E., Li, P.W., Hoskins, R.A., Galle, R.F., et al.: The genome sequence of *Drosophila melanogaster*. Science **287**, 2185–2195 (2000)
2. Allen, J.E., Salzberg, S.L.: JIGSAW: integration of multiple sources of evidence for gene prediction. Bioinformatics **21**, 3596–3603 (2005)
3. Allen, J.E., Majoros, W.H., Pertea, M., Salzberg, S.L.: JIGSAW, GeneZilla, and GlimmerHMM: puzzling out the features of human genes in the ENCODE regions. Genome Biol. **7**, S9 (2007)
4. Altschul, S.F., Gish, W., Miller, W., Myers, E.W., Lipman, D.J.: Basic local alignment search tool. J. Mol. Biol. **215**, 403–410 (1990)
5. Avery, O.T., MacLeod, C.M., McCarty, M.: Studies of the chemical nature of the substance inducing transformation of pneumococcal types. Induction of transformation by a desoxyribonucleic acid fraction isolated from pneumococcus type III. J. Exp. Med. **79**, 137–158 (1944)
6. Baertsch, R., Diekhans, M., Kent, W.J., Haussler, D., Brosius, J.: Retrocopy contributions to the evolution of the human genome. BMC Genomics **9**, 466 (2008)
7. Bartlett, J.M., Stirling, D.: A short history of the polymerase chain reaction. Methods Mol. Biol. **226**, 3–6 (2003)
8. Batzoglou, S., Jaffe, D.B., Stanley, K., Butler, K., Gnerre, S., Mauceli, E., Berger, B., Mesirov, J.P., Lander, E.S.: ARACHNE: a whole-genome shotgun assembler. Genome Res. **12**, 177–189 (2002)
9. Benson, D.A., Karsch-Mizrachi, I., Lipman, D.J., Ostell, J., Sayers, E.W.: Genbank Nucleic Acids Res. **37**, D26–D31 (2009)
10. Bentley, D.R., Balasubramanian, S., Swerdlow, H.P., Smith, G.P., Milton, J., Brown, C.G., Hall, K.P., Evers, D.J., Barnes, C.L., Bignell, H.R., et al.: Accurate whole human genome sequencing using reversible terminator chemistry. Nature **456**, 53–59 (2008)
11. Bergman, C.M., Quesneville, H.: Discovering and detecting transposable elements in genome sequences. Brief. Bioinform. **8**, 382–392 (2007)
12. Bianconi, E., Piovesan, A., Beraudi, A., Casadei, R., Frabetti, F., Vitale, L., Pelleri, M.C., Tassani, S., Piva, F., Perez-Amodio, S., Strippoli, P., Canaider, S.: An estimation of the number of cells in the human body. Ann. Hum. Biol. **40**, 463–471 (2013)
13. Blattner, F.R., Plunkett III, G., Bloch, C.A., Perna, N.T., Burland, V., Riley, M., Collado-Vides, J., Glasner, J.D., Rode, C.K., Mayhew, G.F., Gregor, J., Davis, N.W., Kirkpatrick, H.A.,

Goeden, M.A., Rose, D.J., Mau, B., Shao, Y.: The complete genome sequence of Escherichia coli K-12. Science **277**, 1453–1474 (1997)

14. Boeckmann, B., Bairoch, A., Apweiler, R., Blatter, M.C., Estreicher, A., Gasteiger, E., Martin, M.J., Michoud, K., O'Donovan, C., Phan, I., Pilbout, S., Schneider, M.: The SWISS-PROT protein knowledgebase and its supplement TrEMBL in 2003. Nucleic Acids Res. **31**, 365–370 (2003)

15. Bradnam, K.R., Fass, J.N., Alexandrov, A., Baranay, P., Bechner, M., Birol, I., Boisvert, S., Chapman, J.A., Chapuis, G., Chikhi, R., et al.: Assemblathon 2: evaluating de novo methods of genome assembly in three vertebrate species. Gigascience **2**, 10 (2013)

16. Brady, A., Salzberg, S.L.: Phymm and PhymmBL: metagenomic phylogenetic classification with interpolated Markov models. Nat. Methods **6**, 673–676 (2009)

17. Breitbart, M., Salamon, P., Andresen, B., Mahaffy, J.M., Segall, A.M., Mead, D., Azam, F., Rohwer, F.: Genomic analysis of uncultured marine viral communities. Proc. Natl. Acad. Sci. USA **99**, 14250–14255 (2002)

18. Brenner, S., Johnson, M., Bridgham, J., Golda, G., Lloyd, D.H., Johnson, D., Luo, S., McCurdy, S., Foy, M., Ewan, M., et al.: Gene expression analysis by massively parallel signature sequencing (MPSS) on microbead arrays. Nat. Biotechnol. **18**, 630–634 (2000)

19. Burge, C., Karlin, S.: Prediction of complete gene structures in human genomic DNA. J. Mol. Biol. **268**, 78–94 (1997)

20. Campbell, M.S., Law, M., Holt, C., Stein, J.C., Moghe, G.D., Hufnagel, D.E., Lei, J., Achawanantakun, R., Jiao, D., Lawrence, C.J., et al.: MAKER-p: a tool kit for the rapid creation, management, and quality control of plant genome annotations. Plant Physiol. **164**, 513–524 (2014)

21. Cantarel, B.L., Korf, I., Robb, S.M.C., Parra, G., Ross, E., Moore, B., Holt, C., Sanches Alvarado, A., Yandell, M.: MAKER: an easy-to-use annotation pipeline designed for emerging model organism genomes. Genome Res. **18**, 188–196 (2008)

22. Caspi, A., Pachter, L.: Identification of transposable elements using multiple alignments of related genomes. Genome Res. **16**, 260–270 (2006)

23. Chain, P.S.G., Grafham, D.V., Fulton, R.S., FitzGerald, M.G., Hostetler, J., Muzny, D., Ali, J., Birren, B., Bruce, D.C., Buhay, C., et al.: Genome project standards in a new era of sequencing. Science **326**, 236–237 (2009)

24. Chen, K., Pachter, L.: Bioinformatics for whole-genome shotgun sequencing of microbial communities. PLoS Comput. Biol. **1**, e24 (2005)

25. Clarke, J., Wu, H.-C., Jayasinghe, L., Patel, A., Reid, S., Bayley, H.: Continuouos base identification for single-molecule nanopore DNA sequencing. Nat. Nanotechnol. **4**, 265–270 (2009)

26. Collins, F.S., Green, E.D., Guttmacher, A.E., Guyer, M.S.: A vision for the future of genomics research. Nature **422**, 835–847 (2003)

27. Cunningham, F., Amode, M.R., Barrell, D., Beal, K., Billis, K., Brent, S., Carvalho-Silva, D., Clapham, P., Coates, G., Fitzgerald, S., et al.: Ensembl 2015. Nucleic Acids Res. **43**, D662–D669 (2015)

28. Dahm, R.: Discovering DNA: Friedrich Miescher and the early years of nucleic acid research. Hum. Genet. **122**, 565–581 (2008)

29. Dayhoff, M.O.: Atlas of Protein Sequence and Structure. National Biomedical Research Foundation, Washington (1969)

30. Dayhoff, M.O., Schwartz, R.M., Orcutt, B.C.: A model of evolutionary change in proteins. In: Dayhoff, M.O. (ed.) Atlas of Protein Sequence and Structure, vol. 5, pp. 345–352. Washington, Natl. Biomed. Res. Found (1978)

31. de Brujin, N.G.: A combinatorial problem. Koninklije Nederlandse Akademie v. Wetenschappen **49**, 758–764 (1946)

32. de Filippo, C., Ramazzotti, M., Fontana, P., Cavalieri, D.: Bioinformatic approaches for functional annotation and pathway inference in metagenomics data. Brief. Bioinform. **13**, 696–710 (2012)

33. de la Bastide, M., McCombie, W.R.: Assembling genomic DNA sequences with PHRAP. Curr. Protoc. Bioinform. Chapter 11, Unit 11.4 (2007)

34. Donlin, M.J.: Using the generic genome browser (GBrowse). In: Current Protocols in Bioinformatics, Chapter 9, Unit 9.9 (2009)
35. Earl, D., Bradnam, K., John, J.S., Darling, A., Lin, D., Fass, J., Yu, H.O.K., Buffalo, V., Zerbino, D.R., Diekhans, M., et al.: Assemblathon 1: a competitive assessment of de novo short read assembly methods. Genome Res. **21**, 2224–2241 (2010)
36. Eid, J., Fehr, A., Grey, J., Luong, K., Lyle, J., Otto, G., Peluso, P., Rank, D., Baybayan, P., Bettman, B., et al.: Real-time DNA sequencing from single polymerase molecules. Science **323**, 133–138 (2009)
37. Eilbeck, K., Lewis, S.E., Mungall, C.J., Yandell, M., Stein, L., Durbin, R., Ashburner, M.: The sequence ontology: a tool for the unification of genome annotations. Genome Biol. **6**, R44 (2005)
38. Eilbeck, K., Moore, B., Holt, C., Yandell, M.: Quantitative measures for the management and comparison of annotated genomes. BMC Bioinform. **10**, 67 (2009)
39. El-Metwally, S., Hamza, T., Zakaria, M., Helmy, M.: Next-generation sequencing assembly: four stages of data processing and computational challenges. PLoS One **9**, e1003345 (2013)
40. Elsik, C.G., Mackey, A.J., Reese, J.T., Milshina, N.V., Roos, D.S., Weinstock, G.M.: Creating a honey bee consensus gene set. Genome Biol. **8**, R13 (2007)
41. Engels, R.: Argo Genome Browser. http://www.broadinstitute.organnotationargo
42. Finn, R.D., Tate, J., Mistry, J., Coggill, P.C., Sammut, S.J., Hotz, H.R., Ceric, G., Forslund, K., Eddy, S.R., Sonnhammer, E.L.L.: The Pfam protein families database. Nucleic Acids Res. **36**, D281–D288 (2007)
43. Fleischmann, R., Adams, M., White, O., Clayton, R., Kirkness, E., Kerlavage, A., Bult, C., Tomb, J., Dougherty, B., Merrick, J.: Whole-genome random sequencing and assembly of *Haemophilus influenzae* Rd. Science **269**, 496–512 (1995)
44. Flicek, P., Birney, E.: Sense from sequence reads: methods for alignment and assembly. Nat. Methods **6**, S6–S12 (2009)
45. Generic Feature Format (GFF). http://www.sequenceontology.orggff3.shtml
46. Gilbert, W., Maxam, A.: The nucleotide of the lac operator. Proc. Natl. Acad. Sci. USA **70**, 3581–3584 (1973)
47. Gill, S.R., Pop, M., DeBoy, R.T., Eckburg, P.B., Turnbaugh, P.J., Samuel, B.S., Gordon, J.I., Relman, D.A., Fraser-Liggett, C.M., Nelson, K.E.: Metagenomic analysis of the human distal gut microbiome. Science **312**, 1355–1359 (2006)
48. Gish, W., States, D.J.: Identification of protein coding regions by database similarity search. Nat. Genet. **3**, 266–272 (1993)
49. Glass, E.M., Wilkening, J., Wilke, A., Antonopoulos, D., Meyer, F.: Using the metagenomics RAST server (MG-RAST) for analyzing shotgun metagenomes. Cold Spring Harbor protocols 2010, doi:10.1101/pdb.prot5368 (2010)
50. Gnerre, S., Maccallum, I., Przybylski, D., Ribeiro, F.J., Burton, J.N., Walker, B.J., Sharpe, T., Hall, G., Shea, T.P., Sykes, S., Berlin, A.M., Aird, D., Costello, M., Daza, R., Williams, L., Nicol, R., Gnirke, A., Nusbaum, C., Lander, E.S., Jaffe, D.B.: High-quality draft assemblies of mammalian genomes from massively parallel sequence data. Proc. Natl. Acad. Sci. USA **108**, 1513–1518 (2011)
51. Goffeau, A., Barrell, B.G., Bussey, H., Davis, R.W., Dujon, B., Feldmann, H., Galibert, F., Hoheisel, J.D., Jacq, C., Johnston, M., Louis, E.J., Mewes, H.W., Murakami, Y., Philippsen, P., Tettelin, H., Oliver, S.G.: Life with 6000 genes. Science **274**(546), 563–567 (1996)
52. Grabherr, M.G., Haas, B.J., Yassour, M., Levin, J.Z., Thompson, D.A., Amit, I., Adiconis, X., Fan, L., Raychowdhury, R., Zeng, Q., et al.: Full-length transcriptome assembly from RNA-Seq data without a reference genome. Nat. Biotechnol. **15**, 644–652 (2011)
53. Guttman, M., Garber, M., Levin, J.Z., Donaghey, J., Robinson, J., Adiconis, X., Fan, L., Koziol, M.J., Gnirke, A., Nusbaum, C., Rinn, J.L., Lander, E.S., Regev, A.: *Ab initio* reconstruction of cell type-specific transcriptomes in mouse reveals the conserved multi-exonic structure of lincRNAs. Nat. Biotechnol. **28**, 503–510 (2010)
54. Haas, B.J., Zody, M.C.: Advancing RNA-Seq analysis. Nat. Biotechnol. **28**, 421–423 (2010)

55. Haas, B.J., Delcher, A.L., Mount, S.M., Wortman, J.R., Smith Jr, R.K., Hannick Jr, L.I., Maiti, R., Ronning, C.M., Rusch, D.B., Town, C.D., et al.: Improving the Arabidopsis genome annotation using maximal transcript alignment assemblies. Nucleic Acids Res. **31**, 5654–5666 (2003)
56. Haas, B.J., Salzberg, S.L., Zhu, W., Pertea, M., Allen, J.E., Orvis, J., White, O., Buell, C.R., Wortman, J.R.: Automated eukaryotic gene structure annotation using EVidenceModeler and the program to assemble spliced alignments. Genome Biol. **9**, R7 (2008)
57. Handelsman, J., Rondon, M.R., Brady, S.F., Clardy, J., Goodman, R.M.: Molecular biology access to the chemistry of unknown soil microbes: a new Frontier for natural products. Chem. Biol. **5**, R245–R249 (1998)
58. Hartl, D.L.: Fly meets shotgun: shotgun wins. Nat. Genet. **24**, 327–328 (2000)
59. Havlak, P., Chen, R., Durbin, K.J., Egan, A., Ren, Y., Song, X.Z., Weinstock, G.M., Gibbs, R.A.: The atlas genome assembly system. Genome Res. **14**, 721–732 (2004)
60. Hesper, B., Hogeweg, P.: Bioinformatica: een werkconcept. Kameleon **1**, 28–29 (1970)
61. Hess, M., Sczyrba, A., Egan, R., Kim, T.-W., Chokhawala, H., Schroth, G., Luo, S., Clark, D.S., Chen, F., Zhang, T., et al.: Metagenomic discovery of biomass-degrading genes and genomes from cow rumen. Science **331**, 463–467 (2011)
62. Hoff, K.: The effect of sequencing errors on metagenomic gene prediction. BMC Genomics **10**, 520 (2009)
63. Hoff, K.J., Lingner, T., Meinicke, P., Tech, M.: Orphelia: predicting genes in metagenomic sequencing reads. Nucleic Acids Res. **37**, W101–105 (2009)
64. Holley, R.W., Apgar, J., Everett, G.A., Madison, J.T., Marquisee, M., Merrill, S.H., Penswick, J.R., Zamir, A.: Structure of a ribonucleic acid. Science **147**, 1462–1465 (1965)
65. Holt, C., Yandell, M.: MAKER2: an annotation pipeline and genome-database management tool for second-generation genome projects. BMC Bioinform. **12**, 491 (2011)
66. Huang, X., Madan, A.: CAP3: a DNA sequence assembly program. Genome Res. **9**, 868–877 (1999)
67. Huang, X., Wang, J., Aluru, S., Yang, S.P., Hillier, L.: PCAP: a whole-genome assembly program. Genome Res. **13**, 2164–2170 (2003)
68. Huang, S., Li, R., Zhang, Z., Li, L., Gu, X., Fan, W., Lucas, W.J., Wang, X., Xie, B., Ni, P., et al.: The genome of the cucumber. *Cucumis sativus* L. Nat. Genet. **41**, 1275–1281 (2009)
69. Huson, D.H., Mitra, S., Ruscheweyh, H.J., Weber, N., Schuster, S.C.: Integrative analysis of environmental sequences using MEGAN4. Genome Res. **21**, 1552–1560 (2011)
70. International Human Genome Sequencing Consortium: Finishing the euchromatic sequence of the human genome. Nature **431**, 931–945 (2004)
71. Ju, J., Kim, D.H., Bi, L., Meng, Q., Bai, X., Li, Z., Li, X., Marma, M.S., Shi, S., Wu, J., Edwards, J.R., Romu, A., Turro, N.J.: Four-color DNA sequencing by synthesis using cleavable flourescent nucleotide reversible terminators. Proc. Natl. Acad. Sci. USA **103**, 19635–19640 (2006)
72. Kapustin, Y., Souvorov, A., Tatusova, T., Lipman, D.: Splign: algorithms for computing spliced alignments with identification of paralogs. Biol. Direct **3**, 20 (2008)
73. Kelly, T.J., Smith, H.O.: A restriction enzyme from *Hemophilus influenzae* II. J. Mol. Biol. **51**, 393–409 (1970)
74. Kent, W.J.: BLAT—the BLAST-like alignment tool. Genome Res. **12**, 656–664 (2002)
75. Kim, M., Lee, K.H., Yoon, S.W., Kim, B.S., Chun, J., Yi, H.: Analytical tools and databases for metagenomics in the next-generation sequencing era. Genomics Inform. **11**, 102–113 (2013)
76. Korf, I., Yandell, M., Bedell, J.: BLAST: An Essential Guide to the Basic Local Alignment Search Tool. O'Reilly & Asscociates, Sebastopol (2003)
77. Korf, I.: Gene finding in novel genomes. BMC Bioinform. **5**, 59 (2004)
78. Lander, E.S., Waterman, M.S.: Genomic mapping by fingerprinting random clones: a mathematical analysis. Genomics **2**, 231–239 (1988)
79. Lander, E.S., Linton, L.M., Birren, B., Nusbaum, C., Zody, M.C., Baldwin, J., Devon, K., Dewar, K., Doyle, M., FitzHugh, W., et al.: Initial sequencing and analysis of the human genome. Nature **409**, 745–964 (2001)

80. Langmead, B., Trapnell, C., Pop, M., Salzberg, S.: Ultrafast and memory-efficient alignment of short DNA sequences to the human genome. Genome Biol. **10**, R25 (2009)

81. Lerat, E.: Identifying repeats and transposable elements in sequenced genomes: how to find your way through the dense forest of programs. Hered. (Edinb) **104**, 520–533 (2010)

82. Leung, H.C., Yiu, S.M., Yang, B., Peng, Y., Wang, Y., Liu, Z., Chen, J., Qin, J., Li, R., Chin, F.Y.: A robust and accurate binning algorithm for metagenomic sequences with arbitrary species abundance ratio. Bioinformatics **27**, 1489–1495 (2011)

83. Lewis, S.E., Searle, S.M., Harris, N., Gibson, M., Lyer, V., Richter, J., Wiel, C., Bayrak-taroglir, L., Birney, E., Crosby, M.A.: Apollo: a sequence annotation editor. Genome Biol. **3**, research0082 (2002)

84. Li, R., Fan, W., Tian, G., Zhu, H., He, L., Cai, J., Huang, Q., Cai, Q., Li, B., Bai, Y., et al.: The sequence and De Novo assembly of the giant panda genome. Nature **463**, 311–317 (2010)

85. Li, R., Zhu, H., Ruan, J., Qian, W., Fang, X., Shi, Z., Li, Y., Li, S., Shan, G., Kristiansen, K., Li, S., Yang, H., Wang, J., Wang, J.: De novo assembly of human genomes with massively parallel short read sequencing. Genome Res. **20**, 265–272 (2010)

86. Li, Z., Zhang, Z., Yan, P., Huang, S., Fei, Z., Lin, K.: RNA-Seq improves annotation of protein-coding genes in the cucumber genome. BMC Genomics **12**, 540 (2011)

87. Li, Z., Chen, Y., Mu, D., Yuan, J., Shi, Y., Zhang, H., Gan, J., Li, N., Hu, X., Liu, B., Yang, B., Fan, W.: Comparison of the two major classes of assembly algorithms: overlap-layout-consensus and de-brujin-graph. Brief. Funct. Genomics **11**, 25–37 (2012)

88. Lindblad-Toh, K., Wade, C.M., Mikkelsen, T.S., Karlsson, E.K., Jaffe, D.B., Kamal, M., Clamp, M., Chang, J.L., Kulbokas III, E.J., Zody, M.C.: Genome sequence, comparative, analysis and haplotype structure of the domestic dog. Nature **438**, 803–819 (2005)

89. Liu, B., Gibbons, T., Ghodsi, M., Treangen, T., Pop, M.: Accurate and fast estimation of taxonomic profiles from metagenomic shotgun sequences. BMC Genomics **12** (Suppl 2), S4 (2011)

90. Liu, Q., Mackey, A.J., Roos, D.S., Pereira, F.C.N.: Evigan: a hidden variable model for integrating gene evidence for eukaryotic gene prediction. Bioinformatics **24**, 597–605 (2008)

91. Loftus, B.J., Fung, E., Roncaglia, P., Rowley, D., Amedeo, P., Bruno, D., Vamathevan, J., Miranda, M., Anderson, I.J., Fraser, J.A., et al.: The genome of the basidiomycetous yeast and human pathogen *Cryptococcus neoformans*. Science **307**, 1321–1324 (2005)

92. Lomsadze, A., Ter-Hovhannisyan, V., Chernoff, Y.O., Borodovsky, M.: Gene identification in novel eukaryotic genomes by self-traning algorithm. Nucleic Acids Res. **33**, 6494–6506 (2005)

93. Lorenz, P., Eck, J.: Metagenomics and industrial applications. Nat. Rev. Microbiol. **3**, 510–516 (2005)

94. Margulies, M., Egholm, M., Altman, W.E., Attiya, S., Bader, J.S., Bemben, L.A., Berka, J., Braverman, M.S., Chen, Y.-J., Chen, Z., et al.: Genome Sequencing in microfabricated high-density picolitre reactors. Nature **437**, 376–380 (2005)

95. Maxam, A.M., Gilbert, W.: A new method for sequencing DNA. Proc. Natl. Acad. Sci. USA **74**, 560–564 (1977)

96. McCallum, D., Smith, M.: Computer processing of DNA sequence data. J. Mol. Biol. **116**, 29–30 (1977)

97. McHardy, A.C., Martin, H.G., Tsirigos, A., Hugenholtz, P., Rigoutsos, I.: Accurate phyloge-netic classification of variable-length DNA fragments. Nat. Methods **4**, 63–72 (2007)

98. Miller, J.R., Delcher, A.L., Koren, S., Venter, E., Walenz, B.P., Brownley, A., Johnson, J., Li, K., Mobarry, C., Sutton, G.: Aggressive assembly of pyrosequencing reads with mates. Bioinformatics **24**, 2818–2824 (2008)

99. Miller, J.R., Koren, S., Sutton, G.: Assembly algorithms for next-generation sequencing data. Genomics **95**, 315–327 (2010)

100. Monzoorul Haque, M., Ghosh, T.S., Komanduri, D., Mande, S.S.: SOrt-ITEMS: sequence orthology based approach for improved taxonomic estimation of metagenomic sequences. Bioinformatics **25**, 1722–1730 (2009)

101. Mortazavi, A., Williams, B.A., McCue, K., Schaeffer, L., Wold, B.: Mapping and quantifying mammalian transcriptomes by RNA-Seq. Nat. Methods **5**, 621–628 (2008)
102. Mouse Genome Sequencing Consortium: Initial sequencing and comparative analysis of the mouse genome. Nature **420**, 520–562 (2002)
103. Mulder, N., Apweiler, R.: InterPro and InterProScan: tools for protein sequence classification and comparison. Methods Mol. Biol. **396**, 59–70 (2007)
104. Mullikin, J.C., Ning, Z.: The Phusion assembler. Genome Res. **13**, 81–90 (2003)
105. Myers, E.W.: The fragment assembly string graph. Bioinformatics **21**, ii79–ii85 (2005)
106. Myers, E.W., Sutton, C.G., Delcher, A.L., Dew, I.M., Fasulo, D.P., Flanigan, M.J., Kravitz, S.A., Mobarry, C.M., Reinert, K.H., Remington, K.A., et al.: A whole-genome assembly of Drosophila. Science **287**, 2196–2204 (2000)
107. Nagalakshmi, U., Wang, Z., Waern, K., Shou, C., Raha, D., Gerstein, M., Snyder, M.: The transcriptional landscape of the yeast genome defined by RNA sequencing. Science **320**, 1344–1349 (2008)
108. Namiki, T., Hachiya, T., Tanaka, H., Sakakibara, Y.: MetaVelvet: an extension of Velvet assembler to De Novo metagenome assembly from short sequence reads. Nucleic Acids Res. **40**, e155 (2012)
109. Nene, V., Wortman, J.R., Lawson, D., Haas, B., Kodira, C., Tu, Z.J., Loftus, B., Xi, Z., Megy, K., Grabherr, M., et al.: Genome sequence of Aedes aegypti, a major *arbovirus vector*. Science **316**, 1718–1723 (2007)
110. Noguchi, H., Taniguchi, T., Itoh, T.: MetaGeneAnnotator: detecting species-specific patterns of ribosomal binding site for precise gene prediction in anonymous prokaryotic and phage genomes. DNA Res. **15**, 387–396 (2008)
111. Nygaard, S., Zhang, G., Schiott, M., Li, C., Wurm, Y., Hu, H., Zhou, J., Ji, L., Qiu, F., Rasmussen, M., et al.: The genome of the leaf-cutting ant *Acromyrmex echinatior* suggests key adaptations to advanced social life and fungus farming. Genome Res. **21**, 1339–1348 (2011)
112. Overbeek, R., Begley, T., Butler, R.M., Choudhuri, J.V., Chuang, H.Y., Cohoon, M., de Crecy-Lagard, V., Diaz, N., Disz, T., Edwards, R., et al.: The subsystems approach to genome annoation and its use in the project project to annotate 1000 genomes. Nucleic Acids Res. **33**, 5691–5702 (2005)
113. Pagani, I., Liolios, K., Jansson, J., Chen, I.A., Smirnova, T., Nosrat, B., Markowitz, V.M., Kyrpides, N.C.: The Genomes OnLine Database (GOLD) v. 4: status of genomic and metagenomic projects and their associated metadata. Nucleic Acids Res. **40**, D571–D579 (2011)
114. Park, P.J.: ChIP-seq: advantages and challenges of a maturing technology. Nat. Rev. Genet. **10**, 669–680 (2009)
115. Parra, G., Bradnam, K., Korf, I.: CEGMA: A pipeline to accurately annotate core genes in eukaryotic genomes. Bioinformatics **23**, 1061–1067 (2007)
116. Parra, G., Bradnam, K., Korf, I.: Assessing the gene space in draft genomes. Nucleic Acids Res. **37**, 289–297 (2009)
117. Paszkiewicz, K., Studholme, D.J.: De Novo assembly of short sequence reads. Brief. Bioinform. **11**, 457–472 (2010)
118. Peng, Y., Leung, H.C., Yiu, S.M., Chin, F.Y.: Meta-IDBA: a De Novo assembler for metagenomic data. Bioinformatics **27**, i94–101 (2011)
119. Petrosino, J.F., Highlander, S., Luna, R.A., Gibbs, R.A., Versalovic, J.: Metagenomic pyrosequencing and microbial identification. Clin. Chem. **55**, 856–866 (2009)
120. Pevzner, P.A., Tang, H., Waterman, M.S.: An Eulerian path approach to DNA fragment assembly. Proc. Natl. Acad. Sci. USA **98**, 9748–9753 (2001)
121. Pevzner, P.A., Tang, H., Tesler, G.: De Novo repeat classification and fragment assembly. Genome Res. **14**, 1786–1796 (2004)
122. Pop, M., Phillippy, A., Delcher, A.L., Salzberg, S.L.: Comparative genome assembly. Brief. Bioinform. **5**, 237–248 (2004)
123. Pushkarev, D., Neff, N.F., Quake, S.R.: Single-molecule sequencing of an individual human genome. Nat. Biotechnol. **27**, 847–850 (2009)

124. Rat Genome Sequencing Project Consortium: Genome sequence of the Brown Norway rat yields insights into mammalian evolution. Nature **428**, 493–521 (2004)

125. Rhesus Macaque Genome Sequencing and Analysis Consortium: Evolutionary and biomedical insights from the *rhesus macaque* genome. Science **316**, 222–234 (2007)

126. Rho, M., Tang, H., Ye, Y.: FragGeneScan: predicting genes in short and error-prone reads. Nucleic Acids Res. **38**, e191 (2010)

127. Rondon, M.R., August, P.R., Betterman, A.D., Brady, S.F., Grossman, T.H., Liles, M.R., Loiacono, K.A., Lynch, B.A., MacNeil, I.A., Minor, C., Tiong, C.L., Gilman, M., Osburne, M.S., Clardy, J., Handelsman, J., Goodman, R.M.: Cloning the soil metagenome: a strategy for accessing the genetic and functional diversity of uncultured microorganisms. Appl. Environ. Microbiol. **66**, 2541–2547 (2000)

128. Rosen, G.L., Reichenberger, E.R., Rosenfeld, A.M.: NBC: the naive Bayes classification tool webserver for taxonomic classification of metagenomic reads. Bioinformatics **27**, 127–129 (2011)

129. Rothberg, J.M., Hinz, W., Rearick, T.M., Schultz, J., Mileski, W., Davey, M., Leamon, J.H., Johnson, K., Milgrew, M.J., Edwards, M., et al.: An integrated semiconductor device enabling non-optical genome sequencing. Nature **475**, 348–352 (2011)

130. Rutherford, K., Parkhill, J., Crook, J., Horsnell, T., Rice, P., Rajandream, M.A., Barrell, B.: Artemis: sequence visualization and annotation. Bioinformatics **16**, 944–945 (2000)

131. Salamov, A.A., Solovyev, V.V.: Ab initio gene finding in Drosophila genomic DNA. Genome Res. **10**, 516–522 (2000)

132. Salzberg, S.L., Phillippy, A.M., Zimin, A., Puiu, D., Magoc, T., Koren, S., Treangen, T.J., Schatz, M.C., Delcher, A.L., Roberts, M., Marcais, G., Pop, M., Yorke, J.A.: GAGE: a critical evaluation of genome assemblies and assembly algorithms. Genome Res. **22**, 557–567 (2012)

133. Sanger, F., Air, G.M., Barrell, B.G., Brown, N.L., Coulson, A.R., Fiddes, C.A., Hutchison, C.A., Slocombe, P.M., Smith, M.: Nucleotide sequence of bacteriophage phi X174 DNA. Nature **265**, 687–695 (1977)

134. Sanger, F., Coulson, A.R.: A rapid method for determining sequences in DNA by primed synthesis with DNA polymerase. J. Mol. Biol. **94**, 441–448 (1975)

135. Sanger, F., Niclen, S., Coulson, A.R.: DNA sequencing with chain-terminating inhibitors. Proc. Natl. Acad. Sci. USA **74**, 5463–5467 (1977)

136. Sato, T., Terabe, M., Watanabe, H., Gojobori, T., Hori-Takemoto, C., Miura, K.: Codon and base biases after the initiation codon of the open reading frames in the *Escherichia coli* genome and their influence on the translation efficiency. J. Biochem. **129**, 851–860 (2001)

137. Sayers, E.W., Barrett, T., Benson, D.A., Bryant, S.H., Canese, K., Chetvernin, V., Church, D.M., DiCuccio, M., Edgar, R., et al.: Database resources of the national center for biotechnology information. Nucleic Acids Res. **37**, D5–D15 (2009)

138. Schadt, E.E., Turner, S., Kasarskis, A.: A window into third-generation sequencing. Hum. Mol. Genet. **19**, R227–R240 (2010)

139. Schloss, J.A.: How to get genomes at one ten-thousandth the cost. Nat. Biotechnol. **26**, 1113–1115 (2008)

140. Shendure, J., Porreca, G.J., Reppas, N.B., Lin, X., McCutcheon, J.P., Rosenbaum, A.M., Wang, M.D., Zhang, K., Mitra, R.D., Church, G.M.: Accurate multiplex polony sequencing of an evolved bacterial genome. Science **309**, 1728–1732 (2005)

141. Simpson, J.T., Wong, K., Jackman, S.D., Schein, J.E., Jones, S.J., Birol, I.: ABySS: a parallel assembler for short read sequence data. Genome Res. **19**, 1117–1123 (2009)

142. Simpson, J.T., Durbin, R.: Efficient de novo assembly of large genomes using compressed data structures. Genome Res. **22**, 549–556 (2012)

143. Skinner, M.E., Uzilov, A.V., Stein, L.D., Mungall, C.J., Holmes, I.H.: JBROWSE: a next-generation genome browser. Genome Res. **19**, 1630–1638 (2009)

144. Slater, G.S., Birney, E.: Automated generation of heuristics for biological sequence comparison. BMC Bioinform. **6**, 31 (2005)

145. Smit, A.F.A., Hubley, R., Green, P.: RepeatMasker at http://www.repeatmasker.org

146. Smith, H.O., Wilcox, K.W.: A restriction enzyme from *Hemophilus influeanzae*. I. Purification and general properties. J. Mol. Biol. **51**, 379–391 (1970)
147. Smith, L.M., Sanders, J.Z., Kaiser, R.J., Hughes, P., Dodd, C., Connell, C.R., Heiner, C., Kent, S.B., Hood, L.E.: Flourescence detection in automated DNA sequence analysis. Nature **321**, 674–679 (1986)
148. Smith, C.D., Edgar, R.C., Yandell, M.D., Smith, D.R., Celniker, S.E., Myers, E.W., Karpen, G.H.: Improved repeat identification and masking in Dipterans. Gene **389**, 1–9 (2007)
149. Smith, C.C., Zimin, A., Holt, C., Abouheif, E., Benton, R., Cash, E., Croset, V., Currie, C.R., Elhaik, E., Elsik, C.G., et al.: Draft genome of the globally widespread and invasive Argentine ant (*Linepithema humile*). Proc. Natl. Acad. Sci. USA **108**, 5673–5678 (2011)
150. Staden, R.: Sequence data handling by computer. Nucleic Acids Res. **4**, 4037–4051 (1977)
151. Staden, R., Beal, K.F., Bonfield, J.K.: The Staden package, 1998. Methods Mol. Biol. **132**, 115–130 (2000)
152. Stanke, M., Waack, S.: Gene prediction with a hidden Markov model and a new intron submodel. Bioinformatics **19**, ii215–ii225 (2003)
153. Stanke, M., Steinkamp, R., Waack, S., Morgenstern, B.: AUGUSTUS: a web server for gene finding in eukaryotes. Nucleic Acids Res. **32**, W309–W312 (2004)
154. Suen, G., Teiling, C., Li, L., Holt, C., Abouheif, E., Bornberg-Bauer, E., Bouffard, P., Caldera, E.J., Cash, E., Cavanaugh, A., et al.: The genome sequence of the leaf-cutter ant *Atta cephalotes* reveals insights into its obligate symbiotic lifestile. PLoS Genet. **7**, e1002007 (2011)
155. The Bovine Genome Sequencing and Analysis Consortium: The genome sequence of taurine cattle: a window to ruminant biology and evolution. Science **324**, 522–528 (2009)
156. The Generic Model Organism Database. http://www.gmod.org
157. The Reference Genome Group of the Gene Ontology: Consortium: The gene ontology's reference genome project: a unified framework for functional annotation across species. PLoS Comput. Biol. **5**, e1000431 (2009)
158. The Rice Genome Project: A draft sequence of the rice genome (Oryza sativa L. ssp. indica). Science **296**, 79–92 (2002)
159. The UniProt Consortium: The universal protein resource (UniProt) 2009. Nucleic Acids Res. **37**, D169–D174 (2009)
160. The University of Santa Cruz Genome Browser: http://genome.ucsc.edu
161. The C. elegans Sequencing Consortium: Genome sequence of the nematode C. elegans: a platform for investigating biology. Science **282**, 2012–2018 (1998)
162. Trapnell, C., Pachter, L., Salzberg, S.L.: TopHat: discovering splice junctions with RNA-Seq. Bioinformatics **25**, 1105–1111 (2009)
163. Trapnell, C., Williams, B.A., Pertea, G., Mortazavi, A., Kwan, G., van Baren, M.J., Salzberg, S.L., Wold, B.J., Pachter, L.: Transcript assembly and quantification by RNA-Seq reveals unannotated transcripts and isoform switching during cell differentiation. Nat. Biotechnol. **28**, 511–515 (2010)
164. Treangen, T.J., Salzberg, S.L.: Repetitive DNA and next-generation sequencing: computational challenges and solutions. Nat. Rev. Genet. **13**, 36–46 (2011)
165. Tyson, G.W., Chapman, J., Hugenholtz, P., Allen, E.E., Ram, R.J., Richardson, P.M., Solovyev, V.V., Rubin, E.M., Rokhsar, D.S., Banfield, J.F.: Community structure and metabolism through reconstruction of microbial genomes from the environment. Nature **428**, 37–43 (2004)
166. Valouev, A., Ichikawa, J., Tonthat, T., Stuart, J., Ranade, S., Peckham, H., Zeng, K., Malek, J.A., Costa, G., McKernan, K., Sidow, A., Fire, A., Johnson, S.M.: A high-resolution, nucleosom position map of C. elegans reveals a lack of universal sequence-dictated positioning. Genome Res. **18**, 1051–1063 (2008)
167. van Dijk, E.L., Auger, H., Jaszczyszyn, Y., Thermes, C.: Ten years of next-generation sequencing technology. Trends Genet. **30**, 418–426 (2014)
168. Venter, C.J., Adams, M.D., Myers, E.W., Li, P.W., Mural, R.J., Sutton, G.G., Smith, H.O., Yandell, M., Evans, C.A., Holt, R.A., et al.: The sequence of the human genome. Science **291**, 1304–1351 (2001)

169. Venter, J.C., Remington, K., Heidelberg, J.F., Halpern, A.L., Rusch, D., Eisen, J.A., Wu, D., Paulsen, I., Nelson, K.E., Nelson, W., et al.: Environmental genome sequencing of the Sargasso Sea. Science **304**, 66–74 (2004)
170. Wang, J., Wong, G.K., Ni, P., Han, Y., Huang, X., Zhang, J., Ye, C., Zhang, Y., Hu, J., Zhang, K., et al.: RePS: a sequence assembler that masks exact repeats identified from the shotgun data. Genome Res. **12**, 821–831 (2002)
171. Wang, Z., Gerstein, M., Snyder, M.: RNA-Seq: a revolutionary tool for transcriptomics. Nat. Rev. Genet. **10**, 57–63 (2009)
172. Warren, R.L., Sutton, G.G., Jones, S.J., Holt, R.A.: Assembling millions of short DNA sequences using SSAKE. Bioinformatics **23**, 500–501 (2007)
173. Watson, J.D., Crick, F.H.C.: Molecular structure of nucleic acids. Nature **171**, 737–738 (1953)
174. Whiteford, N., Haslam, N., Weber, G., Prügel-Bennett, A., Essex, J.W., Roach, P.L., Bradley, M., Neylon, C.: An analysis of the feasibility of short read sequencing. Nucleic Acids Res. **33**, e171 (2005)
175. Wold, B., Myers, R.M.: Sequence census methods for functional genomics. Nat. Methods **5**, 19–21 (2008)
176. Worley, K.C., Gibbs, R.A.: Genetics: decoding a national treasure. Nature **463**, 303–304 (2010)
177. Wu, R., Kaiser, A.D.: Structure and base sequence in the cohesive ends of bacteriophage lambda DNA. J. Mol. Biol. **35**, 523–537 (1968)
178. Wu, R., Taylor, E.: Nucleotide sequence analysis of DNA. II. Complete nucleotide sequence of the cohesive ends of bacteriophage lambda DNA. J. Mol. Biol. **57**, 491–511 (1971)
179. Wu, T.D., Nacu, S.: Fast and SNP-tolerant detection of complex variants and splicing in short reads. Bioinformatics **26**, 873–881 (2010)
180. Yandell, M., Ence, D.: A beginner's guide to eukaryotic genome annotation. Nat. Rev. Genet. **13**, 329–342 (2012)
181. Zerbino, D.R., Birney, E.: Velvet: algorithms for de novo short read assembly using de Brujin graphs. Genome Res. **18**, 821–829 (2008)
182. Zhang, W., Chen, J., Yang, Y., Tang, Y., Shang, J., Shen, B.: A practical comparison of De Novo genome assembly software tools for next-generation sequencing technologies. PLoS One **6**, e17915 (2011)

Index

© Springer-Verlag London 2015

M. Axelson-Fisk, *Comparative Gene Finding*, Computational Biology 20,
DOI 10.1007/978-1-4471-6693-1

Printed in the United States
By Bookmasters